Optimum Accelerated Life Testing Models with Time-Varying Stresses

Optimum Accelerated Life Testing Models with Time-Varying Stresses

Preeti Wanti Srivastava
University of Delhi, India

World Scientific

NEW JERSEY • LONDON • SINGAPORE • BEIJING • SHANGHAI • HONG KONG • TAIPEI • CHENNAI • TOKYO

Published by

World Scientific Publishing Europe Ltd.
57 Shelton Street, Covent Garden, London WC2H 9HE
Head office: 5 Toh Tuck Link, Singapore 596224
USA office: 27 Warren Street, Suite 401-402, Hackensack, NJ 07601

Library of Congress Cataloging-in-Publication Data
Names: Srivastava, Preeti Wanti, author.
Title: Optimum accelerated life testing models with time-varying stresses /
Preeti Wanti Srivastava (University of Delhi, India).
Description: [Hackensack?] New Jersey : World Scientific, 2017.
Identifiers: LCCN 2016026719 | ISBN 9789813141254 (hc : alk. paper)
Subjects: LCSH: Accelerated life testing.
Classification: LCC TA169.3 .S75 2017 | DDC 620/.0044--dc23
LC record available at https://lccn.loc.gov/2016026719

British Library Cataloguing-in-Publication Data
A catalogue record for this book is available from the British Library.

Desk Editor: Dr. Sree Meenakshi Sajani

Typeset by Stallion Press
Email: enquiries@stallionpress.com

Printed in Singapore

CONTENTS

Half of all scientific discoveries ever made were made in the 20th century. Creation is accelerating, and even the acceleration appears to be accelerating.

John Template

Preface xxi
Acronyms xxvii
List of Figures xxix
List of Tables xxxi
About the Author xxxvii

1. Different Aspects of ALT Models 1
1.1 Introduction 1
1.1.1 Difference between ALT and ADT 2
1.1.2 Fully and Partially Accelerated Tests 4
1.1.3 Inspection Modes in Accelerated Testing . . 4
1.2 Types of Stress Loadings 4
1.2.1 Constant-Stress Loading 4
1.2.2 Step-Stress Loading 5
1.2.3 Progressive-Stress Loading 5
1.2.4 Cyclic-Stress Loading 5
1.2.5 Random-Stress Loading 6
1.2.6 Ramp-Constant Stress Loading 6
1.2.7 Ramp-Step Stress Loading 6

1.3 Types of Stresses 7
1.3.1 Mechanical Stresses 7
1.3.2 Electrical Stresses 8
1.3.3 Environmental Stresses 9
1.3.4 Usage Rate Acceleration 10
1.4 A Product's Life Distribution 11
1.4.1 Reliability Metrics 11
1.4.2 Life Distribution Models 13
1.4.2.1 Exponential Distribution 13
1.4.2.2 Normal Distribution 13
1.4.2.3 Lognormal Distribution 14
1.4.2.4 Weibull Distribution 15
1.4.2.5 Extreme Value Distribution 15
1.4.2.6 Logistic Distribution 16
1.4.2.7 Log-Logistic Distribution 17
1.4.2.8 Truncated Logistic Distribution 18
1.4.2.9 Burr Type-XII Distribution 19
1.5 Fully ALT Models 20
1.5.1 Life–Stress Relationship 21
1.5.1.1 Arrhenius Relationship 21
1.5.1.2 Inverse Power Law Relationship 22
1.5.1.3 Eyring Relationship 23
1.5.1.4 Temperature–Humidity Relationship 23
1.5.1.5 Temperature–Nonthermal Relationship 23
1.5.1.6 Multivariable Relationship 24
1.6 Constant-Stress ALT Models 24
1.7 Proportional Hazards Based ALT Models 25
1.8 Proportional Odds (PO) Based ALT Models . . . 26
1.9 ALT Models under Time-Varying Stress 27
1.9.1 Step-Stress ALT Model 28
1.9.2 Progressive-Stress ALT Model 30

1.10 Time-Varying Stress ALT Models with Log-Location-Scale Distribution 32
1.11 Relationship Between Constant-Stress, Step-Stress, and Progressive-Stress Loadings 34
1.12 PALT Models 35
1.12.1 Constant-Stress PALT Models 36
1.12.2 Time-varying Stress PALT Models: Step-Stress Set-Up 36
1.12.2.1 Tampered Failure Rate (TFR) Model: Step-Stress Set-Up 37
1.13 Data Types 38
1.14 Data Analysis Methods 42
1.15 Optimum Test Plans 43
1.15.1 Variance Optimality 44
1.15.2 D-Optimality 44
1.15.3 A-Optimality 45
1.16 Likelihood Function Formulation 45
1.16.1 Formulation of Likelihood Function with Time-Censored Data 45
1.16.2 Formulation of Likelihood Function with Failure-Censored Data 46
1.17 Sensitivity Analysis 46
1.18 Design of Fully ALT Models Under Continuous Inspection: A Review 47
1.18.1 Constant-Stress ALT Models 47
1.18.2 Step-Stress ALT Models 49
1.18.3 Progressive-Stress ALT Models 54
1.18.4 Modified-Stress ALT Models 56
1.18.5 ALT Models with Competing Causes of Failure 56
1.19 Design of PALT Models: A Review 58
1.19.1 Constant-Stress PALT Models 58
1.19.2 Step-Stress PALT Models 59
1.20 Design of ALT Models Under Periodic Inspection: A Review 62

2. Optimum Step-Stress Accelerated Life Test Models **63**

2.1 Introduction . . . 63
2.1.1 Notation . . . 64
2.2 An Optimum Simple Log-Logistic Step-Stress ALT Model for Complete Data Set . . . 65
2.2.1 Assumptions . . . 65
2.2.2 Test Procedure . . . 66
2.2.3 Life Distribution under Step-Stress . . . 66
2.2.4 Maximum Likelihood Estimation . . . 67
2.2.5 Fisher Information Matrix . . . 68
2.2.6 Optimality Criterion . . . 71
2.3 An Optimum Simple Log-Logistic Step-Stress ALT Model for Type-I Censored Data Set . . . 72
2.3.1 Assumptions . . . 72
2.3.2 Test Procedure . . . 72
2.3.3 Life Distribution under Step-Stress . . . 74
2.3.4 Maximum likelihood Estimation . . . 74
2.3.5 Fisher Information Matrix . . . 75
2.3.6 Optimum Plan . . . 80
2.4 Testing of Hypotheses . . . 81
2.4.1 For Complete Data Set . . . 81
2.4.2 For Time-Censored Data Set . . . 82
2.5 Numerical Examples . . . 83
2.5.1 For Complete Data Set . . . 83
2.5.2 For Time-Censored Data Set . . . 84
2.6 Sensitivity Analysis . . . 86
2.7 Summary . . . 88

3. Optimum Step-Stress Partially Accelerated Life Test Plans with Type-I and Type-II Censoring **89**

3.1 Introduction . . . 89
3.1.1 Notation . . . 89
3.1.2 Truncated Logistic Distribution under Step-Stress PALT . . . 93

3.2 Optimum Time-Censored Step-Stress PALT Model 95
3.2.1 Assumptions 95
3.2.2 Test Procedure 95
3.2.3 Log-Likelihood Function 96
3.2.4 Fisher Information Matrix 97
3.2.5 Optimal Test Plan 97
3.3 Optimum Failure-Censored Step-Stress PALT Model 97
3.3.1 Case I: Number of Failures Pre-Specified . . 98
3.3.1.1 Assumptions 98
3.3.1.2 Test Procedure 98
3.3.1.3 Log-Likelihood Function 98
3.3.1.4 Fisher Information Matrix 100
3.3.1.5 Optimal Test Plan 100
3.3.2 Case II: Proportion of Units Failing Before Censoring Pre-Specified 100
3.3.2.1 Assumptions 100
3.3.2.2 Test Procedure 101
3.3.2.3 Log-Likelihood Function 101
3.3.2.4 Fisher Information Matrix 102
3.3.2.5 Optimal Test Plan 103
3.4 Numerical Examples 103
3.4.1 Type-I Censoring 103
3.4.1.1 Optimal Plan 103
3.4.1.2 Simulated Data 103
3.4.1.3 MLEs of the Design Parameters 104
3.4.1.4 Confidence Intervals 104
3.4.1.5 Graphical Goodness of Fit 105
3.4.1.6 Coverage Probabilities 106
3.4.2 Type-II Censoring: Case I — Number of Failures Pre-Specified 107
3.4.2.1 Optimal Plan 107
3.4.2.2 Simulated Data 107
3.4.2.3 MLEs of the Design Parameters . 108

3.4.2.4 Confidence Intervals 108
3.4.2.5 Graphical Goodness of Fit 108
3.4.2.6 Coverage Probabilities 110
3.4.3 Type-II Censoring: Case II — Proportion of Censoring Pre-Specified 111
3.4.3.1 Optimal Plan 111
3.4.3.2 Simulated Data 113
3.4.3.3 MLEs of the Design Parameters 113
3.4.3.4 Confidence Intervals 114
3.5 Sensitivity Analysis 115
3.5.1 Type-I Censoring and Type-II Censoring (Case: I — Number of Failures Pre-Specified) 115
3.5.2 Type-II Censoring (Case II: Proportion of Items Failing before Censoring Pre-Specified) 116
3.6 Comparative Study for Type-II Censoring (Case II — Proportion of Items Failing Before Censoring Pre-Specified) 117
3.7 Summary 118

4. Optimum Ramp-Stress Fully Accelerated Life Test Plans Under Type-I Censoring 119
4.1 Introduction 119
4.1.1 Notation 120
4.1.2 Burr Type-XII Life Distribution 122
4.2 Optimum Multi-Level Ramp-Stress ALT Plan Without Stress Upper Bound 123
4.2.1 Assumptions 123
4.2.2 Test Procedure 124
4.2.3 Life Distribution under Multi-Level Ramp-Test without Stress Upper Bound . . 124
4.2.4 Log-Likelihood Function 125
4.2.5 Parameter Estimation 126

4.2.6 Fisher Information Matrix 127
4.2.7 Asymptotic Variance of the MLE of Logarithm of Quantile at Design Stress 128
4.3 Optimum Multi-Objective Ramp-Stress ALT Plan With Stress Upper Bound 129
4.3.1 Assumptions 129
4.3.2 Test Procedure 129
4.3.3 Life Distribution under Ramp-Stress with Stress Upper Bound 130
4.3.4 Log-Likelihood Function 131
4.3.5 Parameter Estimation 132
4.3.6 Fisher Information Matrix 132
4.3.7 Asymptotic Variance of the MLE of Logarithm of Quantile at Design Stress 132
4.4 Formulation of a Multi-Objective Optimization Problem 133
4.4.1 Optimization Problem For Multi-Level Ramp-Stress ALT Plan without Stress Upper Bound 134
4.4.2 Optimization Problem For Ramp-Stress ALT Plan with Stress Upper Bound 135
4.5 Numerical Examples 136
4.5.1 Multi-Level Ramp-Stress ALT Plan without Stress Upper Bound 136
4.5.1.1 Optimal Plan 137
4.5.1.2 Simulated Data 137
4.5.1.3 MLEs of the Design Parameters 138
4.5.1.4 Confidence Intervals 138
4.5.2 Ramp-Stress ALT Plan with Stress Upper Bound 138
4.5.2.1 Optimal Plan 139
4.5.2.2 Simulated Data 139

4.5.2.3 MLEs of the Design Parameters 139
4.5.2.4 Confidence Intervals 140
4.6 Sensitivity Analysis 140
4.7 Comparative Study 141
4.7.1 Multi-Level Ramp-Stress ALT Plan without Stress Upper Bound 141
4.7.2 Ramp-Stress ALT Plan with Stress Upper Bound 143
4.8 Summary 145

5. Optimum Fully Accelerated Life Test Plans with Modified Stress Loading Schemes under Type-I Censoring 147

5.1 Introduction 147
5.1.1 Notation 148
5.1.2 Burr Type-XII Lifetime Distribution 151
5.2 Optimum Multi-Objective Modified Step-Stress ALT Plan 151
5.2.1 Assumptions 152
5.2.2 Test Procedure 152
5.2.3 Life Distribution Under Modified Step-Stress 153
5.2.4 Log-Likelihood Function 156
5.2.5 Fisher Information Matrix 157
5.2.6 Asymptotic Variance of the MLEs of Logarithm of Quantile q at Design Stress 157
5.3 Optimum Multi-Objective Modified Constant-Stress ALT Plan 158
5.3.1 Assumptions 158
5.3.2 Test Procedure 159
5.3.3 Life Distribution under Modified Constant-Stress 160
5.3.4 Log-Likelihood Function 161
5.3.5 Fisher Information Matrix 162

5.3.6 Asymptotic Variance of the MLEs of the Logarithm of Quantile q at Design Stress 163
5.4 Formulation of a Multi-Objective Optimization Problem 163
5.4.1 Modified Step-Stress ALT Plan 164
5.4.2 Modified Constant-Stress ALT Plan 165
5.5 Numerical Examples 166
5.5.1 Modified Step-Stress ALT Plan 166
5.5.1.1 Optimal Plan 166
5.5.1.2 Simulated Data 166
5.5.1.3 MLEs of the Design Parameters 168
5.5.1.4 Confidence Intervals 168
5.5.2 Modified Constant-Stress ALT Plan 168
5.5.2.1 Optimal Plan 168
5.5.2.2 Simulated Data 169
5.5.2.3 MLEs of the Design Parameters 169
5.5.2.4 Confidence Intervals 170
5.6 Sensitivity Analysis 171
5.7 Comparative Study 172
5.7.1 Modified Step-Stress ALT Plan 172
5.7.2 Modified Constant-Stress ALT Plan 174
5.8 Summary 175

6. Optimum Time-Censored Step-Stress Fully ALT with Competing Risks For Failure 177

6.1 Introduction 177
6.1.1 Notation 178
6.2 Assumptions 178
6.3 Test Procedure 179
6.4 Life Distribution Under Step-Stress With Competing Failure Modes 180
6.5 Log-Likelihood 182
6.6 Fisher Information Matrix 183
6.7 Optimal Plan 189

6.8 Testing of Hypotheses 189
6.9 A Numerical Example 190
6.10 Sensitivity Analysis 192
6.11 Summary 193

7. Product Control and Accelerated Life Testing 195
7.1 Introduction 195
7.2 Product Control Metrics 196
7.3 Types of LSPS (RASPs) 201
7.4 LSPs (RASPs) by Attributes 201
7.4.1 Single Sampling RASP 202
7.4.2 Double Sampling Plan 204
7.4.3 Other LSPs By Attributes 206
7.5 LSPs by Variables 208
7.5.1 Variable Sampling Plans for a Process Parameter 209
7.5.1.1 Single Sampling Plan for a Process Parameter 210
7.5.2 Single Sampling by Variables for Proportion Nonconforming 210
7.5.2.1 $\bar{X}$-method, k-method and M-method 211
7.5.3 Variable Repetitive Group Sampling Plan (VRGSP) for Proportion Nonconforming . 216
7.5.4 Other VRASPs for Proportion Nonconforming 218
7.6 LSPs Based on a Cost Model 218
7.6.1 Types of Warranty Policies 219
7.6.2 Warranty Cost Under a Free Replacement Policy 220
7.6.3 Warranty Cost Under a Pro-Rata Replacement Policy 220
7.6.4 Warranty Cost Under a Combination — Free and Pro-Rata Replacement Policy . . 220
7.6.5 Expected Total Cost for the Sampling Plan 221
7.6.6 Optimal LSP 222

7.7 Optimum Accelerated Life Tests Sampling Plans (ALTSPs) 222
7.8 Sensitivity Analysis 222
7.9 Design of ALTSP and PALTSP Models: A Review 223
7.9.1 Constant-Stress ALTSP Models 223
7.9.2 Time-Varying Stress ALTSP Models 224
7.9.3 RASP With Acceleration Factor 224
7.9.4 Constant-Stress PALTSP Models 224
7.9.5 Time-Varying Stress PALTSP Models . . . 224

8. Optimum Time-Censored Ramp-Stress ALTSPs 225

8.1 Introduction 225
8.1.1 Notation 226
8.2 Assumptions 227
8.3 Log-Logistic Life Distribution Under Ramp-Stress Test 228
8.3.1 Log-Lifetime Distribution 230
8.4 Life Test Procedure 230
8.5 Lot Acceptance Sampling Procedure 230
8.6 Optimum Time-Censored Ramp-Stress ALTSP Under Variable SSP for Proportion Nonconforming 231
8.6.1 Log-Likelihood 231
8.6.2 Fisher Information Matrix 232
8.6.3 Asymptotic Variance of the Test Statistic 233
8.6.4 Operating Characteristic (OC) Function 233
8.6.5 Optimal Plans 234
8.7 Optimum Time-Censored Ramp-Stress ALTSP Under Variable SSP for the Process Parameter Based on Median Lifetime 234
8.7.1 Log-Likelihood 235
8.7.2 Asymptotic Variance of the Test Statistic 236

8.7.3 OC Function 236
8.7.4 Optimal Plan 237
8.8 Cost Structure 237
8.8.1 Cost Structure under Variable SSP for Proportion Nonconforming 238
8.8.2 Cost Structure under Variable SSP for the Process Parameter based on Median Lifetime 239
8.9 Design of Optimal Sampling Plans 239
8.9.1 Design of Optimization Problem for Time-Censored Ramp-Stress ALTSP under Variable SSP for Proportion Nonconforming 239
8.9.2 Design of Optimization Problem for Time-Censored Ramp-Stress ALTSP under Variable SSP for the Process Parameter based on Median Lifetime 241
8.10 Numerical Examples 242
8.10.1 Numerical Example on Time-Censored Ramp-Stress ALTSP under Variable SSP for Proportion Nonconforming 242
8.10.2 Numerical Example on Time-Censored Ramp-Stress under Variable SSP for the Process Parameter based on Median Lifetime 252
8.11 Sensitivity Analyses 253
8.12 Summary 255

9. Optimum Time-Censored Step-Stress PALTSP with Warranty Using Tampered Failure Rate Model 259
9.1 Introduction 259
9.1.1 Notation 260
9.2 The Model 261
9.2.1 Assumptions 261
9.2.2 Burr Type-XII Life Distribution 262

9.2.3 Log-Lifetime Distribution 262
9.2.4 Test Procedure 263
9.2.5 Lot Acceptance Sampling Procedure 263
9.2.6 Life Distribution under TFR Model 264
9.3 Parameter Estimation 265
9.4 Fisher Information Matrix 267
9.5 Asymptotic Variance of Test Statistic 267
9.6 Operating Characteristic (OC) Function 268
9.7 Cost Consideration 269
9.7.1 Expected Warranty Cost Per Item 269
9.7.2 Expected Total Cost Per Lot 269
9.8 Formulation of an Optimization Problem 270
9.8.1 Algorithm 271
9.9 Numerical Example 272
9.10 Sensitivity Analysis 278
9.11 Summary 280

10. Optimum Time-Censored Step-Stress PALTSP with Competing Causes of Failure Using Tampered Failure Rate Model 281

10.1 Introduction 281
10.1.1 Notation 281
10.2 Model 282
10.2.1 Assumptions 282
10.2.2 Test Procedure 282
10.2.3 Model Formulation 282
10.3 Lot Acceptance Procedure 285
10.3.1 SSP 285
10.3.2 VRGSP 286
10.4 Log-Likelihood 286
10.5 Fisher Information Matrix 287
10.6 Operating Characteristic (OC) Function 288
10.7 Formulation of an Optimization Problem 289
10.7.1 Bilevel Programming: VRGSP 290
10.7.2 Bilevel Programming: SSP 291
10.8 Illustrative Example 292

10.9 Sensitivity Analyses 292
10.10 Summary 294

Appendix A **297**

A.0 Truncated Distributions and Order Statistics . . . 297
A.01 Truncated Distributions 297
A.02 Order Statistics 298
A.1 Exponential Distribution as a Limiting Case of the Burr Type-XII Distribution 299
A.2 Weibull Distribution as a Limiting Case of the Burr Type-XII Distribution 300
A.3 Calculations of the Elements of the Fisher Information Matrix in 3.2.4 300
A.4 Calculations of Elements of the Fisher Information Matrix Given in 3.3.1.4 308
A.5 Calculations of the Elements of the Fisher Information Matrix Given in Section 3.3.2.4 315
A.6 Justification for the Use of Linear Cumulative Exposure Model in the Burr Type-XII Distribution in Section 4.2.3 328
A.7 Justification for the Use of Linear Cumulative Exposure Model in the Burr Type-XII Distribution in Section 4.3.3 328
A.8 Calculations of the Elements of the Fisher Information Matrix Given in Section 4.2.6 329
A.9 Calculations of the Elements of the Fisher Information Matrix in Section 4.3.6 331
A.10 Justification for the Use of Linear Cumulative Exposure Model in the Burr Type-XII Distribution in Section 5.2.3 337
A.11 Calculations of the Elements of the Fisher Information Matrix in Section 5.2.5 338
A.12 Justification for the Use of Linear Cumulative Exposure Model in the Burr Type-XII Distribution in Section 5.3.3 349

A.13 Calculation of the Elements of the Fisher Information Matrix Given in Section 5.3.5 350

Appendix B **359**

B.1 Calculations of the Elements of the Fisher Information Matrix Given in Section 8.6.2 359

B.2 Calculations of the Asymptotic Variance of the Test Statistic Given in Section 8.6.3 for Time-Censored Case 361

B.3 Calculations of the Elements of the Fisher Information Matrix Given in Section 9.4 362

B.4 Calculation of the Asymptotic Variance of the Test Statistic Given in Section 9.5 366

B.5 Calculations of the Second Partial Derivatives and Elements of the Fisher Information Matrix Given in Section 10.5 367

B.6 Calculation of the Asymptotic Variance of the Estimated Value of Mean Lifetime, $\hat{\mu}$, Given in Sections 10.3.1 and 10.3.2 371

Appendix C **373**

C.1 Bilevel Programming 373

Appendix D **377**

D.1 Tampered Failure Rate (TFR) Model under Constant-Stress Setup 377

References 379

Index 403

PREFACE

Usually a feeling of disappointment follows the book, because what I hoped to write is not what I actually accomplished. However, it becomes a motivation to write the next book.

Anita Desai

Reliability is a highly quantitative engineering discipline. The high rate of technological advances and innovations are driving manufacturers to develop new higher technology products in record time while improving quality, reliability, and productivity. Much of this need is necessitated by the expanding market place and the resulting increased competitors. Reliability test planning is conducted to facilitate manufacturers formalize design, production, and field reliability requirements of their products. A well designed test plan is important in deciding warranty costs, repair, and maintenance strategies of a new product and helps the manufacturer have an edge over his competitors. This is achieved by applying physics based principles, statistical models and methods, and optimization techniques in reliability.

Many modern products are high-reliability products designed to operate without failure for years. Thus, few units will fail or degrade in a test of practical length at normal use conditions. For example, during the design and construction of a communications satellite, there may be only six months available to test components that are expected to be in service for 15 or 20 years. For this reason, reliability tests under accelerated conditions are conducted in manufacturing industries, particularly to obtain timely information on the reliability of products, components, and materials. Generally, information from tests at high levels of stress (usage rate, temperature, voltage, or pressure) is extrapolated, through a physically reasonable model, to obtain estimates of life or long-term performance at normal level of stress. In some cases, stress is increased or otherwise changed during the course of a test. Accelerated Tests (ATs) results are used in the reliability design process to assess or demonstrate component and subsystem reliability, certify components, detect failure modes, compare different manufacturers, and so forth. ATs have become increasingly important because of rapidly changing technologies, more complicated products with more components, and higher customer expectations for better reliability.

Stress under accelerated condition can be applied using **constant-stress**, **step-stress**, **progressive-stress**, **cyclic-stress**, **random-stress**, or combinations of such loadings. The choice of a stress loading depends on how the product or unit is used in service and other practical and theoretical limitations. A test plan is used to decide on optimal values of such experimental variables as the sample size, stress levels, test units' allocations at each stress levels, stress change points etc. It is therefore of paramount importance to have a test plan that helps in accurately estimating reliability measures at operating conditions without waste of time, effort, and money. **ATs** comprise **Accelerated Life Tests (ALTs) and Accelerated Degradation Tests (ADTs)**. An **ALT** can be **Fully ALT** or **Partially Accelerated Life Test (PALT)**. In fully ALT, all the test units are run at accelerated conditions, while in PALT they are run at both normal and accelerated conditions. In ALT, physical failures are observed during the experiment, i.e., when a test unit stops working it is said to have failed. Such failures are called **hard failures**, i.e.,**catastrophic failures.** However, for some highly reliable products even accelerated life testing yield little failure data of units in a feasible amount of time. In such cases, Accelerated Degradation Testing is carried out. Here, failure is defined in terms of performance characteristic of the product exceeding its critical (threshold) value. For example, decreasing light output from a fluorescent light bulb where in failure may be defined as a performance characteristic exceeding a specified level of degradation such as 60% of initial output. Such a failure is called a **soft failure**, i.e., degradation of product performance to an unacceptable level.

The book *Accelerated Testing: Statistical Models, Test Plans, and Data Analyses* by Wayne Nelson (1990) provides an extensive and comprehensive source for background material, practical methodology, basic theory, and examples for fully accelerated testing of nonrepairable systems. It is not only necessary but imperative to examine if the test specimen is able to weather differing stress conditions with the passage of time. The book by Nelson deals with time-varying fully ALT models with respect to step-stress and ramp-stress set-ups only. This book delves exclusively on different aspects of time-varying fully and partially ALT models for nonrepairable systems not only with step-stress and ramp-stress settings but also with different combinations of stress loadings involving one stress factor. Further, Accelerated life test Sampling Plans (ALTSPs) which integrate Accelerated Life Testing (ALT) with quality control technique of Acceptance Sampling Plans (ASPs) have been dealt with in this work. These plans also known as Reliability Acceptance Sampling Plans (RASPs) not only help producers to promote product quality but also prevent consumers from accepting bad products. An **optimum ALTSP** determines optimum total number of test units to be tested under accelerated condition, the number of test units at each stress level, maximum allowable number of non-conforming items, and other experimental variables such that producer's and consumer's interests are safeguarded.

The area covered requires knowledge of statistics of undergraduate level, and will definitely prove to be beneficial for reliability practitioners, post graduate and research students from physics. statistics, and operational research stream.

Chapter 1 deals with the basic concepts of Accelerated Life Testing. It also includes an exhaustive survey of the literature on the design of ALT Models.

Chapter 2 deals with optimum simple step-stress ALT for complete data set as well as for Type-I censored data set. A log-logistic life distribution with median life that is log-linear function of stress and a cumulative exposure model are assumed in both the cases. The models suggested are appropriate for highly reliable components such as insulation system. The optimum test plans are obtained by minimizing the asymptotic variance of the maximum likelihood estimator of the median life at a design stress for some selected values of parameters. It is shown that as censoring time increases optimal stress change time also increases.

Chapter 3 deals with the optimum design of step-stress PALT for the truncated logistic distribution under Type-I and Type-II censoring. Maximum Likelihood method has been used for estimating the acceleration factor and model parameters. The optimal plans have been obtained by minimizing variance of acceleration factor or by using D-optimality criterion. Confidence intervals involving design parameters have also been obtained.

Chapter 4 deals with the formulation of optimum time-censored multi-objective ramp-stress fully ALT plans with multiple linearly increasing stresses as well as with stress upper bound using the Burr Type-XII life distribution. The inverse power law and a cumulative exposure model are assumed. The Burr Type-XII life distribution model can be widely and effectively used in reliability applications because it contains different forms of hazard function. The optimal plans are obtained by minimizing the weighted sum of the asymptotic variances of maximum likelihood estimators of the quantile lifetimes at a design constant-stress.

Chapter 5 considers the optimum design of time-censored multi-objective fully ALT plans with modified step-stress loading and modified constant-stress loading using the Burr type-XII life distribution. Modified constant-stress is also referred to as ramp-constant stress and modified step-stress as ramp-step stress. The inverse power law and cumulative exposure model have been assumed. The optimal plans have been obtained by minimizing the weighted sum of the asymptotic variances of maximum likelihood estimators of the quantile lifetimes at a design constant-stress.

Many products have more than one cause of failure. ATs that yield a mix of failure modes, require valid ways to extrapolate such a data to estimate the product life at a design stress. It is clearly wrong to use data on one failure mode to estimate the life distribution of another failure mode. This has necessitated formulation of optimum simple step-stress ALT with competing causes of failure, involving pre-specified censoring time. The available literature on step-stress accelerated life testing with competing causes of failure deals with exponential distribution that has constant failure rate which is hardly realized in practice. Therefore, Khamis–Higgins model which is an alternative to the Weibull cumulative exposure model for the step-stress accelerated life testing has been used in **Chapter 6**. This chapter is devoted to the design of optimum simple time-censored step-stress model (i.e., two stress levels) with competing causes of failure. The life distribution of each failure cause, which is independent of the others, follows Weibull distribution with a characteristic life that is a log-linear function of the stress, and a cumulative exposure model is assumed. The sum of the asymptotic variances of the maximum likelihood estimators of the log characteristic lives due to all the failure causes is minimized at design stress; and the optimal stress change time is obtained.

Chapters 7–10 deal with application of Accelerated Life Testing methods in the quality control technique of ASPs. An ASP used to determine the acceptability of a product with respect to its life time is called Life Test Sampling Plan (LSP) or RASP. These plans when combined with Accelerated Life Testing for the purpose of inducing early failures in highly reliable items are referred to as ALTSPs. Chapter 7 highlights basics of LSPs and a brief survey of ALTSPs studied in the literature. Chapter 8 focuses on time-censored ALTSP under ramp-stress loading. Chapter 9 discusses time-censored step-stress PALTSP followed by Chapter 10 explaining formulation of time-censored step-stress PALTSP with independent competing causes of failures.

In each chapter, the method developed has been illustrated using numerical examples. In the absence of real life data; simulated data set has been used. However, this should not be a deterrent; as the reliability engineer just needs to replace the simulated data with the experimental data he has obtained and apply the method.

Inferential problems have also been studied, and sensitivity analysis carried out.

A teacher is always a student. I am grateful to my students — Dr. Ruchi Shukla, Dr. Nidhi Jain, Dr. Neha Mittal, and Dr. Deepmala Sharma with whom I worked as a teacher and as a student making this work possible. We are indebted to the staff of World Scientific Press for their editorial work.

Preeti Wanti Srivastava
University of Delhi, India

ACRONYMS

ALT	Accelerated Life Test
ALTSP	Accelerated Life Test Sampling Plan
ANSI	American National Standard Institute
ASP	Acceptance Sampling Plan
ASQ	American Society for Quality
AsVar	Asymptotic Variance
cdf	Cumulative Distribution Function
MIL-STD	Military Standard
ML	Maximum Likelihood
MLE	Maximum Likelihood Estimate
OC	Operating Characteristic
PALT	Partially Accelerated Life Test
PALTSP	Partially Accelerated Life Test Sampling Plan
PD	Percentage Deviation
pdf	Probability Density Function
RASP	Reliability Acceptance Sampling Plan
SSP	Single Sampling Plan
UMP	Uniformly Most Powerful
VRGSP	Variable Repetitive Group Sampling Plan

LIST OF FIGURES

1.1 Constant-stress loading 7
1.2 Step-stress loading 8
1.3 Progressive-stress loading 8
1.4 (a) Sinusoidal cyclic-stress loading. (b) Triangular cyclic-stress loading. (c) Thermal cyclic-stress loading 9
1.5 Random-stress loading 10
1.6 Modified step-stress loading 10
1.7 Modified constant-stress loading 11
1.8 Step-stress profile and the corresponding life distributions (a) Step-stress test. (b) cdfs of constant-stress testing. (c) cdf for step-stress testing. . . 29

2.1 The cumulative exposure log-logistic probability plot for the simulated data 85

3.1 Truncated logistic distribution probability plot for the simulated data under Type-I censoring 105
3.2 Truncated logistic distribution probability plot for the simulated data under Type-II censoring (case I) 109

5.1 Modified step-stress loading 153
5.2 Modified constant-stress ALT 160

8.1 The log-logistic probability plot at low stress for the simulated data ($\alpha = 1.26$, $\gamma_0 = 1.4$, $\gamma_1 = 0.7$, $s_0 = 20$, $\psi_2 = 10$, $\tau = 20$, $k^* = 3.247$) 244
8.2 The log-logistic probability plot at high stress for the simulated data ($\alpha = 1.26$, $\gamma_0 = 1.4$, $\gamma_1 = 0.7$, $s_0 = 20$, $\psi_2 = 10$, $\tau = 20$, $k^* = 3.247$) 245
8.3 OC curve under time-censored variable SSP for proportion nonconforming ($\alpha = 1.26$, $\gamma_0 = 1.4$, $\gamma_1 = 0.7$, $s_0 = 20$, $\psi_2 = 10$, $\tau = 20$, $k^* = 3.247$) 246
8.4 OC curves for $(p_{\tilde{\alpha}}, 1 - \tilde{\alpha})$ as (0.001, 0.95) and $(p_{\tilde{\beta}}, \tilde{\beta})$ as (0.01, 0.05), (0.05, 0.05), (0.10, 0.05) 248
8.5 OC curves for $(p_{\tilde{\alpha}}, 1 - \tilde{\alpha})$ as (0.01, 0.95) and $(p_{\tilde{\beta}}, \tilde{\beta})$ as (0.05, 0.05), (0.10, 0.05), (0.15, 0.05) 248
8.6 OC curves for $(p_{\tilde{\alpha}}, 1 - \tilde{\alpha})$ as (0.001, 0.95) and $(p_{\tilde{\beta}}, \tilde{\beta})$ as (0.01, 0.10), (0.05, 0.10), (0.10, 0.10) 249
8.7 OC curves for $(p_{\tilde{\alpha}}, 1 - \tilde{\alpha})$ as (0.01, 0.95) and $(p_{\tilde{\beta}}, \tilde{\beta})$ as (0.05, 0.10), (0.10, 0.10), (0.15, 0.10) 249
8.8 OC curves for $(p_{\tilde{\alpha}}, 1 - \tilde{\alpha})$ as (0.001, 0.99) and $(p_{\tilde{\beta}}, \tilde{\beta})$ as (0.01, 0.05), (0.05, 0.05), (0.10, 0.05) 250
8.9 OC curves for $(p_{\tilde{\alpha}}, 1 - \tilde{\alpha})$ as (0.01, 0.99) and $(p_{\tilde{\beta}}, \tilde{\beta})$ as (0.05, 0.05), (0.10, 0.05), (0.15, 0.05) 250
8.10 OC curves for $(p_{\tilde{\alpha}}, 1 - \tilde{\alpha})$ as (0.001, 0.99) and $(p_{\tilde{\beta}}, \tilde{\beta})$ as (0.01, 0.10), (0.05, 0.10), (0.10, 0.10) 251
8.11 OC curves for $(p_{\tilde{\alpha}}, 1 - \tilde{\alpha})$ as (0.01, 0.99) and $(p_{\tilde{\beta}}, \tilde{\beta})$ as (0.05, 0.10), (0.10, 0.10), (0.15, 0.10) 251
8.12 UMP-OC curve under time-censored variable SSP for the Process Parameter based on median lifetime ($\alpha = 3.5$, $\gamma_0 = 5$, $\gamma_1 = 1.2$, $s_0 = 25$, $\psi_2 = 12$, $\tau = 24$, $C_s = 25$, $C_{re} = 120$, $C_w = 250$, $C_\tau = 50$) 255

9.1 Log-lifetime distribution probability plot for the simulated data under step-stress PALT 277
9.2 OC curve with $(p_{\tilde{\alpha}}, 1 - \tilde{\alpha})$ as (0.01, 0.99) and $(p_{\tilde{\beta}}, \tilde{\beta})$ as (0.10, 0.01) 278

10.1 OC curve 293

LIST OF TABLES

2.1 Optimal time changing stress $\tau_1^*(\xi, \lambda_1, \alpha)$ 73
2.2 Optimal time changing stress $\tau_2^*(\xi, \lambda_1, T, n_c)$ 81
2.3 Optimal time changing stress $\tau_2^*(\xi, n_c, \lambda_1, \lambda_2)$ 82
2.4 Simple-step stress with two stress levels simulated data (complete data set) 83
2.5 Tests for parameters in the model $\log(\lambda_i) = \beta_0 + \beta_1 x_i, i = 1, 2$ 84
2.6 Simple time-censored step-stress with two stress levels simulated data 85
2.7 Effects of incorrect pre-estimates of β_0, β_1 87
2.8 Effects of incorrect pre-estimates of β_0, β_1 87

3.1 Simulated failure times ($n = 35$, $\beta = 1.1$, $\mu = 0.5$, $\sigma = 3$, $\eta = 6.8$, $\tau^* = 4.41263$, and $n_c = 9$). 103
3.2 Estimated coverage probabilities (in %) of the approximate 95% confidence interval based on 1,000 simulations of Type-I censored samples of size n with ($\beta = 1.1, \mu = 0.5, \sigma = 3$, and $\eta = 6.8$) 106
3.3 Simulated failure times ($n = 35$, $\beta = 1.1$, $\mu = 0.5$, $\sigma = 3$, $y_{(r)} = 7$, $\tau^* = 4.49458$, and $n_c = 7$) 107
3.4 Estimated coverage probabilities (in %) of the approximate 95% confidence interval based on 1,000 simulations of Type-II censored samples case I of size n with ($\beta = 1.1$, $\mu = 0.5$, $\sigma = 3$, and $y_{(r)} = 7$) 110
3.5 Optimum failure-censored step-stress PALT for $n = 36$ 111

3.6 Simulated failure times ($n = 36$, $\beta = 3.5$, $\mu = 3$, $\sigma = 2$, $q_r = 0.8$, $q_l^* = 0.537308$, and $n_c = 7$) 114
3.7 Sensitivity analysis for step-stress PALT plan with Type-I censoring ($n = 35$, $\eta = 6.8$, $\tau^* = 4.41263$, and $\text{AsVar}^* = 0.388484$) 115
3.8 Sensitivity analysis for step-stress PALT plan with Type-II censoring (case I) ($n = 35$, $y_{(r)} = 7$, $\tau^* = 4.49458$, and $\text{AsVar}^* = 0.381725$) 116
3.9 Sensitivity analysis for step-stress PALT plan with Type-II censoring (case II) ($n = 36$, $q_r = 0.8$, and $q_l^* = 0.537308$) 117
3.10 Comparative study of step-stress PALT models with Type-II censoring (case II) 117

4.1 Simulated failure times ($n = 35$, $\gamma_0 = 4.5$, $\gamma_1 = 6$, $s_0 = 30$, $\eta = 100$, $\beta_4 = 9.9$, $k = 1$, $c = 2$, $b = 3$, and $w_1 = w_2 = w_3 = 1/3$) 137
4.2 Simulated failure times ($n = 35$, $\gamma_0 = 0.5$, $\gamma_1 = 5$, $s_d = 20$, $s_h = 30$, $\eta = 19$, $k = 2$, $c = 1$, $n_c = 1$, $b = 3$, and $w_1 = w_2 = w_3 = 1/3$) 139
4.3 Sensitivity analysis for the multi-level ramp-stress ALT plan without stress upper bound ($n = 35$, $s_0 = 30$, $\eta = 100$, $\beta_4 = 9.9$, $k = 1$, $b = 3$, and $w_1 = w_2 = w_3 = 1/3$) 141
4.4 Sensitivity analysis for the ramp-stress ALT plan with stress upper bound $n = 35$, $s_d = 20$, $s_h = 30$, $\eta = 19$, $b = 3$, $w_1 = w_2 = w_3 = 1/3$, and $k = 2$) 141
4.5 Hypothetical Failure time data set for 2-level multi-objective ramp-stress ALT plan ($n = 35$, $\gamma_0 = 4.5$, $\gamma_1 = 6$, $s_0 = 30$, $\eta = 100$, $\beta_4 = 9.9$, $k = 1$, $c = 2$, $b = 3$, and $w_1 = w_2 = w_3 = 1/3$) 142
4.6 Comparative study of ramp-stress ALT models 142
4.7 Comparative study of ramp-stress model with respect to number of stress levels on the basis of changes in parametric values ($n = 35$, $s_0 = 30$, $\eta = 100$, $\beta_4 = 9.9$, $k = 1$, $b = 3$, and $w_1 = w_2 = w_3 = 1/3$) 143

4.8 Comparative study of ramp-stress model with respect to number of stress levels on the basis of equal and unequal weights ($n = 35$, $\gamma_0 = 4.5$, $\gamma_1 = 6$, $s_0 = 30$, $\eta = 100$, $\beta_4 = 9.9$, $k = 1$, $c = 2$, and $b = 3$) 143
4.9 Comparison of a multi-objective plan, and the single-objective plans with respect to changes in parametric values ($n = 35$, $s_d = 20$, $s_h = 30$, $\eta = 19$, $k = 2$, and $n_c = 1$) 144
4.10 Comparison of a multi- objective plan with equal and unequal weights ($n = 35$, $\gamma_0 = 0.5$, $\gamma_1 = 5$, $s_d = 20$, $s_h = 30$, $\eta = 19$, $k = 2$, $c = 1$, $b = 3$, and $n_c = 1$) 144
4.11 Comparison of a multi-objective plan, and the single-objective plans with respect to changes in stress spread ($n = 35$, $\gamma_0 = 0.5$, $\gamma_1 = 5$, $\eta = 19$, $k = 2$, and $c = 1$) 144

5.1 Optimal modified step-stress ALT plan ($n = 35, \gamma_0 = 4.7$, $\gamma_1 = 5.4$, $s_0 = 30$, $s_2 = 60$, $\eta = 10$, $k = 3$, $c = 1$, $b = 3$, and $w_1 = w_2 = w_3 = 1/3$) 167
5.2 Simulated failure times ($n = 35$, $\beta = 9.8$, $\gamma_0 = 4.7$, $\gamma_1 = 5.4$, $s_0 = 30$, $s_2 = 60$, $\eta = 10$, $k = 3$, $c = 1$, $b = 3$, $w_1 = w_2 = w_3 = 1/3$, and $n_c = 7$) 167
5.3 Optimal modified constant-stress ALT plan ($n = 35$, $\gamma_0 = 3.36$, $\gamma_1 = 1.25$, $s_0 = 30$, $s_2 = 60$, $\eta = 10$, $k = 4$, $c = 5$, $b = 3$, and $w_1 = w_2 = w_3 = 1/3$) 169
5.4 Simulated failure times ($n = 35$, $\beta = 3.8$, $\gamma_0 = 3.36$, $\gamma_1 = 1.25$, $s_0 = 30$, $s_2 = 60$, $\eta = 10$, $k = 4$, $c = 1$, $b = 3$, $w_1 = w_2 = w_3 = 1/3$, $n_{c1} = 4$, and $n_{c2} = 1$) 170
5.5 Sensitivity analysis for modified step-stress ALT plan ($n = 35$, $\beta = 9.8$, $s_0 = 30$, $s_2 = 60$, $\eta = 10$, $b = 3$, $w_1 = w_2 = w_3 = 1/3$, and $k = 3$) 171
5.6 Sensitivity analysis for modified constant-stress ALT plan ($n = 35$, $\beta = 3.8$, $s_0 = 30$, $s_2 = 60$, $\eta = 10$, $b = 3$, $w_1 = w_2 = w_3 = 1/3$, and $k = 4$) 172
5.7 Comparative study of modified step-stress ALT models 172

5.8 Comparison of a multi-objective plan, and the single-objective plans with respect to changes in parametric values ($n = 35$, $\beta = 9.8$, $s_0 = 30$, $s_2 = 60$, $\eta = 10$, and $k = 3$) 173
5.9 Comparison of a multi-objective plan with equal and unequal weights ($n = 35$, $\beta = 9.8$, $\gamma_0 = 14.7$, $\gamma_1 = 15.4$, $s_0 = 30$, $s_2 = 60$, $\eta = 10$, $k = 3$, $c = 1$, and $b = 3$) 173
5.10 Comparative study of modified constant-stress ALT models 174
5.11 Comparison of a multi-objective plan, and the single-objective plans with respect to changes in parametric values ($n = 35$, $\beta = 4$, $s_0 = 30$, $s_2 = 60$, $k = 4$, and $\eta = 10$) 174
5.12 Comparison of a multi-objective plan with equal and unequal weights ($n = 35$, $\beta = 4$, $\gamma_0 = 5.66$, $\gamma_1 = 1.95$, $s_0 = 30$, $s_2 = 60$, $\eta = 10$, $k = 4$, $c = 2.9$, and $b = 3$) . . . 175

6.1 Simple step-stress with competing risks of failures simulated data ($n = 30$, $\delta = 2$, $\theta_{11} = 6.08$, $\theta_{12} = 12.03$, $\theta_{21} = 3.05$, $\theta_{22} = 6.02$, $T = 2.8$, $x_0 = 2$, $x_1 = 3$, $x_2 = 5$, $\tau^* = 2.16$) 191
6.2 Tests for parameters in the model $\log \theta_{ij}^{\frac{1}{\delta}} = \alpha_j + \beta_j x_i$; $i = 0, 1, 2$; $j = 1, 2$ 192
6.3 Effects of incorrect pre-estimates of θ_{11}, θ_{12}, θ_{21}, and θ_{22} $\%\Delta\tau^*$ due to $\pm$ 1% to $\pm$5% change in θ_{11}, θ_{12}, θ_{21}, and θ_{22} 193

7.1 $\bar{X}$-method 211
7.2 k-Method (Given n, k) 211
7.3 Decision criteria for acceptance (sample of size n) 213
7.4 M-method (σ known, given n, M) 214

8.1 Simulated Data: Type-I Censored sample on a simple ramp-test using log-logistic model ($\alpha = 1.26$, $\gamma_0 = 1.4$, $\gamma_1 = 0.7$, $s_0 = 20$, $\psi_2 = 10$, $\tau = 20$, $k^* = 3.247$) 244

8.2 Optimal ALTSP under time-censored variable SSP for proportion nonconforming when $p_{\tilde{\alpha}}$ and $p_{\tilde{\beta}}$ change ($\alpha = 1.26$, $\gamma_0 = 1.4$, $\gamma_1 = 0.7$, $s_0 = 20$, $\psi_2 = 10$, $\tau = 20$) 247
8.3 Optimal sample size and the corresponding acceptability constant for selected values of $\tilde{\alpha}$, $\tilde{\beta}$, μ_2/μ_1, $\alpha = 3.5$, $\gamma_0 = 5$, $\gamma_1 = 1.2$, $s_0 = 25$, $\psi_2 = 12$, $\tau = 24$, $C_s = 25$, $C_{re} = 120$, $C_w = 250$, $C_\tau = 50$) 254
8.4 Optimal ALTSP when design parameter estimates change ($\alpha = 1.26$, $\gamma_0 = 1.4$, $\gamma_1 = 0.7$, $s_0 = 20$, $\psi_2 = 10$, $\tau = 20$, $Z^* = 0.5230$, $k^* = 3.247$, $n^* = 35$) 256

9.1 Simulated data: Type-I censored sample on a simple ramp-test using log-lifetime model ($A = 2.17$, $\alpha = 0.24$, $c = 1.6$, $k = 1.2$, $\eta = 1.9$, $\tau^* = 0.57658$, $n_u = 219$, $n_a = 36$, $n_c = 1$) 274
9.2 Optimal ALTSP ($A = 2.17$, $\alpha = 0.24$, $c = 1.6$, $k = 1.2$, $\eta = 1.9$, $1 - \tilde{\alpha} = 0.95$ and $\tilde{\beta} = 0.10$) when $p_{\tilde{\alpha}}$ and $p_{\tilde{\beta}}$ change 278
9.3 Optimal ALTSP ($A = 2.17, \alpha = 0.24, c = 1.6, k = 1.2$ and $\eta = 1.9, 1 - \tilde{\alpha} = 0.99$ and $\tilde{\beta} = 0.01$) when $p_{\tilde{\alpha}}$ and $p_{\tilde{\beta}}$ change 279
9.4 Optimal ALTSP ($A=2.17$, $\alpha=0.24$, $c=1.6$, $k=1.2$) when censoring time changes 279
9.5 Optimal ALTSP ($A = 2.17$, $\alpha = 0.24$, $c = 1.6$, $k = 1.2$, and $\eta = 1.9$) when design parameter estimates change . 280

10.1 Effects of various values of $(p_1, 1 - \tilde{\alpha})$ and $(p_2, \tilde{\beta})$ in optimal ALT plans ($\theta_1 = 25, \theta_2 = 35, A = 1.1, \delta = 1.5$, and $\eta = 2.5$) 293
10.2 Effects of incorrect pre-estimates of θ_1, θ_2, A ($\%\Delta\tau^*$ due to $\pm1\%$ to $\pm5\%$ change in θ_1, θ_2, A) 294

ABOUT THE AUTHOR

Srivastava, Preeti Wanti, is a Professor in the Department of Operational Research, University of Delhi, India. She has a B.Sc. (Hons.) in Statistics, M.Sc. (Statistics), M.Phil. (Statistics), and Ph.D. (Statistics), respectively obtained in 1981, 1989, 1992, and 2000 from the University of Delhi, India. Her research areas are Reliability Theory, Reliability testing and demonstration, Reliability modelling, etc. She has published several research papers in journals of international repute.

CHAPTER 1

DIFFERENT ASPECTS OF ALT MODELS

If your contribution has been vital there will always be somebody to pick up where you left off, and that will be your claim to immortality.

Walter Gropius

1.1 INTRODUCTION

Reliability of a unit is its ability to perform required operations successfully. **Quantitatively** it is the probability that the unit will perform satisfactorily at least for a specified period of time without a major breakdown. The unit may refer to a manufactured product or its components, manufacturing and assembly systems, mechanical, electrical, and electronic systems. Reliability is related to quality of performance of a unit and signifies dependability. The better the performance, higher the reliability and greater the satisfaction to the manufacturer as well as the consumer. It is an important consideration in planning, design and management of such units.

Modern products have become increasingly more sophisticated in the light of rapid technological advances. Very detailed and complex equipment has been researched, designed, developed, and implemented for space exploration, military applications, and commercial uses. In general, each piece of equipment is composed of numerous elementary components and/or subsystems that work as a unit either to achieve specific objectives or to perform a variety of functions. As a consequence, increasing attention has been focused on evaluation of whether a given device successfully performs its intended function.

Studies of operating data for pieces of equipment are used in these evaluations. The requirements of higher-reliability have increased the need for more up-front testing of materials, components, and systems. This is commensurate with the generally accepted modern quality philosophy of producing high-reliability products by improving the design and manufacturing processes, and moving away from reliance on inspections to achieve high reliability.

Many modern products are high-reliability products designed to operate without failure for years. Thus, few units will fail or degrade in a test at normal use conditions. For example, during the design and construction of a communications satellite, there may be only six months available to test components that are expected to be in service for 15 or 20 years. For this reason, **accelerated tests** (ATs) are used widely in manufacturing industries, particularly to obtain timely information on the reliability of products, components, and materials. Generally, information from tests at high levels of stress (usage rate, temperature, voltage, or pressure) is extrapolated, through a physically reasonable model, to obtain estimates of life or long-term performance at normal level of stress. In some cases, stress is increased or otherwise changed during the course of a test. AT results are used in the design-for-reliability processes to assess or demonstrate component and subsystem reliability, certify components, detect failure modes, compare different manufacturers, and so forth. ATs have become increasingly important because of rapidly changing technologies, more complicated products with more components, and higher customer expectations for better reliability.

ATs comprise

(i) **Accelerated Life Tests (ALTs) and**
(ii) **Accelerated Degradation Tests (ADTs)**.

1.1.1 Difference between ALT and ADT

ALT introduced by Chernoff (1962) and Bessler, Chernoff, and Marshall (1962) is undertaken to shorten the test duration and save cost of testing by subjecting all or some of the test units to a more severe environment (increased or decreased stress levels) than the normal operating environment. The test data so obtained at accelerated conditions are extrapolated by means of an appropriate model to estimate the characteristics of life distribution at design condition. In ALT, physical failures are observed during the experiment, i.e., when a test unit stops working it is said to have failed. For example, when the resistance of a resistor deviates too much from its nominal value, causing the oscillator in an electric circuit to stop oscillating or when an incandescent light bulb burns out. Such failures are called **hard failures**, i.e., **catastrophic failures**.

However, for some highly reliable products even accelerated life testing yield little failure data of units in a feasible amount of time. In such cases, Accelerated Degradation Testing is carried out. In an ADT, a performance characteristic of the product whose degradation over time can be related to reliability, is measured, and the degradation data obtained at higher stress level is extrapolated by means of an appropriate model for performance degradation to a certain normal stress level to obtain reliability information about the product. Here, failure is defined in terms of performance characteristic of the product exceeding its critical (threshold) value. For example, decreasing light output from a fluorescent light bulb where in failure may be defined as performance characteristic exceeding a specified level of degradation such as 60% of initial output. Such a failure is called a **soft failure** (Tseng, Hamada, and Chiao, 1995), i.e., degradation of product performance to an unacceptable level. Other examples in which ADTs can be applied include Light Emitting Diode (LED), mechanical products, chemical processes, and materials where the rate of degradation is usually governed by mechanisms such as fatigue, diffusion, oxidation, and shock. For instance, bearings used in automobiles wear mechanically until the loss of their function; batteries with a limited amount of chemicals to react finally run down; materials in aeronautical products may exhibit life reduction with crack growth; storage life of ink cartridges degrades as the chemical oxidizes over time, etc. Stress factors to conduct ADT on these products can be temperature, voltage, acidity, pressure, vibration, load etc. More than one of these factors may be identified to shorten time in an actual test. See for e.g., Tang, Yang, and Xie (2004).

1.1.2 Fully and Partially Accelerated Tests

An accelerated testing can be performed under fully accelerated environmental condition or under partially accelerated environmental condition. In fully accelerated tests, all the test specimens are run at accelerated conditions while in partially accelerated tests, they are run both at normal and accelerated conditions. The term fully ALT has been coined by Bhattacharyya and Soejoeti (1989), and the term Partially Accelerated Life Test (PALT) by Degroot and Goel (1979).

1.1.3 Inspection Modes in Accelerated Testing

Two inspection modes are prevalent in accelerated testing. One is **continuous inspection** that results in exact failure times. The other is the **periodic inspection** in which test units are inspected for failure at pre-determined points in time. The information obtained from a periodic inspection consists of the number of failures in each inspection period, resulting in "**grouped**" or "**interval**" data.

This work focuses on design of ALT models with time-varying stresses for nonrepairable units under continuous inspection.

1.2 TYPES OF STRESS LOADINGS

Stress under accelerated condition can be applied using **constant-stress**, **step-stress**, **progressive-stress**, **cyclic-stress**, **random-stress**, or combinations of such loadings. The choice of a stress loading depends on how the product or unit is used in service and other practical and theoretical limitations (Nelson, 1980; Elsayed, 2012).

1.2.1 Constant-Stress Loading

The constant-stress scheme comprises multiple sub-samples tested at different but fixed stress levels in as many test chambers until a pre-specified time or upon the occurrence of a fixed number of failures.

1.2.2 Step-Stress Loading

In step-stress loading requiring one test chamber, test units are subject to successively higher levels of stress during some pre-specified amount of time, or until the occurrence of fixed number of failures.

1.2.3 Progressive-Stress Loading

In a progressive-stress loading, the stress level is increased continually either until the censoring time or up to the maximum stress level which is maintained until the censoring time. A ramp-stress test results when stress is increased linearly. In particular, a ramp-test with two linearly increasing stresses is a simple ramp-test.

Prot (1948) was the first to propose a ramp-test for fatigue testing under progressive loading. Starr and Endicott (1961), Endicott, Hatch, and Schmer (1965), and David and Montanari (1992) used the ramp-test on capacitors. Solomon, Klein, and Albert (1976) used the ramp-test on insulations, whereas Cacciari, Montanari, and Barry (1992) and Chan (1990) used the ramp-test on integrated circuits. Wang *et al.* (2008) have applied ramped stress breakdown test to estimate lifetime of tunneling magnetoresistance (TMR) heads for the intrinsic and extrinsic failure mode. Aal (2009) has utilized ramp-stress procedures to support both process development as well as continued improvement in the reliability of tunnel oxides used in EEPROM devices of automotive applications. The progressive life test is recommended for rapid acquisition of data for lot acceptance or rejection.

1.2.4 Cyclic-Stress Loading

In cyclic-stress testing, the stress level is changed according to a fixed pattern. For many products, the frequency and length of a cycle affect the lifetime of a product, so they are included in the model as stress variable.

Common examples of such stress are thermal cycling, triangular cycling, and sinusoidal vibration (Zhu and Elsayed, 2011;

Elsayed, 2013). Triangular cyclic-stress involves applying high and low temperatures alternatively over time. In thermal cycling a rest time (soak time) is provided before changing from high to low temperatures whereas in Triangular cyclic-stress temperature is changed immediately. Thermal cycling is widely used in environmental stress testing and is also referred to as ramp-soak stress loading. Automotive engine components experience this type of stress when the engine is ignited in cold weather or when the vehicle is driven through a flooding road. The engine components have to withstand rapidly increasing temperature in the former situation and a sharp temperature drop in the latter. See also Yang (2007).

1.2.5 Random-Stress Loading

In random-stress testing, the level of stress changes at random and has a probabilistic distribution. This type of test is often used for vibration stresses.

1.2.6 Ramp-Constant Stress Loading

This stress loading scheme is combination of ramp-stress and constant-stress loadings, and is also referred to as **Modified constant-stress** loading. The test comprises two test chambers. In the first test chamber, the stress is increased linearly with the constant rate β (say) till stress level s_1 (say) and maintained at the same level till the termination of experiment. In the second test chamber, the stress is increased linearly at the same constant rate as in the first chamber till stress level s_2 ($s_2 > s_1$) and is retained at this level till the end of experiment. Unlike traditional constant-stress loading, this loading scheme prevents sudden change in stress that may result in undesirable failure mode not observed in normal operating condition. See for example Elsayed (2003), Park and Yum (1998), Gao *et al.* (2008) and Shi (2011).

1.2.7 Ramp-Step Stress Loading

This stress loading scheme is combination of ramp-stress and step-stress loadings, and is also referred to as **Modified step-stress** loading. The test consists of one test chamber. The stress is first increased linearly at constant rate β (say) till stress level s_1 (say), maintained at the same level for some time after which it is again increased linearly at the same constant rate, β, till stress level s_2 ($>s_1$) and maintained at the same level till the end of experimentation. Such a gradual change in stress prevents occrrence of undesirable failure mode not encountered under normal operating condition. See for example Elsayed (2003), Park and Yum (1998), Gao *et al.* (2008) and Shi (2011).

Figures 1.1–1.7 depict different forms of stress loadings.

1.3 TYPES OF STRESSES

The type of applied stress depends on the intended operating conditions of the product and the potential cause of failure. See for example, Nelson (1990) and Elsayed (2003).

1.3.1 Mechanical Stresses

The most commonly used acceleration test for mechanical components is fatigue stress. When components are subject to elevated temperature then creep testing (which combines both temperature

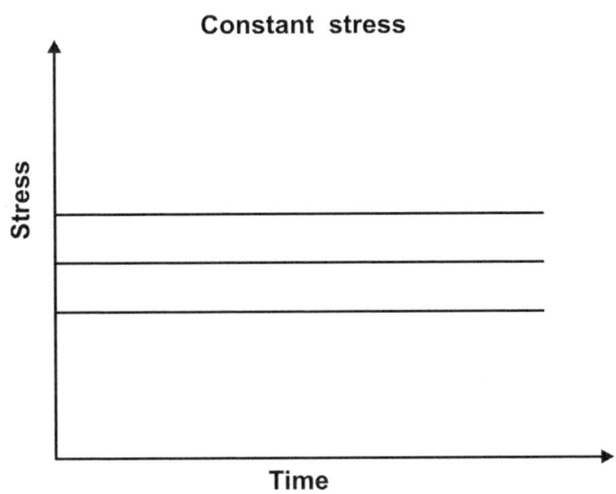

Figure 1.1: Constant-stress loading

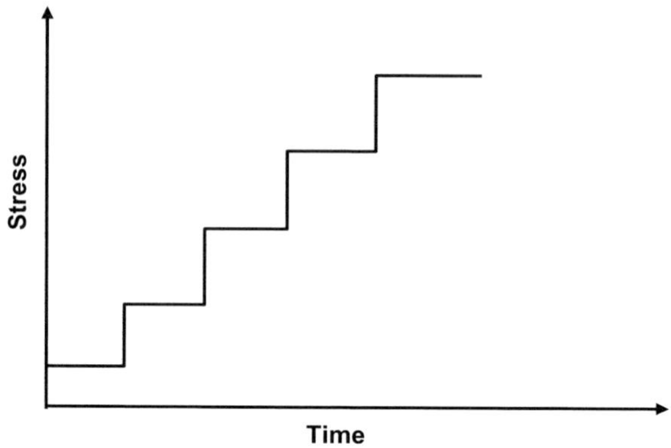

Figure 1.2: Step-stress loading

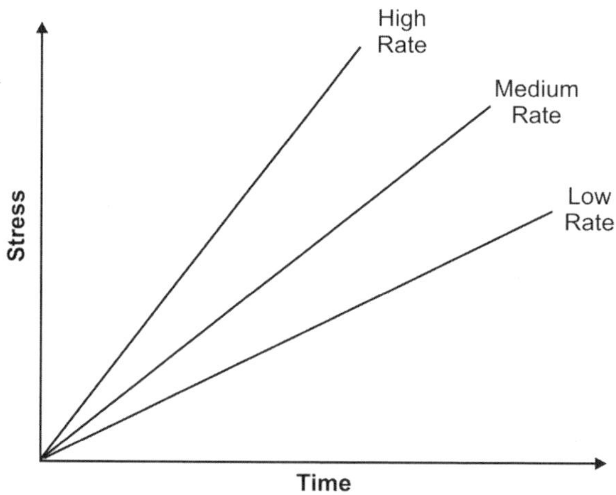

Figure 1.3: Progressive-stress loading

and load and leads to plastic deformation of materials) is applied. Shock and vibration testing is suitable for bearings, shock absorbers, tires, and circuit boards in airplanes and automobiles.

1.3.2 Electrical Stresses

These include power cycling resulting in thermal fatigue, electric field, current density, and electromigration.

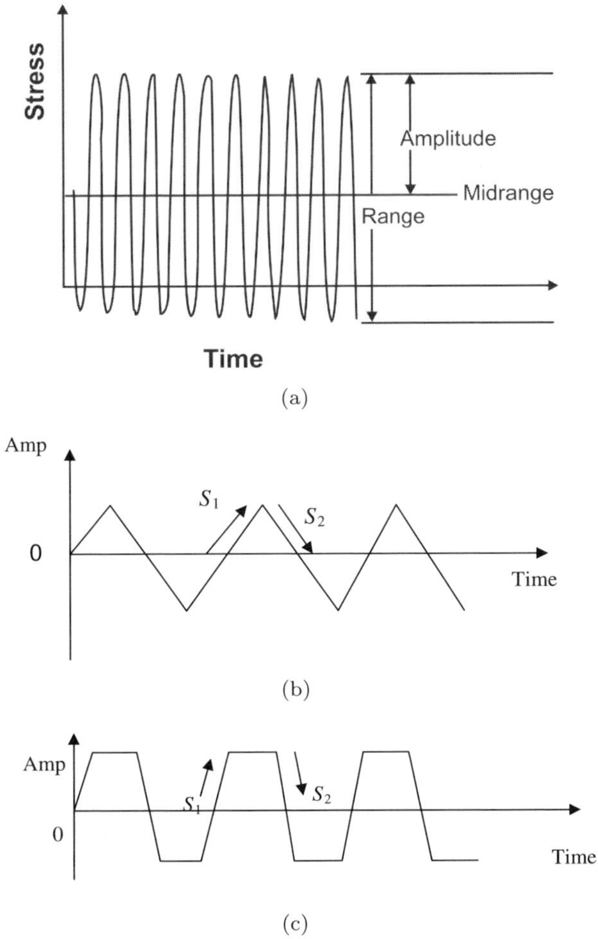

Figure 1.4: (a) Sinusoidal cyclic-stress loading. (b) Triangular cyclic-stress loading. (c) Thermal cyclic-stress loading

1.3.3 Environmental Stresses

Temperature and thermal cycling are commonly used environmental stresses for most products. It is important to use appropriate stress levels that do not induce different failure mechanisms than those under normal conditions. Other environmental stresses include humidity and corrosive environment, ultraviolet light, sulphur dioxide causing corrosion in circuit boards, salt and fine particles, and alpha and gamma rays.

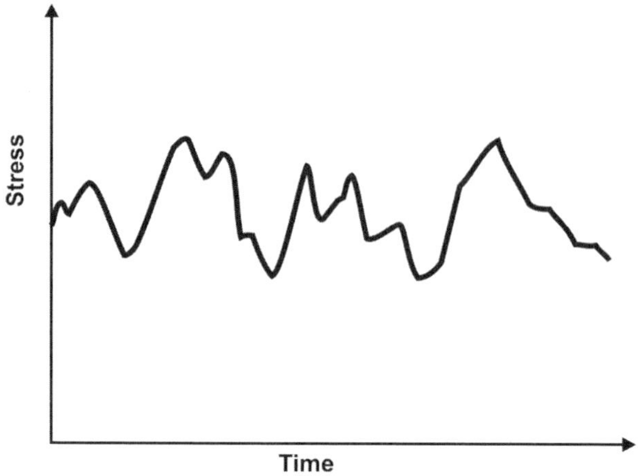

Figure 1.5: Random-stress loading

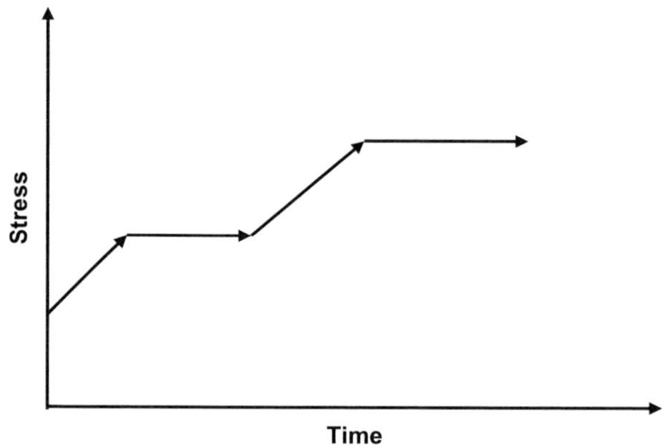

Figure 1.6: Modified step-stress loading

1.3.4 Usage Rate Acceleration

This is appropriate for products that do not operate continuously under normal conditions, so the only alternative to accelerate the life is to run the product more — at a higher usage rate. There are two ways of doing such compressed time testing, namely, faster and reduced off time. In the former, the analyst operates the product faster than at normal conditions. For example, rolling bearings run

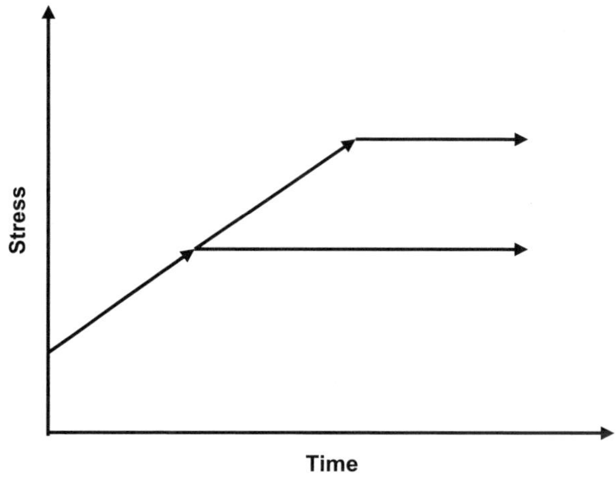

Figure 1.7: Modified constant-stress loading

at about three times their normal speed. The latter is done on products that are off most of the time in actual use. Such products can be accelerated by running them at a greater fraction of time. For example, in most homes, a major appliance (say, washer or dryer) runs an hour or two a day; on test it runs 24 hours a day. Usage rate may be measured by cycles per hour, revolutions per minute, miles per hour, hertz etc.

1.4 A PRODUCT'S LIFE DISTRIBUTION

A life distribution represents scatter in product's life. The use of a correct life distribution model, especially in the presence of a limited data — as typically occurs with modern devices having high reliability, helps in preventing the choice of unnecessary and expensive planned replacements. Significant errors in the extrapolation of accelerated life testing results could occur if the assumed model does not provide a good approximation of the actual failure mechanism.

1.4.1 Reliability Metrics

Let T be a nonnegative random variable representing the lifetime of an item with life distribution $F(t)$, satisfying $F(0) = 0$.

Definition 1.1. The **life distribution** denoted by $F(t)$ is the probability that the item fails before time 't', and is given by

$$F(t) = P[T \leq t], \quad F(0) = 0. \tag{1.1}$$

Definition 1.2. The **reliability** of an item is defined as the probability that it will perform satisfactorily for a specified period of time without a major failure, and is given by

$$\bar{F}(t) = P[T > t] = 1 - F(t). \tag{1.2}$$

It is also known as survival function, survival distribution, and survival probability.

Definition 1.3. Let, $f(\cdot)$ be the probability density function (pdf) of $F(\cdot)$, then the **failure rate** function denoted by $h(t)$ is defined as

$$h(t) = \frac{f(t)}{\bar{F}(t)}, \tag{1.3}$$

for 't' such that $\bar{F}(t) > 0$.

$h(t)$ is a measure of the rate at which failures in the total population occur at time 't'. It is also known as **conditional failure rate function**, or the **hazard rate function**.

It can be easily seen from (1.3) that

$$\bar{F}(t) = \exp\left[-\int_0^t h(u)du\right], \quad 0 \leq t < \infty, \tag{1.4}$$

which is an alternative form of reliability and signifies that one can find $\bar{F}(t)$ from $h(t)$.

Definition 1.4. The **100pth percentile** (or **pth quantile**) of a distribution $F(\cdot)$ is the age ξ_p by which a proportion 'p' of the population fails. It is the solution of

$$p = F(\xi_p). \tag{1.5}$$

It is also called the "100p percent point" or "p fractile". The 50% point is called the **median**, and is usually used as a "typical life". The **characteristic life** is the time at which 63.2% of the units will fail.

Definition 1.5. The **mean lifetime** of a unit is defined as

$$E(X) = \int_0^\infty \bar{F}(x)dx. \tag{1.6}$$

1.4.2 Life Distribution Models

Some of the life distribution models depicted by a product subject to accelerated conditions are briefly described below (Nelson, 1990; Viertl, 1988).

1.4.2.1 *Exponential Distribution*

The population fraction failing by age t is

$$F(t) = 1 - e^{-t/\theta}, \quad t \geq 0, \tag{1.7}$$

where $\theta > 0$ is the mean time to failure, the pdf is

$$f(t) = (1/\theta)e^{-t/\theta}, \quad t \geq 0. \tag{1.8}$$

The 100pth percentile is

$$\xi_p = -\theta \log(1-p), \tag{1.9}$$

where the base of log is e.

The failure rate is

$$h(t) = (1/\theta) = \lambda. \tag{1.10}$$

Only the exponential distribution has a constant failure rate, a key characteristic. In other words, for this continuous distribution only, an old unit and a new unit have the same chance of failing over a future time interval.

1.4.2.2 *Normal Distribution*

The population fraction failing by age t is

$$F(t) = \int_{-\infty}^{t} (2\pi\sigma^2)^{-1/2} \exp\left[-\frac{1}{2}\left(\frac{x-\mu}{\sigma}\right)^2\right] dx, \quad -\infty < t < \infty, \tag{1.11}$$

where μ is the population mean and may have any value, σ is the population standard deviation and must be positive. The pdf is

$$f(t) = (2\pi\sigma^2)^{-1/2} \exp\left[-\frac{1}{2}\left(\frac{t-\mu}{\sigma}\right)^2\right], \quad -\infty < t < \infty. \tag{1.12}$$

The pdf is symmetric about the mean (median) μ, and σ determines the spread.

The range of t is from $-\infty$ to $+\infty$. Life must, of course, be positive. Thus, the distribution fraction below zero must be small for this distribution to be a satisfactory approximation in practice.

The 100pth percentile is

$$\xi_p = \mu + z_p\sigma, \tag{1.13}$$

where z_p is the 100pth standard normal percentile and the median (50th percentile) of the normal distribution is $\xi_P = \mu$, since $z_{0.50} = 0$. The failure rate function of normal distribution increases without limit. Thus, it may describe products with wear out failure.

1.4.2.3 *Lognormal Distribution*

The population fraction failing by age t is

$$F(t) = \Phi\{(\log(t) - \mu)/\sigma\}, \quad t > 0, \sigma > 0, \tag{1.14}$$

where μ is called the log mean and may have any value from $-\infty$ to ∞, σ is called the log standard deviation and must be positive, $\Phi(\cdot)$ denotes the standardized normal distribution function, and the base of log is e. The pdf is

$$f(t) = \exp\{-[\log(t) - \mu]^2/2\sigma^2\}/[t(2\pi)^{1/2}\sigma], \quad t > 0. \tag{1.15}$$

The 100pth percentile is

$$\xi_p = e^{\mu + z_p\sigma}, \tag{1.16}$$

where z_p is the 100pth standard normal percentile, and the median (50th percentile) of the distribution is $\xi_{0.50} = \exp(\mu)$ since $z_{0.50} = 0$.

The failure rate function is given as

$$h(t) = \frac{(1/t(2\pi)^{1/2}\sigma)\exp\{-(\log at)^2/2\sigma^2\}}{1 - \Phi\{\log(at/\sigma)\}}, \tag{1.17}$$

where $a = e^{-\mu}$.

The failure rate function of lognormal distribution has a property seldom seen in products. It is zero at time zero, increases to a maximum, and then decreases to zero with increasing age. Nonetheless, over most of its range and especially over the lower tail, the lognormal distribution fits life data of many products. Often one uses only the lower tail in applications.

1.4.2.4 *Weibull Distribution*

The population fraction failing by age t is

$$F(t) = 1 - \exp[-(t/\alpha)^\beta], \quad t > 0, \tag{1.18}$$

where β is the shape parameter and α is the scale parameter, and are both positive. The pdf is

$$f(t) = (\beta/\alpha^\beta)t^{\beta-1}\exp[-(t/\alpha)^\beta], \quad t > 0. \tag{1.19}$$

The 100pth percentile is

$$\xi_P = \alpha[-\log(1-p)]^{1/\beta}, \tag{1.20}$$

where the base of log is e. The 63.2th percentile of the distribution is always equal to the characteristic life of the distribution, i.e., $\xi_{0.632} \approx \alpha$, and the failure rate function is given by

$$h(t) = (\beta/\alpha^\beta)t^{\beta-1}. \tag{1.21}$$

With a simply increasing or decreasing failure rate, the Weibull distribution flexibly describes failure rate.

1.4.2.5 *Extreme Value Distribution*

The extreme value distribution is used to analyze Weibull data. The natural log of time to failure for a Weibull distribution has an extreme value distribution.

The population fraction failing by age t is

$$F(t) = 1 - \exp[-\exp\{(t-\zeta)/\delta\}], \quad -\infty < t < \infty, \tag{1.22}$$

where ζ is the location parameter and may have value from $-\infty$ to ∞ and δ is the scale parameter and is positive. The pdf is

$$f(t) = \left(\frac{1}{\delta}\right)\exp\left\{\frac{(t-\zeta)}{\delta}\right\}\exp\left[-\exp\left\{\frac{(t-\zeta)}{\delta}\right\}\right], \quad -\infty < t < \infty. \tag{1.23}$$

The 100pth percentile is

$$\xi_p = \zeta + \delta\log(-\log(1-p)). \tag{1.24}$$

The 63.2th percentile of the distribution is equal to the location parameter of the distribution, i.e., $\xi_{0.632} \approx \zeta$.

The failure rate function is

$$h(t) = (1/\delta)\exp[-\exp\{(t-\zeta)/\delta\}]. \tag{1.25}$$

1.4.2.6 *Logistic Distribution*

The Logistic distribution has a shape very close to that of a normal distribution but has heavier tail than the normal.

The population fraction failing by age t is

$$F(t) = \frac{1}{[1+e^{-(t-\mu)/\sigma}]}, \quad -\infty < t < \infty, \tag{1.26}$$

and pdf is

$$f(t) = \frac{1}{\sigma}\frac{e^{-(t-\mu)/\sigma}}{[1+e^{-(t-\mu)/\sigma}]^2}, \quad -\infty < t < \infty, \tag{1.27}$$

where μ is the location parameter and may have value from $-\infty$ to ∞ and σ is the scale parameter which is positive.

The range of t is from $-\infty$ to $+\infty$. Thus, the distribution fraction below zero must be small for this distribution to be a satisfactory approximation in practice as the lifetime of a unit cannot be negative.

The 100pth percentile is

$$\xi_p = \mu - \sigma \log[(1-p)/p]. \tag{1.28}$$

The median (50th percentile) of the logistic distribution is

$$\xi_{0.50} = \mu.$$

The failure rate function is given by

$$h(t) = \frac{1}{\sigma[1 + e^{-(t-\mu)/\sigma}]}. \tag{1.29}$$

Unlike normal distribution which has failure rate function increasing without limit, i.e., increasing failure rate, the $h(t)$ for logistic distribution lies between 0 and $1/\sigma$.

1.4.2.7 *Log-Logistic Distribution*

The log-logistic life distribution has been found appropriate for high-reliability components.

The population fraction failing by age t is

$$F(t) = \frac{\left(\frac{t}{\lambda}\right)^{\alpha}}{1 + \left(\frac{t}{\lambda}\right)^{\alpha}}, \quad t > 0 \tag{1.30}$$

and the pdf is

$$f(t) = \frac{\frac{\alpha}{\lambda}\left(\frac{t}{\lambda}\right)^{\alpha-1}}{\left(1 + \left(\frac{t}{\lambda}\right)^{\alpha}\right)^2}, \quad t > 0, \tag{1.31}$$

where λ is the scale parameter and α is the shape parameter and are positive.

The 100pth percentile is

$$\xi_P = \lambda[p/(1-p)]^{1/\alpha}. \tag{1.32}$$

The median (50th percentile) of the logistic distribution is $\xi_{0.50} = \lambda$.

The failure rate function is given by

$$h(t) = \frac{\frac{\alpha}{\lambda}\left(\frac{t}{\lambda}\right)^{\alpha-1}}{1 + \left(\frac{t}{\lambda}\right)^{\alpha}}, \tag{1.33}$$

which is always decreasing with time if $\alpha \leq 1$; first increasing, then decreasing with time if $\alpha > 1$. The latter property has been observed for insulating material (Chiodo and Mazzanti, 2004, 2006). General properties of the log-logistic distribution are illustrated with some details in Johnson, Kotz, and Balakrishnan (1995) and Cox and Oakes (1984) for what concerns survival analysis.

1.4.2.8 *Truncated Logistic Distribution*

The logistic distribution described in 1.4.2.6 can be used for modeling lifetime data only when the distribution fraction below zero is small, else it can result in modeling negative times-to-failure. This has necessitated the use of the truncated logistic distribution, truncated at point zero for modeling lifetime data. The failure rate of truncated logistic distribution is increasing and more realistically bounded below and above by nonzero finite quantity.

The cumulative distribution function (cdf), pdf, reliability function, and hazard function, respectively, of the truncated logistic distribution, respectively, are

$$F(t) = \frac{1}{A}\left(\frac{1}{1+\exp\{-\frac{t-\mu}{\sigma}\}} - \frac{1}{1+\exp\{\frac{\mu}{\sigma}\}}\right), \quad t, \sigma > 0, -\infty < \mu < \infty \tag{1.34}$$

(Mood, Graybill, and Boes, 1974)

$$f(t) = \frac{\exp\{-\frac{t-\mu}{\sigma}\}}{A\sigma\left(1+\exp\{-\frac{t-\mu}{\sigma}\}\right)^2}, \quad t, \sigma > 0, -\infty < \mu < \infty, \tag{1.35}$$

$$\bar{F}(t) = \frac{\exp\{-\frac{t-\mu}{\sigma}\}}{A\left(1+\exp\{-\frac{t-\mu}{\sigma}\}\right)}, \quad t, \sigma > 0, -\infty < \mu < \infty \tag{1.36}$$

and

$$h(t) = \frac{1}{\sigma\left(1+\exp\{-\frac{t-\mu}{\sigma}\}\right)}, \quad t, \sigma > 0, -\infty < \mu < \infty, \tag{1.37}$$

where $A = \frac{1}{1+\exp\{-\frac{\mu}{\sigma}\}}$ and μ and σ are the location and scale parameters, respectively.

The hazard function (1.37) is an increasing function of t, and is bounded by

$$\frac{1}{\sigma\left(1+\exp\{\frac{\mu}{\sigma}\}\right)}, \quad \text{and} \quad \frac{1}{\sigma}.$$

The 100pth percentile is

$$\xi_p = \mu - \sigma \ln\left(\left(Ap + \frac{1}{1+\exp\{\frac{\mu}{\sigma}\}}\right)^{-1} - 1\right), \tag{1.38}$$

where the 'ln' has natural base. See Appendix A.01 also.

1.4.2.9 *Burr Type-XII Distribution*

The Burr Type-XII distribution has a nonmonotone hazard function which can accommodate many shapes of hazard function.

The pdf and cdf, respectively, of the Burr Type-XII distribution, are

$$g(t) = \left(\frac{kc}{\alpha}\right)\left(\frac{t}{\alpha}\right)^{c-1}\left(1+\left(\frac{t}{\alpha}\right)^{c}\right)^{-(k+1)}, \quad t \geq 0, \tag{1.39}$$

$$G(t) = 1 - \left(1+\left(\frac{t}{\alpha}\right)^{c}\right)^{-k}, \quad t \geq 0. \tag{1.40}$$

The Burr Type-XII distribution is uni-modal, and its mode is $T_{\text{mode}} = \alpha\left(\frac{c-1}{ck+1}\right)^{\frac{1}{c}}$ if $c > 1$. The pdf is L-shaped if $c \leq 1$.

The reliability function and hazard function, respectively, are

$$\bar{F}(t) = \left(1+\left(\frac{t}{\alpha}\right)^{c}\right)^{-k}, \quad t \geq 0, \tag{1.41}$$

$$h(t) = \left(\frac{kc}{\alpha}\right)\left(\frac{t}{\alpha}\right)^{c-1}\left(1+\left(\frac{t}{\alpha}\right)^{c}\right)^{-1}, \quad t \geq 0. \tag{1.42}$$

The exponential life distribution follows from the Burr Type-XII distribution as $k \to \infty$, such that $\alpha = k^{1/c}$, and $c = 1$, and the Weibull life distribution follows from the Burr Type-XII distribution as $k \to \infty$, such that $\alpha = k^{1/c}$ (see (A.1) and (A.2)). The log-logistic distribution is a particular case of this distribution, as for $k = 1$, the distribution reduces to the log-logistic distribution.

Tadikamalla (1980), Lewis (1981), Soliman (2005), and Abd-Elfattah, Hassan, and Nassr (2008) have discussed the statistical and probabilistic properties of the Burr Type-XII distribution, and its relationship to other distributions used in reliability analysis.

This distribution has been found appropriate for life testing experiments under accelerated conditions. As an example of a product lifetime distribution in which the Burr distribution is more appropriate, Nelson (1982, p. 105) has given data on the time to breakdown of an insulating fluid between electrodes at a voltage of 34 KV (minutes). Zimmer, Keats, and Wang (1998) have shown that the Burr Type-XII model is acceptable for these data. Cook and Jhonson (1986) have used the Burr model to obtain better fits to a uranium survey data set.

1.5 FULLY ALT MODELS

In fully accelerated life testing, all the test units are run at accelerated conditions. The term fully ALT has been coined by Bhattacharyya and Soejoeti (1989). The fully ALT model is widely referred to as an ALT model in the literature, and therefore the two terms have been used interchangeably in this work. A fully ALT model or simply an ALT model consists of

(a) a life distribution that represents the scatter in product life, and
(b) relationship between "life" and "stress".

After having specified an appropriate life distribution (see Section 1.4.2) a model that describes a life characteristic of the distribution is formulated. The life characteristic can be any reliability measure such as mean life, quantile life etc., and is expressed as a function of stress. Some of the common life–stress relationships describing life characteristic as a function of stress are described in this section.

1.5.1 Life–Stress Relationship

1.5.1.1 *Arrhenius Relationship*

Arrhenius model is used when the damaging mechanism is temperature sensitive (especially for integrating circuits, LEDs, dielectrics, semiconductors, battery cells, insulating tapes, lubricant, grease, plastic, incandescent bulb filaments). The Arrhenius equation relates the rate of a chemical reaction $\mathcal{R}$ to temperature T. This relationship can be written as

$$\mathcal{R} = A' \exp\left(-\frac{E}{kT}\right), \tag{1.43}$$

where E is the activation energy of the reaction, usually in electron-volts, k is Boltzmann's constant, 8.6171×10^{-5} electron-volts/°C, T is the absolute Kelvin temperature, and A' is a constant that is characteristic of the product failure mechanism and test conditions.

The Arrhenius life relationship is based on the idea that failure occurs after there has been a critical amount $\mathcal{C}$ of chemical reaction. The product is assumed to fail when some critical amount of the chemical has reacted, i.e.,

$$\mathcal{C} = \mathcal{R}(\text{time to failure}).$$

Equivalently,

$$(\text{time to failure}) = \mathcal{C}/\mathcal{R}.$$

This suggests that nominal time, ψ, to failure is inversely proportional to the rate (1.43). This yields the **Arrhenius life relationship**

$$\psi = A \exp\left(\frac{E}{kT}\right), \tag{1.44}$$

where A is a constant that depends on product geometry, specimen size, fabrication, test method, and other factors. ψ represents a quantifiable life measure, such as mean life, characteristic life, median life, etc.

The natural log of (1.44) is the linear function of the inverse of T, i.e.,

$$\log(\psi(T)) = \gamma_0 + \gamma_1/T, \tag{1.45}$$

where γ_0 and γ_1 are characteristic of the product and test method; they are estimated from data. Thus, the log of "nominal life" is a linear function of inverse absolute temperature.

1.5.1.2 *Inverse Power Law Relationship*

It is used for analyzing data for which the accelerated stress is nonthermal in nature. It is frequently used as an accelerating stress for products such as capacitors, transformers, and insulators.

Suppose that the accelerating stress variable V is positive. The inverse power law relationship between "nominal life" ψ of a product and V is

$$\psi(V) = A/V^{\gamma_1}, \tag{1.46}$$

where A and γ_1 are parameters characteristic of the product, specimen geometry and fabrication, the test method, etc.

Equivalent forms are

$$\psi(V) = (A'/V)^{\gamma_1} \quad \text{and} \quad \psi(V) = A''(V_0/V)^{\gamma_1},$$

where V_0 is a specified (standard) level of stress.

The natural log of (1.46) is

$$\log[\psi(V)] = \gamma_0 + \gamma_1[-\log(V)]. \tag{1.47}$$

Thus, the natural log of "typical life", $\log(\psi)$, is a linear function of the transformed stress

$$x = -\log(V) \quad \text{with} \quad \gamma_0 = \log(A).$$

1.5.1.3 *Eyring Relationship*

It was formulated from quantum mechanics principles and is most often used when thermal stress (temperature) is the acceleration variable. However, the Eyring relationship is also often used for stress factors other than temperature, such as humidity.

The Eyring relationship for "nominal" life ψ as a function of absolute temperature T is

$$\psi(T) = (A/T)\exp\left(\frac{B}{kT}\right), \tag{1.48}$$

where A and B are constants of the product and test method, and k is Boltzmann's constant. For the small range of absolute temperature in most applications, (A/T) is essentially constant, and (1.48) is close to the Arrhenius relationship (1.44).

1.5.1.4 *Temperature–Humidity Relationship*

It is a two-stress relationship and is used for predicting the life at use conditions when temperature and humidity are the accelerated stresses in a test.

The model is given by

$$\psi(Y,T) = A\exp\left(\frac{\phi}{T} + \frac{v}{Y}\right), \tag{1.49}$$

where A, ϕ, and ν are constants of the product and test method, T is temperature (in absolute units) and Y is the relative humidity.

1.5.1.5 *Temperature–Nonthermal Relationship*

It is another two-stress model. It is used for analyzing the data for which the temperature and nonthermal are accelerated stresses.

The model is given by

$$\psi(Y,T) = \frac{C}{Y^n e^{B/T}}, \tag{1.50}$$

where B, C, and n are constants of the product and test method, T is temperature (in absolute units), and Y is the nonthermal stress, for example, voltage.

1.5.1.6 *Multivariable Relationship*

In most practical applications however, life is a function of more than one or two stress variables, say temperature and voltage. In addition, there are many applications where the life of a product as a function of stress and of some engineering variable other than stress is sought. A multivariable relationship is therefore needed in order to analyze such data.

Such a relationship is the general log-linear relationship, which describes a life characteristic as a function of 'n' stresses. Mathematically, the model is given by,

$$\ln \psi(x_1, x_2, \ldots, x_n) = \exp\left(\beta_0 + \sum_{i=1}^{n} \beta_i x_i\right), \tag{1.51}$$

where β_0 and β_i, $i = 1, 2, \ldots, n$, are unknown parameters, and $x_1, x_2, \ldots, x_n$ are stress variables.

Remark 1.1. A great variety of other stress relationships, for example, exponential relationship, exponential–power relationship, quadratic and polynomial relationship, and relationships involving more than one accelerating variables exist in the literature (Nelson, 1990; Yang, 2007).

Remark 1.2. Similarly Arrhenius-exponential, Arrhenius-Lognormal, Power-exponential, Power-Lognormal, and other models with different life distributions and stress–life relationships have been developed (Nelson, 1990; Yang, 2007).

1.6 CONSTANT-STRESS ALT MODELS

When the stress is constant, the stress level applied to a sample of units does not vary with time. Each unit is tested under the same accelerated stress, say, temperature for a fixed duration of time. For example, 10 units are tested at 310 K for 100 hours, 10 different units are tested at 320 K for 100 hours and 10 different units are tested at 330 K for 100 hours.

Example 1.1 (Arrhenius–Weibull Model). At absolute temperature T, using (1.18) and (1.45), the population fraction failing by age t is

$$\begin{aligned} F(t,T) &= 1 - \exp\{-[t/\alpha(T)]^{\beta}\} \\ &= 1 - \exp\{-[t\exp[-\gamma_0 - \gamma_1/T]]^{\beta}\}, \end{aligned} \tag{1.52}$$

where the shape parameter β is a constant independent of temperature.

Example 1.2 (Power–Weibull Model). At stress level V, using (1.18) and (1.46) with $A = e^{\gamma_0}$, the population fraction failing by age t is

$$\begin{aligned} F(t,V) &= 1 - \exp\{-[t/\alpha(V)]^{\beta}\} \\ &= 1 - \exp\{-[te^{-\gamma_0}V^{\gamma_1}]^{\beta}\}. \end{aligned} \tag{1.53}$$

1.7 PROPORTIONAL HAZARDS BASED ALT MODELS

The Cox (or proportional hazards) model introduced by Cox (1972) can be used as an accelerated life testing model (Nelson, 1990). A brief description of the model is as follows:

Let $x_1, \ldots, x_J$ denote the (possibly transformed and centred) stress variables, and let $h_0(t)$ denote the hazard function of the unknown life distribution at $x_1 = 0$, $x_2 = 0, \ldots, x_J = 0$.

The Cox model for the hazard function of the distribution at variable values $x_1, \ldots, x_J$ is

$$h(t; x_1, \ldots, x_J) = h_0(t)\exp(\gamma_1 x_1 + \gamma_2 x_2 + \cdots + \gamma_J x_J). \tag{1.54}$$

The hazard function $h_0(t)$ and the coefficients $\gamma_1, \gamma_2, \ldots, \gamma_J$ are estimated from the data.

The corresponding reliability function and pdf, respectively are:

$$\bar{F}(t; x_1, \ldots, x_J) = \bar{F}_0(t)\exp(\gamma_1 x_1 + \gamma_2 x_2 + \cdots + \gamma_J x_J) \tag{1.55}$$

and

$$f(t; x_1, \ldots, x_J) = h(t; x_1, \ldots, x_J)\bar{F}(t; x_1, \ldots, x_J). \qquad (1.56)$$

The reliability function at $x_1 = 0$, $x_2 = 0, \ldots, x_J = 0$ is:

$$\bar{F}_0(t) = \exp\left(-\int_0^t h_0(u)du\right). \qquad (1.57)$$

1.8 PROPORTIONAL ODDS (PO) BASED ALT MODELS

The assumptions of proportional hazards model are not always valid for all failure data. Brass (1971) has observed that the ratio of the death rates, or hazard rates, of two populations under different stress levels (for example, one population for smokers and the other for nonsmokers) is not constant with age, or time, but follows a more complicated course, in particular moving closer to unity for older people (see Zhang, 2007 also). Brass (1971) has proposed a more realistic model based on PO which is described briefly in this section.

Define the odds function $\theta(t)$ as the odds of failure of a unit at time t, i.e.,

$$\theta(t) = \frac{F(t)}{(1 - F(t))}. \qquad (1.58)$$

The odds function $\theta(t)$ is related to the hazard rate $h(t)$ at time t by

$$h(t) = \frac{\theta'(t)}{\theta(t) + 1}, \qquad (1.59)$$

where $\theta'(t)$ is the derivative of $\theta(t)$ with respect to t,

$$\text{or } \theta(t) = \exp\left(\int_0^t h(u)du\right) - 1. \qquad (1.60)$$

Then, the basic PO model is given by:

$$\theta(t; x_1, \ldots, x_J) = \theta_0(t)\exp(\gamma_1 x_1 + \gamma_2 x_2 + \cdots + \gamma_J x_J), \qquad (1.61)$$

or equivalently by

$$\frac{F(t)}{(1-F(t))} = \exp(\gamma_1 x_1 + \gamma_2 x_2 + \cdots + \gamma_J x_J)\frac{F_0(t)}{(1-F_0(t))}, \tag{1.62}$$

where $\theta_0(t)$ is a baseline odds function and $F_0(t)$ is the baseline cdf. The baseline odds function can be any monotone increasing function of time t with the property of $\theta_0(0) = 0$.

This model can be represented in terms of hazard rate:

$$h(t; x_1, \ldots, x_J) = \frac{\exp(\gamma_1 x_1 + \gamma_2 x_2 + \cdots + \gamma_J x_J)h_0(t)}{1-(1-\exp(\gamma_1 x_1 + \gamma_2 x_2 + \cdots + \gamma_J x_J))F_0(t)}. \tag{1.63}$$

For two failure time samples with stress levels x_1 and x_2, and $\underset{\sim}{\gamma}'$ denoting the transpose of the vector of unknown parameters, the difference between the respective log odds functions:

$$\log\theta(t; x_1) - \log\theta(t; x_2) = \underset{\sim}{\gamma}'(x_1 - x_2), \tag{1.64}$$

is independent of the baseline odds function, $\theta_0(t)$, and time, t, implying that one odds function is constantly proportional to the other. Since the odds ratios under different stress levels are proportional to each other, the model is referred to POs model.

1.9 ALT MODELS UNDER TIME-VARYING STRESS

In some areas of reliability testing, stress on individual units is varied during the test. The primary motivation for using varying stress models is to turn up the stress on the surviving units in order to force all or most of the units to fail more quickly than with constant stress model.

When the stress is time-dependent, the stress applied to a sample of units varies with time. Time-dependent stresses can be applied in a variety of ways. For example, if temperature is the stress type, each unit may be tested at 310 K for 10 hours then increased to 320 K for 10 hours then increased to 330 K for 10 hours over the duration of the test. Alternatively, the units may be placed in a test chamber where the temperature starts at 310 K and increases by 5° every 10 minutes until the temperature reaches 330 K.

Some common types of time-dependent stress profiles include step-stress, ramp-stress, and various profiles in which the application of the stress is a continuous function of time (see Nelson, 1990; Miller and Nelson, 1983; Nachalas, 1986).

Some models are described below.

1.9.1 Step-Stress ALT Model

To model a data from step-stress test, the life distribution under step-stress needs to be related to the distribution under a constant stress. Such a model known as "**Cumulative Exposure Model**" was put forward by Nelson (1980).

The model assumes that the remaining life of units depend only on the current cumulative fraction failed and current stress-regardless of how the fraction got accumulated.

The model explained in Figure 1.8(a) depicts a step-stress pattern with four steps. Figure 1.8(b) depicts the four cdfs for the constant stresses (S_1, S_2, S_3, S_4). The arrows show that the specimens first follow the cdf for S_1 up to the first hold time t_1. When the stress increases from S_1 to S_2, the unfailed specimens continue along the cdf for S_2, starting at the accumulated fraction failed. Similarly, for the remaining stresses, the unfailed specimens continue along the next cdf, starting at the accumulated fraction failed. The cdf for life under the step-stress pattern appears in Figure 1.8(c) and consists of the segments of the cdfs for the constant stresses.

Mathematical Formulation: Let $\tilde{F}(\cdot)$ be the cdf of time to failure under a particular step-stress pattern. Mathematically, the model means that if the cdf under constant stresses $S_1 < S_2 < \cdots < S_k$ are $F(t, S_1), F(t, S_2), \ldots, F(t, S_k)$, then under step-stress

$$\tilde{S}(t) = S_i, t_{i-1} \le t < t_i, \quad t_0 = 0, \, i = 1, 2, \ldots, k,$$

the cdf is

$$\tilde{F}(t) = F[(t - t_{i-1}) + l_{i-1}, S_i],$$
$$t_{i-1} \le t < t_i, \;\; t_0 = 0, \, i = 1, 2, \ldots, k, \tag{1.65}$$

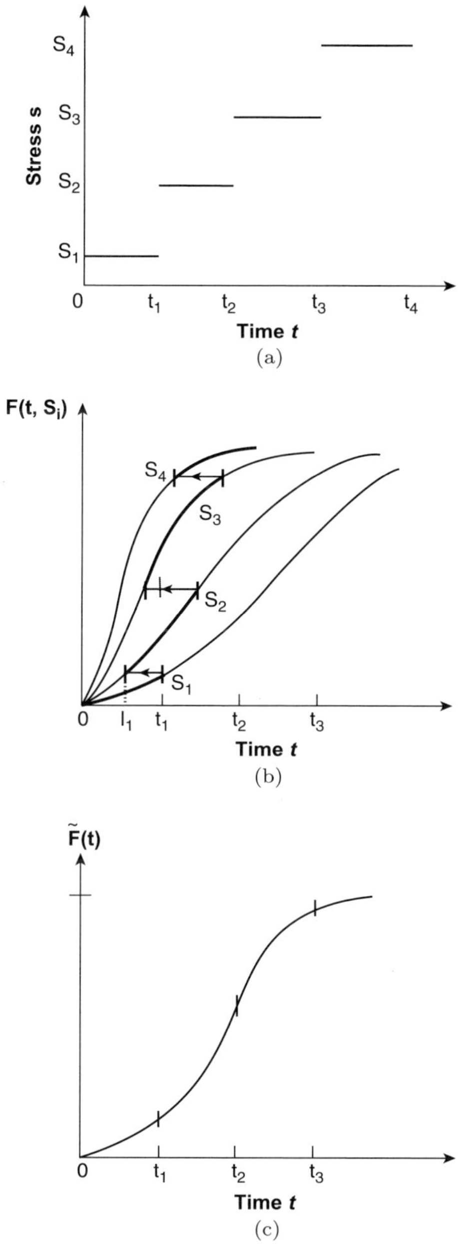

Figure 1.8: Step-stress profile and the corresponding life distributions (a) Step-stress test. (b) cdfs of constant-stress testing. (c) cdf for step-stress testing

where l_{i-1} $(i = 1, 2, \ldots, k)$ is determined by

$$F(l_{i-1}, S_i) = F[(t_{i-1} - t_{i-2}) + l_{i-2}, S_{i-1}], \quad i = 1, 2, \ldots, k, l_0 = 0. \tag{1.66}$$

Example 1.3 (Power–Weibull Step-Stress Model). Using (1.18) and (1.46) with $A = V_0^{\gamma_1}$, the cdf for the fraction of units failing by time t for the constant stress S_i is

$$F(t, V_i) = 1 - \exp[-\{t(V_i/V_0)^{\gamma_1}\}^{\beta}]. \tag{1.67}$$

Then, for step i,

$$\begin{aligned} l_{i-1} &= (t_{i-1} - t_{i-2} + l_{i-2})(V_{i-1}/V_i)^{\gamma_1}, \\ \tilde{F}(t) &= 1 - \exp[-\{(t - t_{i-1} + l_{i-1})(V_i/V_0)^{\gamma_1}\}^{\beta}], \quad t_{i-1} \le t < t_i. \end{aligned} \tag{1.68}$$

Thus, $\tilde{F}(\cdot)$ consists of segments of Weibull distributions.

1.9.2 Progressive-Stress ALT Model

Consider the life distribution $F(t; V, x)$ for a failure mode that depends on constant stress V and possibly other (constant) variables x only through the scale parameter $\alpha(V, x)$ that is a function of V, x, and coefficients, namely,

$$F(t; V, x) = G\left(\frac{t}{\alpha(V, x)}\right), \tag{1.69}$$

where $G(.)$ is the assumed cdf with the scale parameter set equal to one.

$$\Rightarrow F(t_I) = G(\varepsilon), \tag{1.70}$$

where step i is at stress level V_i with scale parameter $\alpha_i \equiv \alpha(V_i, x)$ for a time $\Delta_i = t_i - t_{i-1}$, $t_0 = 0$, $t_I = \Delta_1 + \Delta_2 + \cdots + \Delta_I$, and the cumulative exposure, ε, for the failure mode is

$$\varepsilon = \frac{\Delta_1}{\alpha(V_1, x)} + \frac{\Delta_2}{\alpha(V_2, x)} + \cdots + \frac{\Delta_I}{\alpha(V_I, x)}. \tag{1.71}$$

The fraction failed after any step-stress pattern is independent of the order of the steps.

When the stress $V(t)$ is a function of time, the distribution scale parameter $\alpha(V, x)$ is a function of time; namely, $\alpha(t) = \alpha(V(t), x)$. The corresponding cumulative exposure, $\varepsilon(t)$, which appears as a sum in (1.71), becomes the integral

$$\varepsilon(t) = \int_0^t \frac{dt}{\alpha(V(t), x)}. \tag{1.72}$$

This is the limit of step-stress approximation to $V(t)$ which approaches $V(t)$ as all intervals $\Delta_i \to 0$. $\varepsilon(t)$ is a function of $V(t)$, x, and the model parameters. Thus, the population fraction failed by time t under $V(t)$ is

$$F_0(t; V(t), x) = G(\varepsilon(t)), \tag{1.73}$$

where $G(\cdot)$ is the assumed cumulative distribution with the scale parameter set equal to one. **This is a generalized formulation of the cumulative damage model, where stress can be any function of time, for example, a ramp-stress, a monotonically increasing stress, ramp-constant stress, ramp-step stress, triangular cyclic stress, thermal cyclic stress, etc.**

When $V(t) = kt$, where k is the rate of increase of stress, we have **ramp-stress** model.

Example 1.4 (Power–Weibull Ramp-Stress Model). Using (1.18) and (1.46) with $A = V_0^{\gamma_1}$ and $V(t) = kt$, the cumulative exposure (1.72) at time t is

$$\varepsilon(t) = \int_0^t \frac{du}{\alpha(V(u))}, \tag{1.74}$$

where $\alpha(V(t)) = \left(\frac{V_0}{V(t)}\right)^{\gamma_1}$

$$\Rightarrow \varepsilon(t) = \left(\frac{k}{V_0}\right)^{\gamma_1} \left(\frac{t^{\gamma_1+1}}{\gamma_1 + 1}\right). \tag{1.75}$$

The population fraction failed by time t under $V(t)$ is

$$F_0(t, V(t)) = 1 - \exp\{-[\varepsilon(t)]^{\beta}\} \text{ (using (1.73))} \quad (1.76)$$

$$= 1 - \exp\left\{ - \left[\left(\frac{k}{V_0} \right)^{\gamma_1} \left(\frac{t^{\gamma_1 + 1}}{\gamma_1 + 1} \right) \right]^{\beta} \right\}, \text{ (using (1.75))} \quad (1.77)$$

which is a Weibull distribution with shape and scale parameters

$$\beta' = (\gamma_1 + 1)\beta \quad \text{and} \quad \alpha' = (\gamma_1 + 1) \left(\frac{V_0}{k} \right)^{\gamma_1},$$

respectively.

1.10 TIME-VARYING STRESS ALT MODELS WITH LOG-LOCATION-SCALE DISTRIBUTION

Let V_L, V_U, and V_H be initial low level, pre-specified use level, and highest possible level of the original stress. Further let $s = \log\left(\frac{V_H}{V}\right)$ and $\xi = \frac{s - s_u}{s_H - s_u}$ be stress and standardized stress, respectively. Assume that at any level of stress the failure time, T, follows a log-location-scale distribution with cdf

$$P[T \leq t] = \Phi\left(\frac{\log t - \mu}{\sigma}\right), \quad (1.78)$$

where the location parameter is $\mu(\xi) = \gamma_0 + \gamma_1 \xi$, the scale parameter σ is constant, and ξ is a standardized stress. Also assume that Nelson's (1990) cumulative exposure model holds. This model implies that the distribution of the remaining life of a test unit depends only on the cumulative exposure it has received no matter how it was exposed. In a multiple step-stress test, we define the standardized log time at stress level i to be

$$z_i = \frac{\log(t - \delta_{i-1}) - \mu(\xi_i)}{\sigma}, \quad i = 2, 3, \ldots, h \quad (1.79)$$

and

$$z_\eta = \frac{\log(\eta - \delta_{h-1}) - \mu(\xi_h)}{\sigma}, \tag{1.80}$$

where $\mu(\xi) = \gamma_0 + \gamma_1\xi$, $\tau_{i-1} \leq t < \tau_i$, τ_i is the time at which the stress level change from s_i to s_{i+1}, $\tau_0 = 0$, η is the censoring time and h is the total number of stress levels in the experiment. The time shift induced by the change in stress levels from the beginning of the test to the beginning of step i is

$$\delta_{i-1} = \tau_{i-1} - \sum_{j=1}^{i-1}(\tau_j - \tau_{j-1})e^{\gamma_1(\xi_i - \xi_j)}. \tag{1.81}$$

Under the cumulative exposure model, we have

$$z_i(\tau_i) = z_{i+1}(\tau_i), \quad i = 2, 3, \ldots, \tag{1.82}$$

When $h \to \infty$, under the limiting process described at the beginning

$$z(t) = \frac{\log(w(t)) - \gamma_0}{\sigma}, \tag{1.83}$$

where

$$w(t) = \int_0^t \exp(-\gamma_1\xi(x))dx \tag{1.84}$$

is the cumulative exposure at time t, and $\xi(t)$ must be integrable. This is a weak condition that is met in both step-stress and ramp-stress test plans, whether the changes in stress are monotone or not.

Thus, the cdf and pdf of T when $h \to \infty$ are

$$F(t; \gamma_0, \gamma_1, \sigma) = \Phi\left(\frac{\log w(t)) - \gamma_0}{\sigma}\right) \tag{1.85}$$

and

$$f(t; \gamma_0, \gamma_1, \sigma) = \frac{e^{-\gamma_1\xi(t)}}{w(t)\sigma}\phi\left(\frac{\log(w(t)) - \gamma_0}{\sigma}\right), \tag{1.86}$$

respectively. The failure probabilities at the highest level of stress $\xi_H = 1$, and the use level of stress $\xi_U = 0$ can be expressed as

$$p_H = \Phi\left(\frac{\log(\eta) - \gamma_0 - \gamma_1}{\sigma}\right) \tag{1.87}$$

and

$$p_U = \Phi\left(\frac{\log(\eta) - \gamma_0}{\sigma}\right), \tag{1.88}$$

respectively. Some commonly used log-location-scale distributions are lognormal, Weibull and log-logistic distributions. See Hong, Ma, and Meeker (2010) for reference.

1.11 RELATIONSHIP BETWEEN CONSTANT-STRESS, STEP-STRESS, AND PROGRESSIVE-STRESS LOADINGS

Constant-stress and step-stress are particular cases of progressive-stress. (1.73) is valid for constant-stress as well as step-stress. If $V(t)$ becomes a constant V, then $G(\varepsilon(t))$ in (1.73) reduces to the cdf under constant-stress.

If stress $V(t)$ becomes a step-stress, $\tilde{V}(t)$, then

$$\begin{aligned}
G(\varepsilon(t)) &= G\left(\int_0^t \frac{du}{\alpha(\tilde{V}(u))}\right) \\
&= G\left(\frac{t - t_{i-1}}{\alpha(t_i)} + \sum_{j=1}^{i-1} \frac{t_j - t_{j-1}}{\alpha(t_j)}\right), \\
&\qquad t_{i-1} \le t < t_i,\ t_0 = 0,\ i = 1, 2, \ldots, k \\
&= G\left(\frac{t - t_{i-1}}{\alpha_i} + \sum_{j=1}^{i-1} \frac{t_j - t_{j-1}}{\alpha_j}\right) \\
&= G\left(\frac{t - t_{i-1}}{\alpha_i} + \frac{l_{i-1}}{\alpha_i}\right),
\end{aligned} \tag{1.89}$$

where

$$\frac{l_{i-1}}{\alpha_i} = \sum_{j=1}^{i-1} \frac{t_j - t_{j-1}}{\alpha_j}.$$

Since argument in $G(.)$ is of same form as in (1.65), that is,

$$\begin{aligned}\frac{l_{i-1}}{\alpha_i} &= \sum_{j=1}^{i-1} \frac{t_j - t_{j-1}}{\alpha_j} \\ &= \sum_{j=1}^{i-1} \frac{t_{i-j} - t_{i-j-1}}{\alpha_{i-j}} \\ &= \frac{t_{i-1} - t_{i-2}}{\alpha_{i-1}} + \sum_{j=1}^{i-2} \frac{t_{i-j} - t_{i-j-1}}{\alpha_{i-j}} \\ &= \frac{t_{i-1} - t_{i-2}}{\alpha_{i-1}} + \frac{l_{i-2}}{\alpha_{i-1}}, \end{aligned} \tag{1.90}$$

therefore, step-stress loading is also a particular case of progressive-stress loading.

Example 1.5 (Constant-stress as a particular case of progressive-stress using Weibull distribution). If $V(t)$ becomes a constant V, then (1.74) reduces to $\varepsilon(t) = \frac{t}{\alpha}$, and (1.76) reduces to (1.18), that is, Weibull distribution under constant-stress.

Example 1.6 (Step-stress as a particular case of progressive-stress using Weibull distribution). If stress $V(t)$ becomes a step-stress, $\tilde{V}(t)$, then

$$\begin{aligned}F_0(t, \tilde{V}(t)) &= 1 - \exp\{-[\varepsilon(t)]^{\beta}\} \\ &= 1 - \exp\left[-\left\{\frac{t - t_{i-1}}{\alpha_i} + \sum_{j=1}^{i-1} \frac{t_j - t_{j-1}}{\alpha_j}\right\}^{\beta}\right], \\ & \qquad t_{i-1} \le t < t_i, \;\; t_0 = 0, \;\; i = 1, 2, \ldots, k \end{aligned} \tag{1.91}$$

which is nothing but (1.68) with $\alpha_i = \left(\frac{V_0}{V_i}\right)^{\gamma_1}$.

1.12 PALT MODELS

In partially accelerated life testing, the test units are run at both normal and accelerated conditions. If life–stress relationship as envisaged in fully ALT models (see Section 1.5) is not known or cannot be assumed to be known the accelerated data cannot be extrapolated to normal operating condition, and PALT is then the reasonable scheme to extrapolate accelerated data to normal operating condition.

A PALT model consists of a life distribution and an acceleration factor. The acceleration factor provides a quantitative estimate of the relationship between the test condition and the field condition and is a necessary relationship to properly design and execute an AT.

Definition 1.6. An acceleration factor is defined as the ratio of a reliability measure, say mean life, at use condition to that at accelerated condition.

Thus, acceleration factor k (say) means that the unit under consideration runs k times longer at normal operating condition than at accelerated condition. The two PALT models existing in the literature are constant-stress PALT and step-stress PALT.

1.12.1 Constant-Stress PALT Models

The constant-stress PALT runs each item at either use or accelerated condition only.

$n\pi$ items randomly chosen among n test items are allocated to use condition, and the $n(1-\pi)$ remaining items to accelerated condition, where π denotes the proportion of items allocated at use condition. Each test item is run until a pre-specified time or upon the occurrence of a fixed number of failures.

To describe the model for this constant-stress PALT, let T denote the lifetime of a test item tested at use condition and Y denotes the lifetime of a test item at accelerated condition, then $Y = \alpha T$, where α is the acceleration factor. The lifetimes T and Y are mutually independent.

Example 1.7. Let T follow exponential distribution with hazard rate λ, then the pdf of T is $\lambda e^{-\lambda t}$, $(t, \lambda) > 0$, and that of Y is $\lambda\alpha^{-1}e^{-\lambda\alpha^{-1}y}$, $(y, \lambda) > 0$.

1.12.2 Time-varying Stress PALT Models: Step-Stress Set-Up

The step-stress PALT permits the test to be changed from use to accelerated condition at a pre-determined time.

Let the lifetime of an item under use condition be denoted by the random variable T with life distribution $F(t;\theta)$. Here the value of the parameter θ is unknown and is to be estimated. Suppose the item has not failed by some time 'τ', then it is switched to the higher level of stress and the test is continued until the item fails. It is assumed that the effect of this switch is to multiply the remaining lifetime of the item by unknown acceleration factor α (>1). In general, α will be a function of the higher stress levels that are chosen. If only one higher stress is used, then α is regarded as a constant. Furthermore, since the effect of switching to the higher stress level will typically be to shorten the life of the test item, usually α will be less than one.

To describe the model for this step-stress PALT, let Y denote the total lifetime of a test item. Then, Y is defined by the relation

$$Y = \begin{cases} T, & T \leq \tau \\ \tau + \alpha(T-\tau), & T > \tau. \end{cases} \tag{1.92}$$

Since switching to the higher stress level can be regarded as tampering with the ordinary life test, Y is called a ***tampered random variable***, τ is called the ***tampering point*** or ***stress change point***, and α is called the ***tampering coefficient or acceleration factor***. See for example, Degroot and Goel (1979).

Example 1.8. Let lifetime of an item at use condition be exponential with hazard rate λ, then the pdf of the total lifetime Y can be written in the form

$$g_Y(y) = \begin{cases} \lambda e^{-\lambda y}, & 0 < y \leq \tau \\ \lambda\alpha^{-1} e^{-\lambda(\tau+\alpha^{-1}(y-\tau))}, & y > \tau. \\ 0 \end{cases} \tag{1.93}$$

1.12.2.1 *Tampered Failure Rate (TFR) Model: Step-Stress Set-up*

Bhattacharyya and Soejoeti (1989) have proposed the TFR model in step-stress setting. Their model assumes that a change in the stress has a multiplicative effect on the failure rate function over the remaining life. In the special setting of a two-step PALT, and assuming that the initial life distribution belongs to a two-parameter

Weibull family, they have studied maximum likelihood estimation and derived the Fisher information matrix. They have also proposed a set of assumptions under which the extension of the model to a fully accelerated situation becomes possible. Madi (1993) has generalized the TFR model from a two-step or simple step-stress setting to a k-step or multiple step-stress setting, where by continuously raising the stress level on the unfailed units over a specified number of k of time intervals. Maximum likelihood estimation for the multiple step-stress setting is investigated when the initial life distribution belongs to a two-parameter Weibull family.

The proposed **TFR** model is defined as

$$h^*(y) = \begin{cases} h_1(y) & \text{if } y \leq t \\ \alpha h_1(y) & \text{if } y > t. \end{cases} \tag{1.94}$$

The factor α will typically depend on low and high stress levels and possibly also on t but it is plausible to assume that it does not depend on y. Also, because higher stress level is the single accelerated stress, one can leave α as a parameter of the model without specifying a regression structure on the stress variable.

From the relation (1.4), it follows that the reliability function corresponding to (1.94) is

$$\bar{F}^*(y) = \begin{cases} \bar{F}_1(y) & \text{if } y \leq t \\ \bar{F}_1(t)\left(\dfrac{\bar{F}_1(y)}{\bar{F}_1(t)}\right)^{\alpha} & \text{if } y > t. \end{cases} \tag{1.95}$$

Srivastava and Sharma (2014c) have formulated tampered failure rate model under constant-stress setup, see Appendix D.

1.13 DATA TYPES

The analysis of life tests under use and accelerated conditions relies extensively on data. Specifically, analysis relies on life and stress data or time-to-failure data at a specific stress level.

These common types of life data from single test or design condition are:

(i) **Complete Data**: It consists of the exact life (failure age) of each sample unit. For example, if five units are tested and they all failed, one would then have complete information as to the time-to-failure for each unit in the sample.

(ii) **Censored Data**: Censoring is very common in life tests. It usually applies when the distribution of exact lifetimes are known for only a portion of the products and the remainder of the lifetimes are known only to exceed certain values under a life test. These censoring schemes are described below briefly:

Type-I (Time) Censoring: Here, all the n test units are activated at time $t = 0$. The experiment is terminated when all the units have failed or until time t_0 whichever is early. After the experiment, only the lifetimes of those units that have failed before t_0 will be known exactly. The information in the data set obtained consists of s ($\leq n$) observed ordered lifetimes:

$$t_{(1)} < t_{(2)} < \cdots < t_{(s)}.$$

In addition, one knows that $(n - s)$ components have survived the time t_0, and this information should be utilized.

Since the number s of components that fail before t_0 is obviously stochastic, there is a chance that none or relatively few of the components will fail before t_0 which is the weakness of the design. This censoring scheme is also called Type-I one-stage right censoring.

Type-I left censoring can be realized by starting the observation of failures at a pre-fixed time $\mathrm{T}_L > 0$.

Type-II (Failure) Censoring: Here all the n test units are put to test at time $t = 0$, and the experiment is terminated at the occurrence of rth failure $(0 < r < n)$. The information obtained through the experiment consists of the data set

$$t_{(1)} < t_{(2)} < \cdots < t_{(r)}$$

in addition to the fact that $n - r$ components have survived the time $t_{(r)}$.

In this case, the number r is nonstochastic, and the time $t_{(r)}$ to complete the experiment is stochastic. A weakness of this scheme is therefore that one cannot know beforehand how long the experiment will last. This censoring scheme is also called Type-II one-stage right censoring.

Similarly, Type-II left censoring can be realized by assuming that the observation of failures begins at the time of the $(r+1)$th failure. The exact failure times of the r units known to have failed before this starting time are unknown.

Hybrid censoring: is a combination of Type-I censoring and Type-II censoring. The experiment terminates at the time t_0 or at the occurrence of rth failure whichever occurs first.

Progressive censoring: The conventional Type-I and Type-II censoring schemes do not have the flexibility of allowing removal of units at points other than the terminal point of the experiment. Progressive censoring, however, allows for the removal of items at various points during experimentation, and has also been called multi-censoring or multi-stage censoring.

Progressive Type-I right censoring: Consider the following Type-I censoring scheme:

m censoring times, $T_1, \ldots, T_m$, are fixed such that at these times, $R_1, \ldots, R_m$ surviving units are to be randomly removed (censored) from the test, respectively. Here, the R_is are fixed, with the provision that there are R_i surviving items at time T_i, $i = 1, 2, \ldots, m-1$, i.e., the number of items removed at time T_i is

$$R_i^{\text{obs}} = \min(R_i, \text{number of items remaining at } T_i).$$

Furthermore, the experiment terminates at time T_m with R_m^{obs} being the number of surviving units at that time. This type of censoring is referred to as progressive Type-I right censoring, and is a generalization of conventional Type-I one-stage right censoring, where $R_i = 0$, $i = 1, 2, \ldots, m-1$ so that no items are removed from experimentation until some pre-specified time T_m.

Progressive Type-II right censoring: Under this censoring scheme, n units are placed on life test at time zero. Immediately following the first failure, R_1 surviving units are to be removed from the test at random. Then, immediately following the second observed failure, R_2 surviving units are to be removed from the test at random. This process continues until, at the time of the mth observed failure, the remaining

$$R_m = n - R_1 - R_2 - \cdots - R_{m-1} - m$$

units are all to be removed from the experiment. In this censoring scheme, $R_1, R_2, \ldots, R_m$ (and therefore m) are all pre-fixed. The resulting m ordered values which are obtained as a consequence of this type of censoring are appropriately referred to as progressively Type-II right censored order statistics. If $R_1 = R_2 = \cdots = R_{m-1} = 0$ so that $R_m = n - m$, this scheme reduces to conventional Type-II one-stage right censoring, where just the first m usual order statistics are observed.

Both the Type-I and Type-II progressive censoring schemes described above can be generalized in order to accommodate censoring on the left as well. In the case of progressive Type-II censoring, it can be assumed that the observation of failures begins at the time of the $(r + 1)$th failure, at which time R_{r+1} surviving units are removed from the sample. The exact failure times of the r units known to have failed before this starting time are unknown. Similarly, progressive Type-I censoring can be generalized by starting the observation of failures at a pre-fixed time $T_L > 0$.

Interval Censoring: Interval censored data reflects uncertainty as to the exact times the units failed within an interval. This type of data frequently comes from tests or situations where the objects of interest are not constantly monitored. If one is running a test on five units and inspecting them every 100 hours, one only knows that a unit failed or did not fail between inspections. More specifically, if one inspects a certain unit at 100 hours and find it operating, and then performs another inspection at 200 hours to find that the unit is no longer operating, then it is known that a failure occurred in the interval between 100 and 200 hours. In other words, the only information that is available is that the unit failed in a certain interval of time.

(iii) **Grouped Data**: Data can also be entered individually or into groups. Grouped data analysis is used for tests in which groups of units possess the same time-to-failure or in which groups of units were suspended at the same time. Entering redundant data into groups is highly recommended. Grouped data speeds data entry by the user and significantly speeds up the calculations.

(iv) **Multiple Censored Data**: Most data censored on right have different running times intermixed with failure times. Such data are called multiple censored. Multiple censored data arise when units go on test at different times. Thus, they have different running times when the data are recorded. Such data may be time censored (running times differ from failure times) or failure censored (running times equal failure times).

(v) **Competing Modes**: A mix of competing failure modes occurs when sample units fail from different causes. Data on a particular failure mode consist of the failure times of units failing by that mode. Such data for a mode are multiple censored.

See for example, Hφyland and Rausand (1994) and Aggarwala (2001a, 2001b).

1.14 DATA ANALYSIS METHODS

Some specific data analysis methods in ALT are estimation confidence intervals, and hypothesis tests.

There are several methods of estimation for fitting an accelerated life testing model to the data. Applicable methods include maximum likelihood estimation, linear estimation based on order statistics, and graphical estimation. Ordinary least squares estimation does not apply to censored data. These methods are compared by Hahn and Nelson (1974). ML estimation is preferred for the reasons outlined by Meeker and Nelson (1975) and Kielpinski and Nelson (1975).

Confidence intervals are essential. By their width, confidence intervals for parameters indicate the uncertainty in estimates.

Hypothesis tests indicate whether observed differences are statistically significant. Goodness of fit is employed to determine how well the observed sample data fits some proposed models.

Probability plot is the most commonly used graphical goodness of fit test.

Mann, Schafer, and Singpurwalla (1974) and Lawless (1982) have provided general theory and applications of lifetime data analysis.

1.15 OPTIMUM TEST PLANS

A detailed test plan is usually designed before conducting a test under accelerated condition. A test plan is used to decide on the appropriate stress levels, appropriate stress change times, optimum number of test units to be allocated at different stress levels, and optimum duration of the test. It is therefore not only important but imperative to have a test plan that helps in accurately estimating reliability measures at operating conditions without waste of time, effort, and money.

The optimal test plans are considered as minimizing or maximizing the selected objective functions that are purely based on the Fisher information matrix. The Fisher information is a way of measuring the amount of information that an observable random variable X carries about an unknown parameter θ upon which the likelihood function $L(\theta)$ depends. The likelihood function is the joint probability of the data, the Xs, conditional on the value of θ, or a function of θ. Fisher information is the negative of the expectation of the second derivative of the log of $L(\theta)$ with respect to θ. Thus, for k parameters $\theta_1, \theta_2, \ldots, \theta_k$, the true Fisher information matrix F is

$$F = \begin{bmatrix} E\{-\partial^2 L/\partial^2\theta_1\} & E\{-\partial^2 L/\partial^2\theta_1\theta_2\} & \cdots & E\{-\partial^2 L/\partial^2\theta_1\theta_k\} \\ E\{-\partial^2 L/\partial^2\theta_2\theta_1\} & E\{-\partial^2 L/\partial^2\theta_2\} & \cdots & E\{-\partial^2 L/\partial^2\theta_2\theta_k\} \\ \cdot & \cdot & \cdots & \cdot \\ \cdot & \cdot & \cdots & \cdot \\ \cdot & \cdot & \cdots & \cdot \\ E\{-\partial^2 L/\partial^2\theta_k\theta_1\} & E\{-\partial^2 L/\partial^2\theta_k\theta_2\} & \cdots & E\{-\partial^2 L/\partial^2\theta_k\} \end{bmatrix}, \tag{1.96}$$

where

$$L \equiv \log L(\boldsymbol{\theta}), \quad \boldsymbol{\theta} = (\theta_1, \theta_2, \ldots, \theta_k).$$

An optimum test plan can be obtained using variance optimality, D-optimality or A-optimality criterion (see e.g., Gouno, Sen, and Balakrishnan, 2004; Balakrishnan and Han, 2009). These optimality criteria are briefly described below.

1.15.1 Variance Optimality

The Variance optimality criterion is based on minimizing the asymptotic (large sample) variance of the ML estimators of the mean or of a quantile life at a specified design stress. It is assumed that the high test stress, x_H, has been specified.

Example 1.9. For a constant-stress test, the optimum plan — the low test stress x_L and the proportion of test units allocated at the lower stress is found by minimizing the asymptotic variance of the maximum likelihood estimator of a stated mean or quantile life at design stress.

Example 1.10. For a simple time step-stress test, the variance is minimized by the optimum choice of the stress change time of x_L.

Example 1.11. For a ramp stress test, the optimum plan — low stress rate and the proportion of test units allocated at low stress rate is found by minimizing the asymptotic variance of the maximum likelihood estimator of a stated mean or quantile life at design stress. The higher stress rate is assumed to be given because of technical consideration.

1.15.2 *D*-Optimality

The D-optimality criterion is based on the determinant of the Fisher information matrix which is the same as the reciprocal of the asymptotic variance–covariance matrix. The volume of the asymptotic joint confidence region of parameters, say, (α, β) is proportional to the square root of the determinant of the inverse of the Fisher information matrix, $|F^{-1}|^{1/2}$, at a fixed confidence level. In other words, it is inversely proportional to $|F|^{1/2}$. Consequently, a smaller value of the determinant would correspond to a higher (joint) precision of the estimators of α, β. Thus, the D-optimality criterion is the one which maximizes the $|F|/n^2$.

Example 1.12. For a simple step-stress test, the D-optimal stress change time of x_L is the one which maximizes the $|F|/n^2$.

1.15.3 *A*-Optimality

The A-optimality criterion is also known as trace criterion. It is based on the sum of the diagonal entries of Fisher's information matrix. Thus, the A-optimality criterion is the one which maximizes trace $(F)/n$.

Example 1.13. For a simple step-stress test, the A-optimal stress change time of x_L is the one which maximizes the objective function defined by trace $(F)/n$.

1.16 LIKELIHOOD FUNCTION FORMULATION

This work deals with optimal design of ALT models with time-varying stresses based on complete, time-censored and failure censored data sets (see Section 1.13). The construction of likelihood function with these data sets is briefly explained in this section.

Let the failure times $t_1, t_2, \ldots, t_n$ of n units be independent and identically distributed. The complete data likelihood is given by:

$$L(\boldsymbol{\theta}; t_1, t_2, \ldots, t_n) = \prod_{i=1}^{n} f(t_i, \boldsymbol{\theta}), \qquad (1.97)$$

where $\boldsymbol{\theta} = (\theta_1, \theta_2, \ldots, \theta_k)$.

1.16.1 Formulation of Likelihood Function with Time-Censored Data

Let η_i be censoring time and δ_i denote the censoring indicator variable corresponding to the ith unit, $i = 1, 2, \ldots, n$.

$$\delta_i = \begin{cases} 0, & \text{if } t_i > \eta_i \\ 1, & \text{if } t_i \leq \eta_i \end{cases}, \quad i = 1, 2, \ldots, n. \qquad (1.98)$$

Let the observed part of $t_1, t_2, \ldots, t_n$ be denoted by $x_1, x_2, \ldots, x_{n_1}$, and censored part by $y_1, y_2, \ldots, y_{n_2}$, where $y_i > \eta_i$. Then, the likelihood function based on time-censored data is:

$$\begin{aligned} L^* &\equiv L^*(\boldsymbol{\theta}; x_1, x_2, \ldots, x_{n_1}) \\ &= \int L(\boldsymbol{\theta}; x_1, x_2, \ldots, x_{n_1}, y_1, y_2, \ldots, y_{n_2}) dy_1 dy_2 \ldots dy_{n_2} \\ &= \prod_{i=1}^{n_1} f(x_i, \boldsymbol{\theta}) \prod_{j=1}^{n_2} \left(\int_{y_j > \eta_j} f(y_j, \boldsymbol{\theta}) dy_j \right) \\ &= \prod_{i=1}^{n_1} f(x_i, \boldsymbol{\theta}) \prod_{j=1}^{n_2} (1 - F(\eta_j)), \end{aligned} \tag{1.99}$$

where $n_1 + n_2 = n$.

Alternatively, L^* can be written as:

$$L^* = \prod_{i=1}^{n} (f(t_i, \boldsymbol{\theta}))^{\delta_i} (1 - F(\eta_i))^{1-\delta_i}. \tag{1.100}$$

1.16.2 Formulation of Likelihood Function with Failure-Censored Data

Let $t_{(r)}$ be the time to rth failure at which the experiment is terminated (see Section 1.13). Then, the likelihood function is based on the order statistics $t_{(1)}, t_{(2)}, \ldots, t_{(r)}$ and is given by (see David, 1981):

$$L^{**} = \frac{n!}{(n-r)!} \prod_{i=1}^{r} (f(t_{(i)}, \boldsymbol{\theta})) \prod_{j=1}^{n_2} (1 - F(t_{(r)})), \tag{1.101}$$

where $r + n_2 = n$.

1.17 SENSITIVITY ANALYSIS

The use of an optimum test plan requires information about the design parameters. Incorrect choice of these gives a poor estimate of the mean or quantile life at design stress. The sensitivity analysis identifies the sensitive parameters which need to be estimated with special care for the purpose of minimizing the risk of obtaining erroneous solutions. If a small change in any parameter value does not result in relatively large change in the optimum plan, then the proposed test plan is robust, and the initial values have a small effect on optimal values.

1.18 DESIGN OF FULLY ALT MODELS UNDER CONTINUOUS INSPECTION: A REVIEW

Accelerated testing under continuous inspection results in exact failures (see Section 1.1.3). Several life test models under fully accelerated environmental conditions have been devised in the literature keeping in view the physical behavior of the material subject to the elevated stress level under various life distribution models and different types of censoring schemes. A fully ALT model is commonly referred to as an ALT model in the literature. A brief description of these models follows.

1.18.1 Constant-Stress ALT Models

In the literature, much research has been focused on the design of constant-stress ALT models.

Chernoff (1962) has developed locally optimum ALT plans for the exponential distribution to estimate the failure rate at a specified design stress, and has called his plans "locally optimum" because they depend on true (unknown) parameters values.

For the simple linear regression model, Gaylor and Sweeny (1965) have summarized the literature.

Little and Jebe (1969) have considered least squares estimation of the mean of a normal distribution at a design stress. They have given the optimum plans for successive testing with complete data.

Mann (1972) has considered linear estimation with order statistics to estimate a percentile of an extreme value (or Weibull) distribution at a design stress. The optimum plans for Type-II censored data have also been obtained.

Herzberg and Cox (1972) briefly survey key papers on the optimum design for multiple linear regression models.

Kielpinski and Nelson (1975); Nelson and Kielpinski (1976) have considered ML estimation of the median of a normal (or lognormal) distribution at a design stress. They have given the optimum plans, and best "standard" plans for simultaneous testing with Type-I censored data.

Meeker and Nelson (1978) have given the optimum plans for simultaneous testing for Weibull distribution with Type-I censored data.

Yang (1994) has formulated a constrained nonlinear program to obtain an optimal test plan. The solution for four stress levels has been presented. His approach has been used by Yang and Jin (1994) with three stress levels, where the middle stress is the average of the low and high stresses.

Bai and Yun (1996) have considered optimal design of constant-stress ALT plans for products of unequal size for Weibull life distribution.

Most of the work has been summarized in Nelson (1990, 2005a, 2005b) and Meeker and Escobar (1993, 1998).

Tang (1999) has imposed constraints that limit the probability of having probability plots with best-fit lines crossing at the lower tails. In this way, the stress levels, and their corresponding sample allocations are best suited to infer whether the shape parameters are indeed different at different stress levels. Motivated by the fact that the mid stress level is meant for validating the assumed stress–life model, Tang, Tan, and Ong (2002) have considered the test plans in which the mid stress level will have the least influence on the slope of the stress–life relationship plot. Although these plans are more robust, variances of the estimates at design stress for these compromised plans may be much higher than those under the two stress level optimum plans.

Tang and Yang (2002) have proposed a graphical approach for planning multiple constant-stress levels ALT so that the uncertainty involved for some estimate of interest is not worse than that of a statistical optimum plan by margin determined by the experimenter before the test.

Tang and Xu (2005) have considered a multiple objective framework for planning ALT.

Watkins and John (2008) have considered the problem of constant-stress ALT based on Type-II censored data using Weibull distribution for estimating the MLEs of model parameters.

Liu and Tang (2009) have proposed a sequential constant-stress ALT scheme and its Bayesian inference.

Attia, Aly, and Bleed (2011) have carried out estimation and planning of constant-stress ALT for Generalized Logistic Distribution under Type-I censoring.

1.18.2 Step-Stress ALT Models

Amongst time-varying stress loadings, step-stress ALT plans have been studied mostly extensive.

Nelson (1980) obtained the ML estimators for the parameters of the Weibull distribution under the inverse power law using the break down time data of an electrical insulation.

Miller and Nelson (1983) have obtained the optimum simple step-stress ALT plans for the case where tests units have exponentially distributed life, and are observed continuously until all test units are run to fail.

Bai, Kim, and Lee (1989) have extended the results of Miller and Nelson (1983) to the case where prescribed censoring time is involved.

Bai and Chun (1991) have obtained optimum simple step-stress accelerated life-tests with competing causes of failure.

Bai and Kim (1993) have presented an optimal simple step-stress ALT for the Weibull distribution under Type-I censoring. A log-linear relationship and the cumulative exposure model have been assumed, and nomographs have been used to find the optimal plan.

Madi (1993) has generalized from the simple (2-step) step-stress setting to the multiple (k-step, $k \geq 2$) step-stress setting and got the MLEs of the parameters for the multiple step-stress setting when the initial life distribution belongs to a two-parameter Weibull family.

Tang *et al.* (1996) have analyzed data from a step-stress ALT using linear cumulative exposure model, and three-parameter

Weibull distribution. Khamis and Higgins (1996) have obtained the optimal three-step, step-stress test using exponential distribution. Khamis (1997) has developed an optimum design for the multiple-step, multiple-stress exponential model. Khamis and Higgins (1998) have proposed an optimum design for the multiple-step, multiple-stress model using an alternative to the Weibull cumulative exposure model which is based on a time transformation of the exponential cumulative exposure model. They have called this model Khamis–Higgins (KH) model. They have also used the proposed model to obtain an optimum design, and study inferential problem for two-step ALT with known shape parameter (Khamis and Higgins, 1999).

Watkins (2001) has considered simple step-stress models with Type-II censored exponential data.

Yeo and Tang (1999) have derived a simple step-stress accelerated life testing for an optimum hold time under low stress, and an optimum low stress level by taking the target acceleration factor into consideration. These studies are based on the assumption that the failure time follows exponential distribution because of its simplicity. Tang (2003) has studied how to plan multiple-steps step-stress ALT and how to analyze data from step-stress ALT.

Bai and Chung (1998) have provided the optimal designs of simple step-stress ALTs for lognormal lifetime distributions.

Xiong (1998) has presented the inferences of parameters in the simple step-stress model in accelerated life testing with type-II censored exponential data. Xiong and Milliken (1999) have considered step-stress accelerated life testing with random stress change times for exponential data.

Nonparametric approaches for step-stress testing have been proposed by Shaked and Singpurwalla (1983) and Schmoyer (1991).

Gouno and Balakrishnan (2001) provide a concise review of step-stress ALTs. Gouno (2000, 2001) presents a Bayesian model in order to estimate the assumed constant failure rate under each stress level and to obtain graphical estimates (least square estimator) of the failure rate under use conditions and of the activation energy E_a in the Arrhenius model. These estimates are compared with maximum likelihood estimates, and are shown to give more coherent estimation.

Lu and Rudy (2002) have dealt with the Weibull cumulative exposure (CE) model under the inverse power law in the simple SSALT.

McSorley, Lu, and Li (2002) have studied maximum likelihood estimators of parameters in Weibull cumulative exposure model with log-linear function of stress using three step-stress ALT data.

Alhadeed and Yang (2002) have obtained the optimal design for the lognormal simple step-stress accelerated life testing model using the KH model, assuming constant shape parameter, and no censoring.

Teng and Yeo (2002) have treated general step-stress experiments with failure-censored structure and exponential failures, while assuming a log-linear life–stress relationship. They have further applied a least-squares approach for estimating the model parameters and presented illustrative numerical examples.

Xiong and Ji (2004) have discussed the analysis of grouped and censored data from step-stress life test.

Alhadeed and Yang (2005) have obtained optimal simple step-stress plan for cumulative exposure model using lognormal distribution.

Elsayed and Zhang (2006) have obtained the optimum multiple-stress-type ALT plans based on the PO model with simple step-stress loading and cumulative exposure model.

Wang (2006) has derived the unbiased estimators of the mean life and the failure rate at a design stress based on censored data in step-stress ALT using exponential life distribution.

Xiong, Zhu, and Ji (2006) have studied a variation of a simple step-stress ALT in which the stress change time is random, and the test is subject to Type-II censoring.

Elsayed and Zhang (2007) have also designed the optimum multiple-stress-type ALT plans based on the proportional hazards model.

Xu and Fei (2007) have described the methods and the guidelines for planning the two-factor step-stress ALT experiment for model in which there is no interaction between the factors.

Li and Fard (2007) have developed a step-stress accelerated life testing for two stress variables considering the censored Weibull distributed failure data. Balakrishnan, Leiva, and López (2007) have investigated the simple step-stress model for an exponential distribution under Type-II censoring using the ML estimators. Gouno (2007) has obtained optimum step-stress for temperature accelerated life testing.

Srivastava and Shukla (2008a) have obtained optimum simple step-stress ALT model with log-logistic distribution when all the test units are run to fail. Srivastava and Shukla (2008b) have obtained optimum simple step-stress ALT model for Type-I censored log-logistic data set.

Ma and Meeker (2008) have provided an approach to calculate the large-sample approximate variance in step-stress test plans for a general log-location-scale distribution.

Al-Ghamdi and Hassan (2009) and Al-Masri and Al-Haj Ebrahem (2009) have obtained an optimum simple step-stress accelerated life testing model using the Lomax distribution and log-logistic distribution, respectively.

Fard and Li (2009) have obtained an optimum simple step-stress accelerated life testing model using the Weibull distributed Type-I censored data. Balakrishnan, Xie, and Kundu (2009) have considered a simple step-stress model from an exponential distribution when there is time constraint on the duration of the experiment.

Balakrishnan, Xie, and Zhang (2009) have studied inferential problems for a simple step-stress model with Type-I censoring and lognormally distributed lifetimes.

Balakrishnan, Beutner, and Kateri (2009) have developed order restricted maximum likelihood estimation for multiple step-stress models with exponentially distributed lifetimes for both Type-I and Type-II censored data. Balakrishnan (2009) has reviewed various developments on exact inferential methods for the parameters of exponential step-stress model based on different forms of censored data.

The multi-sample version of a simple step-stress experiment under Type-II censoring has been discussed by Kateri, Kamps, and Balakrishnan (2009).

Lee and Pan (2010) have analyzed step-stress ALT data using generalized linear model.

Kateri, Kamps, and Balakrishnan (2011) have discussed the optimal allocation of change points in step-stress ALT using exponential life distribution and Type-II censored data. Kateri, Kamps, and Balakrishnan (2012) have proposed sequential order statistics approach to step-stress testing.

Saxena *et al.* (2012) have studied optimum step-stress ALT for Rayleigh distribution using complete data set.

Aly and Bleed (2013) have estimated and derived optimum test plan for time-censored step-stress ALT using generalized logistic distribution. Kamal, Zarrin, and Arif-ul-Islam (2013) have obtained optimal step-stress accelerated life testing plan for two parameter Pareto distribution.

The simple step-stress model for a two-parameter exponential distribution, when both the parameters are unknown and the data are Type-II censored have been considered by Mitra *et al.* (2013). It is assumed that under two different stress levels, the scale parameter only changes but the location parameter remains unchanged.

Han and Ng (2013) have compared the optimal k-level constant-stress and step-stress ALTs for the exponential failure data under complete sampling and Type-I censoring. The objective is to quantify the advantage of using the step-stress testing relative to the constant-stress one.

Zhang *et al.* (2014) have obtained reliability information for a white organic light-emitting diode (OLED) by conducting two constant and one step-stress test using Weibull life distribution.

Hassan, Assar, and Shelbaia (2014) have studied time-censored multiple step-stress ALT for Weibull Poisson distribution.

Rezk, Helbawy, and Dayian (2014) have obtained optimal test plans for step-stress ALTs when the lifetime of a product follows Kumaraswamy Weibull distribution with both Type-I and Type-II censored data.

Step-stress ALTs using data types other than Type-I and Type-II have also been studied in the literature. Gouno, Sen, and

Balakrishnan (2004) have obtained an optimal step-stress test using progressive Type-I censored exponential data. Wu, Lin, and Chen (2006) have studied a k-stage step-stress ALT under progressive Type-I censoring with grouped data. The ML method has been used to obtain the estimators of the model parameters. Fan, Wang, and Balakrishnan (2008) have studied a k-stage step-stress ALT with m-stress variables based on progressive Type-I group censored data. Balakrishnan and Han (2009) have obtained an optimal k-step-stress accelerated life testing with equal step duration by modifying the model considered by Gouno, Sen, and Balakrishnan (2004). Ling, Xu, and Li (2009) have considered a simple step-stress ALT model based on progressive Type-I hybrid censoring data. The authors have provided the progressive Type-I hybrid censoring scheme and statistical method in synthetic accelerated stresses so as to decrease the lifetime and reduce the test cost. Wang (2010) has derived the exact confidence intervals for the exponential step-stress accelerated life testing model as well as the approximate confidence intervals for the k-step exponential step-stress accelerated life testing model under progressive Type-II censoring. A Monte Carlo simulation study has been carried out to examine the performance of these confidence intervals.

1.18.3 Progressive-Stress ALT Models

Prot (1948) was the first to propose fatigue testing using ramp-stress. See 1.2.3 also.

Allen (1959) has suggested a statistical model for units having exponential lives under progressive stress, and presented methods of estimating the parameters. Yurkowski, Schafer, and Finkelstein (1967) surveyed early statistical methods. Nelson (1990, Chapter 10) has given some practical situations in which ramp-stress ALT has been used. He has also presented theory to calculate the ML estimator of a quantile associated with a ramp-stress ALT using the Weibull distribution under a particular cumulative exposure model.

Nilsson (1985) has considered the problem of estimating the parameters of the Weibull distribution under the ramp-test.

Yin and Sheng (1987) have derived the life distribution of an item under progressive-stress and studied the properties of the MLEs of the parameters under the ramp-test when the test items have exponential life distribution.

Fei, Leng, and Shu (1991) have used the progressive-stress ALT for solid tantalum electrolytic capacitor.

Bai, Chung, and Cha (1992) have considered the design of the ramp-tests with two stress rates for the Weibull distribution under Type-I censoring, and have determined low stress rate minimizing the asymptotic variance of the MLE of a specified quantile at design stress.

Bai and Chun (1993) have developed nonparametric inferences for ramp-stress tests under random censoring.

Lin and Fei (1987) and Fei and Xun (1996) have discussed the statistical analysis for progressive-stress accelerated life testing.

Bai, Chun, and Cha (1997) have obtained an optimum time-censored ramp test for items with the Weibull life distribution when there is stress upper bound. Wang and Fei (2004) have derived the ML estimates of model parameters of Weibull distribution for TFR model in progressive-stress ALT.

Tang (2006) has investigated Bayes analysis on progressive-stress ALT using Weibull distribution.

Srivastava and Shukla (2009) have obtained optimum simple ramp-stress test using log-logistic distribution and Type-I censoring.

Hong, Ma, and Meeker (2010) have proposed a new optimum ramp-stress test plan by using a new approach for computing approximate large-sample variances of ML estimators of a quantile of a general log-location-scale distribution with censoring, and time-varying stress.

Srivastava and Mittal (2012a) have obtained optimum multi-level ramp-stress ALT plan with multiple-objectives for Burr Type-XII distribution under Type-I censoring. The optimal test plan chooses the stress rates and proportion of units allocated to each stress by minimizing the weighted sum of the asymptotic variances of the maximum likelihood estimator of quantile lifetimes at design constant stress. Srivastava and Mittal (2012b) have also obtained optimum multi-objective ramp-stress ALT with stress upper bound for Burr Type-XII distribution.

1.18.4 Modified-Stress ALT Models

Park and Yum (1998) are the first to propose the optimum ALT plan under modified stress loading methods.

Fei (2000) has studied the statistical analysis of combined data from the progressive-stress and constant-stress life tests under power-Weibull model. Wang (2001) has presented models and statistical analysis for the mixed ALT. Gao *et al.* (2008) have discussed the statistical inference for the mixed ALT with progressive censoring using exponential distribution. In the tests discussed by Wang (2001), and Gao *et al.* (2008); n test units were put on progressive-stress for some time, and then surviving test units on constant-stress.

Srivastava and Mittal (2013a) have formulated optimum multi-objective modified constant-stress ALT plan for the Burr Type-XII distribution with Type-I censoring. Srivastava and Mittal (2014) have designed optimum multi-objective modified step-stress ALT plan for the Burr Type-XII distribution with Type-I censoring. The optimal test plan consists in determining the optimum low stress level and optimal time at which stress starts linearly increasing from low stress by minimizing the weighted sum of the asymptotic variances of the maximum likelihood estimator of quantile lifetimes at a design constant stress.

1.18.5 ALT Models with Competing Causes of Failure

The presence of competing risks in accelerated life testing (ALT) has been focus of attention of many research workers in the field of reliability.

Nelson (1975) has done graphical analysis of ALT data with a mix of failure modes under the assumption that each of p independent failure cause follows an independent lognormal distribution.

McCool (1978) has presented a technique for calculating interval estimates for Weibull parameters of a primary failure mode, when a secondary failure mode having the same (but unknown) Weibull shape parameter is acting.

David and Moeschberger (1971, 1978) have provided an expression for the likelihood of competing risk data with censoring and under fixed experimental conditions. In particular, they have discussed the large sample properties of ML estimators under the Weibull and lognormal distributions.

Herman and Patell (1971) have also discussed the maximum likelihood estimation under competing causes of failure.

Klein and Basu (1981, 1982a) have worked on analysis of ALT when more than one failure mode is acting. Assuming independence among competing failure modes for each stress level, the authors obtain maximum likelihood estimators when the lifetimes are exponential or Weibull, with common or different shape parameter, and data is Type-I, Type-II, or progressively censored.

Further, Klein, and Basu (1982b) derive estimates of model parameters for various censoring schemes when a general stress function is used. A dependent competing risk model is proposed by considering a bivariate Weibull distribution as the joint survival function of two competing risks.

Nelson (1990, Chapter 7) and Craiu and Lee (2005) have discussed the analysis of ALTs under competing causes of failure and have given examples of products which have multiple causes of failure, e.g., semiconductor devices, ball bearing assemblies, and insulation systems.

Bai and Chun (1991) have focused on optimum simple step-stress ALTs for products with competing causes of failure where the life distribution of each failure cause is assumed to be independently and exponentially distributed.

Kim and Bai (2002) have analyzed ALT data when there are two competing risks and they assumed a mixture of two Weibull distributions, and modeled the location parameters as linear functions of stress.

Pascual (2007) has considered the problem of planning ALT when the respective times to failure of competing risks are independently distributed Weibull with a common shape parameter that is assumed to be known. Balakrishnan and Han (2008) and Han and Balakrishnan (2010) have considered an exponential SSALT with competing risks using respectively Type-I and Type-II censored data. Donghoon and Balakrishnan (2010) have studied inferential problem for exponential distribution under time constraint.

Liu and Qiu (2011) have developed a method for planning multiple-step step-stress ALT with independent competing risks using time-censored data.

Shi *et al.* (2013) have considered a constant-stress ALT with competing risks for failure from exponential life distribution under progressive Type-II hybrid censoring.

1.19 DESIGN OF PALT MODELS: A REVIEW

In the literature, PALT models under constant-stress scheme and step-stress scheme have been devised under continuous inspection mode. These are highlighted below briefly:

1.19.1 Constant-Stress PALT Models

Extensive research has been conducted on constant-stress PALT.

Bhattacharyya and Soejoeti (1989) have estimated the parameters of the Weibull distribution and acceleration factor using maximum likelihood method.

Bai and Chung (1992) have considered optimal designs for constant PALTs under Type-I censoring.

Bai, Chun, and Chung (1993) have used the maximum likelihood method to estimate the scale parameter and the acceleration factor for the lognormally distributed lifetime using constant-stress PALT and Type-I censored data. Abdel-Ghani (1998) has considered the estimation problem in constant-stress PALT for the Weibull distribution.

Abdel-Ghaly, El-Khodary, and Ismail (2003a) have obtained the maximum likelihood estimates and developed optimum test plans for simple constant-stress PALT with Type-I censoring for items having Pareto distribution as a lifetime model. Such plans minimize the generalized asymptotic variance of the MLEs of the model parameters.

Ismail (2004) has used maximum likelihood and Bayesian methods for estimating the acceleration factor and the parameters of Pareto distribution of the second kind for constant-stress PALT with Type-I and Type-II censoring. Abdel-Ghaly, Amin, and Omar (2008) have considered the estimation problem of parameters of the Burr Type-XII distribution for constant-stress PALT for Type-I and Type-II censored samples.

Ismail (2009) has planned constant-stress PALT for the Weibull life distribution with Type-II censoring.

Abdel-Hamid (2009) has used the constant-stress PALT and estimated the parameters of the Burr Type-XII life distribution and the acceleration factor under progressive Type-II censoring.

Srivastava and Mittal (2013b, 2013c) have devised optimum constant-stress PALT model for the truncated logistic life distribution under time and failure censoring, respectively.

1.19.2 Step-Stress PALT Models

Goel (1971) has considered the estimation problem of the acceleration factor using both maximum likelihood and Bayesian methods for items having exponential distribution and uniform distribution. The estimates of the parameters of the lifetime distributions were obtained in the case of step-stress PALT under complete sampling.

DeGroot and Goel (1979) have used the Bayesian approach with different loss functions to estimate the parameters of the exponential distribution and the acceleration factor for step-stress PALT in the case of complete sampling.

Bhattacharyya and Soejoeti (1989) have indicated that step-stress PALT is practical for many problems of life testing where the test process requires a long time if the test is simply carried out under the use condition.

Bai and Chung (1992) have considered optimal designs for step-stress PALTs under Type-I censoring.

Bai, Chun, and Chung (1993) have used the maximum likelihood method to estimate the scale parameter and the acceleration factor for the lognormally distributed lifetime step-stress PALT, using Type-I censored data.

Abdel-Ghaly, Attia, and Abdel-Ghani (1996) have estimated the parameter of the Weibull distribution and acceleration factor using maximum likelihood method in time-censored step-stress PALT. Abdel-Ghaly, Attia, and Abdel-Ghani (1997) have used the Bayesian approach for estimating the Weibull distribution parameters given that the shape parameter is known. They have studied the estimation problem in step-stress PALT under both Type-I and Type-II censored data.

Madi (1997) has applied the Gibbs sampling approach to the PALT.

Abdel-Ghani (1998) has considered the estimation problem of the parameters of Weibull distribution and the acceleration factor for step-stress PALT using maximum likelihood method and Bayesian approach under Type-I and Type-II censored data.

Abdel-Ghaly, El- Khodary, and Ismail (2002) have studied both the estimation and optimal design problems for the Pareto distribution under step-stress PALT with Type-I censoring. They have derived the maximum likelihood estimates of the model parameters.

Abdel-Ghaly, Attia, and Abdel-Ghani (2002) have used the maximum likelihood method to estimate the acceleration factor and the parameters of Weibull distribution in step-stress PALT using Type-I and Type-II censored data.

Tahir (2003) has considered a two-stage sequential procedure for estimating the failure rate in the step-stress PALT model under the squared error loss. In the first stage, the units are subjected to normal stress until time t, where t is determined as the stopping time which minimizes the expected loss plus cost of running the test. In the second stage, the stress is raised to a higher level for those units that did not fail by time t and is held constant until they all fail. The accumulated data are then used to estimate the failure rate with the Bayes estimator. It is assumed that the lifetimes of the test units are independent and exponentially distributed random variables with a common failure rate.

Abdel-Ghaly, El-Khodary, and Ismail (2003b) have obtained the maximum likelihood estimates and developed optimum test plans for simple step-stress PALT with Type-II censoring for items having Pareto distribution as a lifetime model. Such plans minimize the generalized asymptotic variance of the MLEs of the model parameters.

Ismail (2004) has used maximum likelihood and Bayesian methods for estimating the acceleration factor and the parameters of Pareto distribution of the second kind for step-stress PALT and constant-stress PALT with Type-I and Type-II censoring. Also, he has considered the problem of optimal designs for the two types of PALT. Abdel-Ghani (2004) has estimated the parameters of log-logistic distribution under step-stress PALT.

Abdel-Ghaly, Amin, and Omar (2008) have considered the estimation problem of parameters of the Burr Type-XII distribution for step-stress PALT for Type-I and Type-II censored samples.

Abd-Elfattah, Hassan, and Nassr (2008) have estimated the parameters of the Burr Type-XII life distribution and acceleration factor using the maximum likelihood method in time-censored step-stress PALT.

Aly and Ismail (2008) have obtained optimum simple step-stress PALT using Weibull life distribution model and Type-I censoring scheme.

Ismail and Sarhan (2009) have discussed the step-stress PALT under progressive Type-II censoring with random removals. The removals from the test are assumed to have binomial distribution. The lifetimes of the test units are considered to be exponentially distributed.

Ismail and Aly (2009) have described the optimum plan for failure step-stress PALT assuming Weibull distribution as lifetime model.

Ismail and Aly (2010) have discussed optimal test plans for failure-step-stress partial ALT under Weibull distribution.

Ismail (2010) has applied the Bayesian approach to the estimation problem in the case of step-stress PALT with two stress levels and Type-I censoring assuming the two-parameter Gompertz distribution as a lifetime model. Srivastava and Mittal (2010, 2012c) have studied step-stress PALT using truncated logistic distribution with time-censored and failure censored data.

Ismail (2011) has discussed the point and interval estimations of two-parameter Gompertz distribution under step-stress PALT with Type-II censoring.

Hassan and Al-Thobety (2012) have discussed the failure step-stress PALT and statistical inferences for the model parameters and acceleration factor by assuming that the lifetime of the test items follows inverse Weibull distribution.

Lakshmi and Sankari (2012) have studied time step-stress PALT for the Gompertz distribution using Type-I censoring. They have critically reviewed the physiological aspects and clinical sequelae of energy deficiency and hypoestrogenism associated with the female athlete triad in exercising women.

1.20 DESIGN OF ALT MODELS UNDER PERIODIC INSPECTION: A REVIEW

In contrast to continuous inspection; in periodic inspection the test units are inspected for failure at pre-determined points in time. The information obtained from a periodic inspection consists of the number of failures in each inspection period, resulting in "grouped" or "interval" data, and the number of units that survive until the censoring time.

In the literature, optimum ALTs under periodic inspection have been formulated using constant-stress loadings. Yum and Choi (1989) have obtained optimum ALT under periodic inspection for exponentially distributed lifetimes, Seo and Yum (1991) for Weibull distribution, Ahmad *et al.* (1994) for Rayleigh failure law, Islam and Ahmad (1994) for Weibull distribution, Ahmad and Islam (1996) for Burr Type-XII distribution, Ahmad, Islam, and Salam (2006) for exponentiated Weibull distribution, and Ahmad, Quadri, and Choudhary (2008) for Burr Type-III distribution.

CHAPTER 2

OPTIMUM STEP-STRESS ACCELERATED LIFE TEST MODELS

2.1 INTRODUCTION

In the present chapter, optimum test plan when stress is applied in step-stress setup is dealt with. The step-stress scheme allows the stress setting of a unit to be changed at a pre-specified time or upon the occurrence of a fixed number of failures. A test unit starts at a specified low stress. If the unit does not fail at a specified time, stress on it is raised and held for a specified time. Stress is repeatedly increased and held, until the unit fails or is censored. A simple step-stress ALT uses only two stress-levels.

The literature on step-stress accelerated life testing is reviewed in Section 1.18.2.

In this chapter, optimum design of log-logistic cumulative exposure models in step-stress accelerated life testing, both when all the test units are run to fail (complete data set) and when censoring time is pre-specified are described.

The cumulative exposure model (Nelson, 1990; Miller and Nelson, 1983) for simple step-stress testing with stresses x_1 and x_2 is

$$F(t) = \begin{cases} F_1(t), & t < \tau \\ F_2(t - \tau + \tau'), & t \geq \tau \end{cases}, \tag{2.1}$$

where $F_i(t)$ is the cumulative distribution function (cdf) of the failure time at stress x_i, τ is the time to changing stress, and τ' is the solution of

$$F_1(\tau) = F_2(\tau'). \tag{2.2}$$

See Section 1.9.1 also.

The log-logistic cdf is

$$F(t) = \frac{\left(\frac{t}{\lambda}\right)^{\alpha}}{1 + \left(\frac{t}{\lambda}\right)^{\alpha}}, \quad t \geq 0, \ \ \alpha > 0, \ \ \lambda > 0. \tag{2.3}$$

The scale parameter λ is the median life time of the distribution and α is the shape parameter.

The hazard rate of log-logistic distribution is given by

$$h(t) = \frac{\frac{\alpha}{\lambda}\left(\frac{t}{\lambda}\right)^{\alpha-1}}{1 + \left(\frac{t}{\lambda}\right)^{\alpha}}, \tag{2.4}$$

which is always decreasing with time if $\alpha \leq 1$; first increasing, then decreasing with time if $\alpha > 1$. The latter property has been observed for insulating material. The log-logistic distribution has been found appropriate for high reliability component. See Section 1.4.2.7 also.

2.1.1 Notation

x_0, x_1, x_2	design, low and high stress levels
ξ	extrapolation factor, where $\xi \equiv \frac{(x_1 - x_0)}{(x_2 - x_1)}$
n	number of test units
t_{ij}	failure time of the test unit j subject to stress x_i, $i = 1, 2,\ j = 1, 2, \ldots, n_i$
λ_i	median life at stress x_i, $i = 0, 1, 2$
$n_1,\ n_2$	number of failure times observed (before, after) the stress change; $n_1 + n_2 = n$
$F_i(t)$	cdf of log-logistic distribution with median life λ_i
$F_0(t)$	cdf of a test unit under simple step-stress

τ_1	stress change time for complete data set
τ_1^*	optimal time of changing stress for complete data set
τ_2	stress change time for censored data set
τ_2^*	optimal time of changing stress for censored data set
α	shape parameter
β_0, β_1	parameters of log-linear relationship between stress x_i and the median life λ_i, $i = 0, 1, 2$.

2.2 AN OPTIMUM SIMPLE LOG-LOGISTIC STEP-STRESS ALT MODEL FOR COMPLETE DATA SET

Srivastava and Shukla (2008a) have formulated optimum simple log-logistic step-stress ALT model when all the test units are run to fail. They have also discussed the inferential procedures pertaining to the design parameters and carried out sensitivity analysis. The methodology used is explained in this section.

2.2.1 Assumptions

1. Two stress levels x_1 and x_2 $(x_1 < x_2)$ are used.
2. For any level of stress, the life distribution of a test unit is log-logistic.
3. The median life of a test unit is log-linear function of stress, i.e.,

$$\log \lambda(x_i) = \beta_0 + \beta_1 x_i, \quad i = 0, 1, 2, \tag{2.5}$$

   where β_0 and $\beta_1(<0)$ are unknown parameters depending on the nature of the product and the method of test. The median life of a test unit at lower stress is longer than at higher stress, i.e., $\lambda_0 > \lambda_1 > \lambda_2$.
4. A cumulative exposure model holds, i.e., the remaining life of a test unit depends only on the cumulative exposure it has seen.
5. The lifetimes of test units are independent and identically distributed.

This is quite a valid assumption for nonrepairable units. For a nonrepairable unit, there is no repair once the unit fails, and so, the unit is discarded. Because, the failure of one unit does not affect the performance of any other similar unit, the assumption that different units have lifetimes that are independent is reasonable. Also if many copies of a unit were produced by the same manufacturing system, then it is reasonable to assume that the system lifetimes have the same distribution. These two assumptions can be combined into one statement that says lifetimes are independent and identically distributed.

6. The shape parameter α is constant and known.

2.2.2 Test Procedure

1. All n units are initially placed on low stress x_1 and run until τ_1 when the stress is changed to x_2. At x_2 testing continues until all remaining units fail.
2. n_i failure times t_{ij}, $j = 1, 2, \ldots, n_i$, of test units are observed while testing at stress x_i, $i = 1, 2$.

2.2.3 Life Distribution under Step-Stress

From the assumptions of cumulative exposure model and the log-logistic life distribution,

$$\tau_1' = \left(\frac{\lambda_2}{\lambda_1}\right)\tau_1 \quad \text{(using (2.1)–(2.3))}$$

Thus, the pdf of a test unit is

$$f_0(t) = \begin{cases} f_1(t), & t < \tau_1 \\ f_2(t), & t \geq \tau_1, \end{cases} \tag{2.6}$$

where

$$f_1(t) = \frac{\frac{\alpha}{\lambda_1}\left(\frac{t}{\lambda_1}\right)^{\alpha-1}}{\left(1 + \left(\frac{t}{\lambda_1}\right)^{\alpha}\right)^2}, \tag{2.7}$$

and

$$f_2(t) = \frac{\frac{\alpha}{\lambda_2}\left(\frac{t}{\lambda_2}\right)^{\alpha-1}}{\left(1 + \frac{t^\alpha - \tau_1^\alpha}{\lambda_2^\alpha} + \frac{\tau_1^\alpha}{\lambda_1^\alpha}\right)^2}. \tag{2.8}$$

2.2.4 Maximum Likelihood Estimation

Under the assumption of the cumulative exposure model, the likelihood function from observations t_{ij}, $i = 1, 2$, $j = 1, 2, \ldots, n_i$, using (2.6), is

$$\begin{aligned} L(\lambda_1, \lambda_2) &\\ &= \prod_{j=1}^{n_1} f_0(t_{1j}) \prod_{j=1}^{n_2} f_0(t_{2j}) \\ &= \prod_{j=1}^{n_1}\left[\frac{\frac{\alpha}{\lambda_1}\left(\frac{t_{1j}}{\lambda_1}\right)^{\alpha-1}}{\left(1+\left(\frac{t_{1j}}{\lambda_1}\right)^\alpha\right)^2}\right] \prod_{j=1}^{n_2}\left[\frac{\frac{\alpha}{\lambda_2}\left(\frac{t_{2j}}{\lambda_2}\right)^{\alpha-1}}{\left(1+\left(\frac{t_{2j}^\alpha - \tau_1^\alpha}{\lambda_2}\right) + \left(\frac{\tau_1}{\lambda_1}\right)^\alpha\right)^2}\right] \end{aligned} \tag{2.9}$$

By substituting for $\lambda_i, i = 1, 2$ in (2.9); the log-likelihood function is a function of unknown parameters β_0 and β_1, and is given by

$$\begin{aligned} \log L(\beta_0, \beta_1) &\\ &= \sum_{j=1}^{n_1}[\log\alpha - (\beta_0 + \beta_1 x_1) + (\alpha - 1)\log t_{1j} \\ &\quad - (\alpha - 1)(\beta_0 + \beta_1 x_1) - 2\log(1 + t_{1j}^\alpha e^{-\alpha\beta_0 - \alpha\beta_1 x_1})] \\ &\quad + \sum_{j=1}^{n_2}[\log\alpha - (\beta_0 + \beta_1 x_2) + (\alpha - 1)\log t_{2j} \\ &\quad - (\alpha - 1)(\beta_0 + \beta_1 x_2) - 2\log(1 + (t_{2j}^\alpha - \tau_1^\alpha)e^{-\alpha\beta_0 - \alpha\beta_1 x_2} \\ &\quad + \tau_1^\alpha e^{-\alpha\beta_0 - \alpha\beta_1 x_1})]. \end{aligned} \tag{2.10}$$

The maximum likelihood estimates $\hat{\beta}_0, \hat{\beta}_1$ are the parameter values that maximize $\log L(\beta_0, \beta_1)$.

2.2.5 Fisher Information Matrix

The Fisher information matrix is obtained by taking the expected values of the second partial and mixed partial derivatives of $\log L(\beta_0, \beta_1)$ with respect to β_0, β_1.

The first-order partial derivatives are

$$\frac{\partial \log L}{\partial \beta_0} = -\alpha n + 2\alpha \sum_{j=1}^{n_1} \left[\frac{e^{-\alpha\beta_0 - \alpha\beta_1 x_1} t_{1j}^{\alpha}}{1 + e^{-\alpha\beta_0 - \alpha\beta_1 x_1} t_{1j}^{\alpha}} \right]$$
$$+ 2\alpha \sum_{j=1}^{n_2} \left[\frac{e^{-\alpha\beta_0 - \alpha\beta_1 x 2} \tau_1^{\alpha} + e^{-\alpha\beta_0 - \alpha\beta_1 x 2}(t_{2j}^{\alpha} - \tau_1^{\alpha})}{1 + e^{-\alpha\beta_0 - \alpha\beta_1 x_1} \tau_1^{\alpha} + e^{-\alpha\beta_0 - \alpha\beta_1 x_2}(t_{2j}^{\alpha} - \tau_1^{\alpha})} \right], \tag{2.11}$$

$$\frac{\partial \log L}{\partial \beta_1} = -\alpha(n_1 x_1 + n_2 x_2) + 2\alpha \sum_{j=1}^{n_1} \left[\frac{e^{-\alpha\beta_0 - \alpha\beta_1 x_1} t_{1j}^{\alpha} x_1}{1 + e^{-\alpha\beta_0 - \alpha\beta_1 x_1} t_{1j}^{\alpha}} \right]$$
$$+ 2\alpha \sum_{j=1}^{n2} \left[\frac{x_1 e^{-\alpha\beta_0 - \alpha\beta_1 x_1} \tau_1^{\alpha} + x_2 e^{-\alpha\beta_0 - \alpha\beta_1 x_2}(t_{2j}^{\alpha} - \tau_1^{\alpha})}{1 + e^{-\alpha\beta_0 - \alpha\beta_1 x_1} \tau_1^{\alpha} + e^{-\alpha\beta_0 - \alpha\beta_1 x_2}(t_{2j}^{\alpha} - \tau_1^{\alpha})} \right]. \tag{2.12}$$

The second-order partial derivatives are

$$\frac{\partial^2 \log L}{\partial \beta_0^2} = -(B_1 + B_2 + B_3), \tag{2.13}$$

$$\frac{\partial^2 \log L}{\partial \beta_0 \partial \beta_1} = -(x_1(B_1 + B_3) + x_2 B_2), \tag{2.14}$$

$$\frac{\partial^2 \log L}{\partial \beta_1 \partial \beta_0} = \frac{\partial^2 \log L}{\partial \beta_0 \partial \beta_1}, \tag{2.15}$$

$$\frac{\partial^2 \log L}{\partial \beta_1^2} = -(x_1^2(B_1 + B_3) + x_2^2 B_2 + (x_1 - x_2)^2 B_4), \tag{2.16}$$

where

$$B_1 = 2\sum_{j=1}^{n_1} \left(\frac{\alpha^2 t_{1j}^{\alpha}}{(1 + e^{-\alpha\beta_0 - \alpha\beta_1 x_1} t_{1j}^{\alpha})^2} \right),$$

$$B_2 = 2\sum_{j=1}^{n_2}\left(\frac{\alpha^2(t_{2j}^{\alpha} - \tau_1^{\alpha})e^{-\alpha\beta_0-\alpha\beta_1 x_2}}{(1+(t_{2j}^{\alpha} - \tau_1^{\alpha})e^{-\alpha\beta_0-\alpha\beta_1 x_2} + \tau_1^{\alpha}e^{-\alpha\beta_0-\alpha\beta_1 x_1})^2}\right),$$

$$B_3 = 2\sum_{j=1}^{n_2}\left(\frac{\alpha^2\tau_1^{\alpha}e^{-\alpha\beta_0-\alpha\beta_1 x_1}}{(1+(t_{2j}^{\alpha} - \tau_1^{\alpha})e^{-\alpha\beta_0-\alpha\beta_1 x_2} + \tau_1^{\alpha}e^{-\alpha\beta_0-\alpha\beta_1 x_1})^2}\right)$$

and

$$B_4 = 2\sum_{j=1}^{n_2}\left(\frac{\alpha^2(t_{2j}^{\alpha} - \tau_1^{\alpha})e^{-\alpha\beta_0-\alpha\beta_1 x_2}\tau_1^{\alpha}e^{-\alpha\beta_0-\alpha\beta_1 x_1}}{(1+(t_{2j}^{\alpha} - \tau_1^{\alpha})e^{-\alpha\beta_0-\alpha\beta_1 x_2} + \tau_1^{\alpha}e^{-\alpha\beta_0-\alpha\beta_1 x_1})^2}\right).$$

The Fisher information matrix F is

$$F = \begin{bmatrix} M_{11} & M_{12} \\ M_{21} & M_{22} \end{bmatrix}.$$

The elements of the Fisher information matrix for an observation are the negative expectations of these second partial derivatives. These elements have been obtained using the following results:

(i) $n_1 \sim \text{Bin}(n, p_1)$, where

$$\begin{aligned} E[n_1] &= np_1, \\ p_1 &= P[0 < T_1 < \tau_1] \\ &= \int_0^{\tau_1} f_1(t)dt \quad \text{(using (2.6))} \\ &= \frac{\frac{\tau_1^{\alpha}}{\lambda_1^{\alpha}}}{\left(1 + \frac{\tau_1^{\alpha}}{\lambda_1^{\alpha}}\right)}. \end{aligned}$$

(ii) $n_2 \sim \text{Bin}(n, p_2)$, where

$$E[n_2] = np_2,$$

where

$$\begin{aligned} p_2 &= P[\tau_1 < T_2 < \infty] \\ &= \int_{\tau_1}^{\infty} f_2(t)dt \quad \text{(using (2.6))} \\ &= \frac{1}{\left(1+\frac{\tau_1^\alpha}{\lambda_2^\alpha}\right)}. \end{aligned}$$

(iii)

$$E\left[\sum_{i=1}^{m} g(X_i)\right] = E[m]E[g(X)],$$

where m is a nonnegative integer valued random variable and $g(.)$ is a function of i.i.d. random variables.

Thus,

$$\begin{aligned} M_{11} &= E\left[\frac{\partial^2 \log L}{\partial \beta_0^2}\right] \\ &= 2np_1 \int_0^{\tau_1} \frac{\alpha^2 \frac{t_1^\alpha}{\lambda_1^\alpha}}{\left(1+\frac{t_1^\alpha}{\lambda_1^\alpha}\right)^2} \left(\frac{f(t_1)}{p_1}\right) dt_1 \\ &\quad + 2np_2 \int_{\tau_1}^{\infty} \frac{\alpha^2 \left(\frac{t_2^\alpha - \tau_1^\alpha}{\lambda_2^\alpha}\right)}{\left(1+\frac{\tau_1^\alpha}{\lambda_1^\alpha}+\left(\frac{t_2^\alpha-\tau_1^\alpha}{\lambda_2^\alpha}\right)\right)^2} \left(\frac{f(t_2)}{p_2}\right) dt_2 \\ &\quad + 2np_2 \int_{\tau_1}^{\infty} \frac{\alpha^2 \left(\frac{\tau_1^\alpha}{\lambda_1^\alpha}\right)}{\left(1+\frac{\tau_1^\alpha}{\lambda_1^\alpha}+\left(\frac{t_2^\alpha-\tau_1^\alpha}{\lambda_2^\alpha}\right)\right)^2} \left(\frac{f(t_2)}{p_2}\right) dt_2 \\ &= \frac{2}{6} n\alpha^2, \end{aligned}$$

$$\begin{aligned} M_{22} &= E\left[\frac{\partial^2 \log L}{\partial \beta_1^2}\right] \\ &= 2np_1\alpha^2 \int_0^{\tau_1} \frac{x_1^2 \frac{t_1^\alpha}{\lambda_1^\alpha}}{\left(1+\frac{t_1^\alpha}{\lambda_1^\alpha}\right)^2} \left(\frac{f(t_1)}{p_1}\right) dt_1 + 2np_2\alpha^2 \end{aligned}$$

$$\times \int_{\tau_1}^{\infty} \frac{\left(x_1^2\left(\frac{\tau_1^\alpha}{\lambda_1^\alpha}\right) + \frac{x_2^2}{\lambda_1^\alpha}\left(t_2^\alpha - \tau_1^\alpha\right) + (x_1 - x_2)^2 \frac{\tau_1^\alpha}{\lambda_1^\alpha}\left(\frac{t_2^\alpha - \tau_1^\alpha}{\lambda_2^\alpha}\right)\right)}{\left(1 + \frac{\tau_1^\alpha}{\lambda_1^\alpha} + \frac{t_2^\alpha - \tau_1^\alpha}{\lambda_2^\alpha}\right)^2}$$

$$\times \left(\frac{f(t_2)}{p_2}\right) dt_2$$

$$= \frac{2}{6} n\alpha^2 [x_1^2 A_1 + x_2^2 A_2 + (x_1 - x_2)^2 A_3],$$

$$M_{12} = M_{21} = E\left[\frac{\partial^2 \log L}{\partial \beta_0 \partial \beta_1}\right]$$

$$= 2np_1\alpha^2 \int_0^{\tau_1} \frac{x_1 \frac{t_1^\alpha}{\lambda_1^\alpha}}{\left(1 + \frac{t_1^\alpha}{\lambda_1^\alpha}\right)^2} \left(\frac{f(t_1)}{p_1}\right) dt_1$$

$$+ 2np_2\alpha^2 \int_{\tau_1}^{\infty} \frac{x_1\left(\frac{\tau_1^\alpha}{\lambda_1^\alpha}\right) + x_2 \frac{\left(t_2^\alpha - \tau_1^\alpha\right)}{\lambda_2^\alpha}}{\left(1 + \frac{\tau_1^\alpha}{\lambda_1^\alpha} + \frac{t_2^\alpha - \tau_1^\alpha}{\lambda_2^\alpha}\right)^2} \left(\frac{f(t_2)}{p_2}\right) dt_2$$

$$= \frac{2}{6} n\alpha^2 [x_1 A_1 + x_2 A_2].$$

Hence, the Fisher information matrix F is

$$F = \frac{2}{6} n\alpha^2 \begin{bmatrix} 1 & [x_1 A_1 + x_2 A_2] \\ [x_1 A_1 + x_2 A_2] & [x_1^2 A_1 + x_2^2 A_2 + (x_1 - x_2)^2 A_3] \end{bmatrix}.$$

2.2.6 Optimality Criterion

The asymptotic variance $\mathrm{Var}(\tau)$ of the MLE of the median life at the design stress is then given by

$$\begin{aligned} n\mathrm{Var}(\tau) &= n.\mathrm{AsVar}(log\hat{\lambda}_0) \\ &= n\mathrm{AsVar}(\hat{\beta}_0 + \hat{\beta}_1 x_0) \\ &= n[\mathrm{Var}(\hat{\beta}_0) + x_0^2 \mathrm{Var}(\hat{\beta}_1) + 2x_0 \mathrm{Cov}(\hat{\beta}_0, \hat{\beta}_1)] \end{aligned}$$

$$= \frac{3}{\alpha^2 A_2 \left(1 - A_2 + \frac{\tau_1^*}{\lambda_1^\alpha}\right)} \left[\xi^2 + 2\xi A_2 + \left(1 + \frac{\tau_1^*}{\lambda_1^\alpha}\right) A_2\right], \tag{2.17}$$

where ^ implies the MLE.

The optimal τ_1 is found by minimizing $n\mathrm{Var}(\tau_1)$ using NMinimize option of *Mathematica 5.0.*

Table 2.1 gives optimal $\tau_1(\tau_1^*)$.

2.3 AN OPTIMUM SIMPLE LOG-LOGISTIC STEP-STRESS ALT MODEL FOR TYPE-I CENSORED DATA SET

In this section, the approach used by Srivastava and Shukla (2008b) to obtain optimum simple step-stress ALT model for Type-I censored log-logistic data set is brought forth.

2.3.1 Assumptions

1. Testing is done at stresses x_1 and x_2, where $x_1 < x_2$,
2. For any level of stress, the life distribution of a test unit follows the log-logistic distribution,
3. The median life of a test unit is log-linear function of stress, i.e.,

$$\log \lambda(x) = \beta_0 + \beta_1 x, \tag{2.18}$$

   where β_0 and $\beta_1(<0)$ are unknown parameters depending on the nature of the product and the method of test. Therefore, the median life of a test unit at lower stress is longer than at higher stress, i.e.,

$$\lambda_0 > \lambda_1 > \lambda_2,$$

4. A cumulative exposure model holds. That is, the remaining life of a test unit depends only on the cumulative exposure it has seen.

2.3.2 Test Procedure

1. All n units are initially placed on low stress x_1 and run until time τ_2, when the stress is changed to x_2, and the test is continued until all units fail or until a pre-specified censoring time T.

Table 2.1: Optimal time changing stress $\tau_1^*(\xi, \lambda_1, \alpha)$

	ξ				
λ_1	**0.2**	**0.4**	**0.6**	**0.8**	**1.0**
(a) ($\alpha = 0.5$)					
3	27.000	11.690	7.315	5.362	4.229
4	36.001	15.587	9.753	7.150	5.719
5	45.000	19.484	12.192	8.937	7.149
6	53.998	23.381	14.630	10.725	8.580
7	63.000	27.278	17.069	12.513	10.010
8	72.000	31.175	19.507	14.300	11.439
9	81.000	35.072	21.946	16.088	12.869
(b) ($\alpha = 1.5$)					
3	6.240	4.720	4.037	3.640	3.798
4	8.320	6.294	5.383	4.854	4.506
5	10.400	7.868	6.729	6.068	5.633
6	12.480	9.441	8.075	7.281	6.759
7	14.560	11.015	9.421	8.495	7.886
8	16.640	12.589	10.767	9.709	9.012
9	18.720	14.162	12.113	10.922	10.139
(c) ($\alpha = 2$)					
3	5.196	4.215	3.748	3.468	3.280
4	6.928	5.620	4.998	4.625	4.374
5	8.660	7.025	6.248	5.781	5.467
6	10.392	8.430	7.497	6.937	6.561
7	12.124	9.835	8.747	8.094	7.654
8	13.856	11.240	9.996	9.250	8.748
9	15.588	12.645	11.246	10.406	9.841
(d) ($\alpha = 2.5$)					
3	4.655	3.937	3.585	3.369	3.222
4	6.207	5.250	4.780	4.492	4.296
5	7.759	6.563	5.975	5.615	5.370
6	9.311	7.875	7.170	5.739	6.449
7	10.862	9.188	8.366	7.862	7.519
8	12.414	10.501	9.561	8.985	8.593
9	13.966	11.813	10.756	10.108	9.667

2. n_i failure times t_{ij}, $j = 1, 2, \ldots, n_i$ of test units are observed while testing at stress x_i, $i = 1, 2$, and n_c units are censored.

2.3.3 Life Distribution under Step-Stress

From the assumptions of cumulative exposure model and log-logistic life distribution,

$$\tau_2' = \left(\frac{\lambda_2}{\lambda_1}\right)\tau_2 \quad \text{(using (2.1)–(2.3))}.$$

Thus, the pdf of a test unit is

$$f_0(t) = \begin{cases} f_1(t), & t < \tau_2 \\ f_2(t), & t \geq \tau_2, \end{cases} \tag{2.19}$$

where

$$f_1(t) = \frac{\frac{\alpha}{\lambda_1}\left(\frac{t}{\lambda_1}\right)^{\alpha-1}}{\left(1+\left(\frac{t}{\lambda_1}\right)^{\alpha}\right)^2}, \tag{2.20}$$

and

$$f_2(t) = \frac{\frac{\alpha}{\lambda_2}\left(\frac{t}{\lambda_2}\right)^{\alpha-1}}{\left(1+\frac{t^\alpha-\tau_2^\alpha}{\lambda_2^\alpha}+\frac{\tau_2^\alpha}{\lambda_1^\alpha}\right)^2}. \tag{2.21}$$

2.3.4 Maximum likelihood Estimation

The likelihood function from observations t_{ij}, $i = 1, 2, = 1, 2, \ldots, n_i$, can be written as

$$L(\lambda_1, \lambda_2) = \prod_{j=1}^{n_1} f_0(t_{1j}) \prod_{j=1}^{n_2} f_0(t_{2j})(1 - F_0(T))^{n_c},$$

where

$$F_0(t) = \begin{cases} \dfrac{\left(\frac{t}{\lambda_1}\right)^{\alpha}}{1+\left(\frac{t}{\lambda_1}\right)^{\alpha}}, & t < \tau_2 \\[2ex] \dfrac{\frac{t^\alpha-\tau_2^\alpha}{\lambda_2^\alpha}+\frac{\tau_2^\alpha}{\lambda_1^\alpha}}{1+\frac{t^\alpha-\tau_2^\alpha}{\lambda_2^\alpha}+\frac{\tau_2^\alpha}{\lambda_1^\alpha}}, & t \geq \tau_2. \end{cases}$$

Thus,

$$L(\lambda_1,\lambda_2)=\prod_{j=1}^{n_1}\left[\frac{\left(\frac{\alpha}{\lambda_1}\right)\left(\frac{t_{1j}}{\lambda_1}\right)^{\alpha-1}}{\left(1+\left(\frac{t_{1j}}{\lambda_1}\right)^{\alpha}\right)^2}\right]\prod_{j=1}^{n_2}\left[\frac{\left(\frac{\alpha}{\lambda_2}\right)\left(\frac{t_{2j}}{\lambda_2}\right)^{\alpha-1}}{\left(1+\frac{t_{2j}^{\alpha}-\tau_2^{\alpha}}{\lambda_2^{\alpha}}+\frac{\tau_2^{\alpha}}{\lambda_1^{\alpha}}\right)^2}\right]$$
$$\times\prod_{j=1}^{n_c}\left[\frac{1}{1+\left(\frac{T^{\alpha}-\tau_2^{\alpha}}{\lambda_2^{\alpha}}\right)+\frac{\tau_2^{\alpha}}{\lambda_1^{\alpha}}}\right]. \tag{2.22}$$

By substituting for λ_i, $i=1,2$, in (2.18); the log-likelihood function is a function of unknown parameters β_0 and β_1.

$$\begin{aligned}\log L(\beta_0,\beta_1)=&\sum_{j=1}^{n_1}[\log\alpha-(\beta_0+\beta_1x_1)+(\alpha-1)\log t_{1j}\\&-(\alpha-1)(\beta_0+\beta_1x_1)-2\log(1+t_{1j}^{\alpha}e^{-\alpha\beta_0-\alpha\beta_1x_1})]\\&+\sum_{j=1}^{n_2}[\log\alpha-(\beta_0+\beta_1x_2)+(\alpha-1)\log t_{2j}^{\alpha}\\&-(\alpha-1)(\beta_0+\beta_1x_2)\\&-2\log(1+(t_{2j}^{\alpha}-\tau_2^{\alpha})e^{-\alpha\beta_0-\alpha\beta_1x_2}+\tau_2^{\alpha}e^{-\alpha\beta_0-\alpha\beta_1x_1})]\\&+n_c[-\log(1+(T^{\alpha}-\tau_2^{\alpha})e^{-\alpha\beta_0-\alpha\beta_1x_2}\\&+\tau_2^{\alpha}e^{-\alpha\beta_0-\alpha\beta_1x_1})].\end{aligned} \tag{2.23}$$

The maximum likelihood estimates $\hat{\beta}_0,\hat{\beta}_1$ are the parameter values that maximize log likelihood function given in (2.23).

2.3.5 Fisher Information Matrix

The Fisher information matrix is obtained by taking the expected values of the second partial and mixed partial derivatives of $\log L(\beta_0,\beta_1)$ with respect to β_0,β_1.

The first partial derivatives are

$$\begin{aligned}\frac{\partial\log L}{\partial\beta_0}=&\frac{\alpha n_c(e^{-\alpha\beta_0-\alpha\beta_1x_1}\tau_2^{\alpha}+e^{-\alpha\beta_0-\alpha\beta_1x_2}(T^{\alpha}-\tau_2^{\alpha}))}{1+e^{-\alpha\beta_0-\alpha\beta_1x_1}\tau_2^{\alpha}+e^{-\alpha\beta_0-\alpha\beta_1x_2}(T^{\alpha}-\tau_2^{\alpha})}\\&+\sum_{j=1}^{n_1}\left(-\alpha+\frac{2\alpha e^{-\alpha\beta_0-\alpha\beta_1x_1}t_{1j}^{\alpha}}{1+e^{-\alpha\beta_0-\alpha\beta_1x_1}t_{1j}^{\alpha}}\right)\end{aligned}$$

$$+\sum_{j=1}^{n_2}\left(-\alpha+2\alpha\left(\frac{e^{-\alpha\beta_0-\alpha\beta_1 x_1}\tau_2^\alpha + e^{-\alpha\beta_0-\alpha\beta_1 x_2}(t_{2j}^\alpha-\tau_2^\alpha)}{1+e^{-\alpha\beta_0-\alpha\beta_1 x_1}\tau_2^\alpha + e^{-\alpha\beta_0-\alpha\beta_1 x_2}(t_{2j}^\alpha-\tau_2^\alpha)}\right)\right), \tag{2.24}$$

$$\frac{\partial \log L}{\partial \beta_1} = \frac{n_c\alpha(x_1 e^{-\alpha\beta_0-\alpha\beta_1 x_1}\tau_2^\alpha + x_2 e^{-\alpha\beta_0-\alpha\beta_1 x_2}(T^\alpha-\tau_2^\alpha))}{(1+e^{-\alpha\beta_0-\alpha\beta_1 x_1}\tau_2^\alpha + e^{-\alpha\beta_0-\alpha\beta_1 x_2}(T^\alpha-\tau_2^\alpha))}$$
$$+\sum_{j=1}^{n_1}\left(-x_1-(\alpha-1)x_1+\frac{2\alpha e^{-\alpha\beta_0-\alpha\beta_1 x_1}x_1 t_{1j}^\alpha}{1+e^{-\alpha\beta_0-\alpha\beta_1 x_1}t_{1j}^\alpha}\right)+\sum_{j=1}^{n_2}$$
$$\times\left(-x_2-(\alpha-1)x_2+2\alpha\left(\frac{x_1\tau_2^\alpha e^{-\alpha\beta_0-\alpha\beta_1 x_1} + x_2 e^{-\alpha\beta_0-\alpha\beta_1 x_2}(t_{2j}^\alpha-\tau_2^\alpha)}{1+e^{-\alpha\beta_0-\alpha\beta_1 x_1}\tau_2^\alpha + e^{-\alpha\beta_0-\alpha\beta_1 x_2}(t_{2j}^\alpha-\tau_2^\alpha)}\right)\right). \tag{2.25}$$

These two expressions, when set equal to zero, yield the likelihood equations. The parameter values that are the solutions to these equations are the maximum likelihood estimates.

The second partial derivatives are

$$\frac{\partial^2 \log L}{\partial \beta_0^2}$$
$$= -n_c\alpha^2\left(\frac{e^{-\alpha\beta_0-\alpha\beta_1 x_2}(T^\alpha-\tau_2^\alpha)+\tau_2^\alpha e^{-\alpha\beta_0-\alpha\beta_1 x_1}}{(1+e^{-\alpha\beta_0-\alpha\beta_1 x_2}(T^\alpha-\tau_2^\alpha)+\tau_2^\alpha e^{-\alpha\beta_0-\alpha\beta_1 x_1})^2}\right)$$
$$-2\sum_{j=1}^{n_1}\left(\frac{\alpha^2 e^{-\alpha\beta_0-\alpha\beta_1 x_1}t_{1j}^\alpha}{(1+e^{-\alpha\beta_0-\alpha\beta_1 x_1}t_{1j}^\alpha)^2}\right)$$
$$-2\sum_{j=1}^{n_2}\left(\frac{\alpha^2\tau_2^\alpha e^{-\alpha\beta_0-\alpha\beta_1 x_1}+\alpha^2 e^{-\alpha\beta_0-\alpha\beta_1 x_2}(t_{2j}^\alpha-\tau_2^\alpha)}{(1+\tau_2^\alpha e^{-\alpha\beta_0-\alpha\beta_1 x_1}+e^{-\alpha\beta_0-\alpha\beta_1 x_2}(t_{2j}^\alpha-\tau_2^\alpha))^2}\right), \tag{2.26}$$

$$\frac{\partial^2 \log L}{\partial \beta_1^2}$$
$$= -n_c\alpha^2 \left(\frac{e^{-\alpha\beta_0-\alpha\beta_1 x_1} x_1^2 \tau_2^\alpha + x_2^2 e^{-\alpha\beta_0-\alpha\beta_1 x_2}(T^\alpha - \tau_2^\alpha)}{(1 + e^{-\alpha\beta_0-\alpha\beta_1 x_2}(T^\alpha - \tau_2^\alpha) + e^{-\alpha\beta_0-\alpha\beta_1 x_1}\tau_2^\alpha)^2} \right)$$
$$-2\sum_{j=1}^{n_1} \left(\frac{\alpha^2 x_1^2 e^{-\alpha\beta_0-\alpha\beta_1 x_1} t_{1j}^\alpha}{(1 + e^{-\alpha\beta_0-\alpha\beta_1 x_1} t_{1j}^\alpha)^2} \right) - 2\sum_{j=1}^{n_2}$$
$$\times \left(\frac{\begin{array}{c}\alpha^2 x_1^2 e^{-\alpha\beta_0-\alpha\beta_1 x_1}\tau_2^\alpha + \alpha^2 x_2^2 e^{-\alpha\beta_0-\alpha\beta_1 x_2}(t_{2j}^\alpha - \tau_2^\alpha) \\ +(x_1 - x_2)^2\alpha^2 \times e^{-\alpha\beta_0-\alpha\beta_1 x_1}\tau_2^\alpha e^{-\alpha\beta_0-\alpha\beta_1 x_2}(t_{2j}^\alpha - \tau_2^\alpha)\end{array}}{(1 + e^{-\alpha\beta_0-\alpha\beta_1 x_1}\tau_2^\alpha + e^{-\alpha\beta_0-\alpha\beta_1 x_2}(t_{2j}^\alpha - \tau_2^\alpha))^2} \right), \tag{2.27}$$

$$\frac{\partial^2 \log L}{\partial \beta_0 \partial \beta_1}$$
$$= -n_c\alpha^2 \left(\frac{x_2 e^{-\alpha\beta_0-\alpha\beta_1 x_2}(T^\alpha - \tau_2^\alpha) + x_1 \tau_2^\alpha e^{-\alpha\beta_0-\alpha\beta_1 x_1}}{(1 + e^{-\alpha\beta_0-\alpha\beta_1 x_2}(T^\alpha - \tau_2^\alpha) + \tau_2^\alpha e^{-\alpha\beta_0-\alpha\beta_1 x_1})^2} \right)$$
$$-2\sum_{j=1}^{n_1} \left(\frac{\alpha^2 x_1 \tau_2^\alpha e^{-\alpha\beta_0-\alpha\beta_1 x_1} t_{1j}^\alpha}{(1 + e^{-\alpha\beta_0-\alpha\beta_1 x_1} t_{1j}^\alpha)^2} \right)$$
$$-2\sum_{j=1}^{n_2} \left(\frac{\alpha^2 x_1 \tau_2^\alpha e^{-\alpha\beta_0-\alpha\beta_1 x_1} + \alpha^2 x_2 e^{-\alpha\beta_0-\alpha\beta_1 x_2}(t_{2j}^\alpha - \tau_2^\alpha)}{(1 + \tau_2^\alpha e^{-\alpha\beta_0-\alpha\beta_1 x_1} + e^{-\alpha\beta_0-\alpha\beta_1 x_2}(t_{2j}^\alpha - \tau_2^\alpha))^2} \right). \tag{2.28}$$

The Fisher information matrix F is

$$F = \begin{bmatrix} F_{11} & F_{12} \\ F_{21} & F_{22} \end{bmatrix}.$$

The elements of the Fisher information matrix for an observation are the negative expectations of these second partial derivatives. These elements have been obtained using following results:

(i) $n_1 \sim \text{Bin}(n, p_1)$,

$$\begin{aligned} \text{where } p_1 &= P[0 < T_1 < \tau_2] \\ &= \int_0^{\tau_2} f_1(t)dt. \end{aligned}$$

$$\begin{aligned} E[n_1] &= n \int_0^{\tau_2} \frac{\frac{\alpha}{\lambda_1}\left(\frac{t_1}{\lambda_1}\right)^{\alpha-1}}{\left(1+\left(\frac{t_1}{\lambda_1}\right)^{\alpha}\right)^2} \\ &= n \frac{\frac{\tau_2^\alpha}{\lambda_1^\alpha}}{\left(1+\frac{\tau_2^\alpha}{\lambda_1^\alpha}\right)}. \end{aligned}$$

(ii) $n_2 \sim \text{Bin}(n, p_2)$,

$$\begin{aligned} \text{where } p_2 &= P[\tau_2 < T_1 < T] \\ &= \int_{\tau_2}^{T} f_2(t)dt. \end{aligned}$$

$$\begin{aligned} E[n_2] &= n \int_{\tau_2}^{T} \frac{\frac{\alpha}{\lambda_2}\left(\frac{t_2}{\lambda_2}\right)^{\alpha-1}}{\left(1+\left(\frac{t_2^\alpha-\tau_2^\alpha}{\lambda_2^\alpha}\right)+\frac{\tau_2^\alpha}{\lambda_1^\alpha}\right)^2} dt_2 \\ &= n \left(\frac{1}{\left(1+\frac{\tau_2^\alpha}{\lambda_1^\alpha}\right)} - \frac{1}{\left(1+\frac{T^\alpha-\tau_2^\alpha}{\lambda_2^\alpha}+\frac{\tau_2^\alpha}{\lambda_1^\alpha}\right)} \right). \end{aligned}$$

(iii)
$$\begin{aligned} E[n_c] &= E[n] - E[n_1] - E[n_2] \\ &= n - E[n_1] - E[n_2]. \end{aligned}$$

(iv)
$$E\left[\sum_{i=1}^{m} g(X_i)\right] = E[m]E[g(X)],$$

where m is a nonnegative integer valued random variable and $g(.)$ is a function of i.i.d. random variables.

Thus,

$$
\begin{aligned}
F_{11} &= E\left[\frac{-\partial^2 \log L}{\partial \beta_0^2}\right] \\
&= \alpha^2 \left[\frac{\frac{(T^\alpha - \tau_2^\alpha)}{\lambda_2^\alpha} + \frac{\tau_2^\alpha}{\lambda_1^\alpha}}{\left(1 + \frac{T^\alpha - \tau_2^\alpha}{\lambda_2^2} + \frac{\tau_2^\alpha}{\lambda_1^\alpha}\right)^2}\right] E[n_c] + 2\alpha^2 E\left[\frac{\frac{T_{1j}^\alpha}{\lambda_1^\alpha}}{\left(1 + \frac{T_{1j}^\alpha}{\lambda_1^\alpha}\right)^2}\right] E[n_1] \\
&\quad + 2\alpha^2 E\left[\frac{\frac{\tau_2^\alpha}{\lambda_1^\alpha} + \frac{T_{2j}^\alpha - \tau_2^\alpha}{\lambda_2^\alpha}}{\left(1 + \frac{\tau_2^\alpha}{\lambda_1^\alpha} + \frac{T_{2j}^\alpha - \tau_2^\alpha}{\lambda_2^\alpha}\right)^2}\right] E[n_2] \\
&= I_1, \\
F_{22} &= E\left[\frac{-\partial^2 \log L}{\partial \beta_1^2}\right] \\
&= \alpha^2 \left\{\frac{\frac{T^\alpha - \tau_2^\alpha}{\lambda_2^\alpha} x_2^2 + \frac{\tau_2^\alpha}{\lambda_1^\alpha} x_1^2}{\left(1 + \frac{T^\alpha - \tau^\alpha}{\lambda_2^\alpha} + \frac{\tau^\alpha}{\lambda_1^\alpha}\right)^2}\right\} E[n_c] + 2\alpha^2 x_1^2 E[n_1] E\left[\frac{\frac{T_{1j}^\alpha}{\lambda_1^\alpha}}{\left(1 + \frac{T_{1j}^\alpha}{\lambda_1^\alpha}\right)^2}\right] \\
&\quad + 2\alpha^2 E[n_2] E\left[\frac{x_1^2 \frac{\tau_2^\alpha}{\lambda_1^\alpha} + x_2^2 \frac{T_{2j}^\alpha - \tau_2^\alpha}{\lambda_2^\alpha} + (x_1 - x_2)^2 \left(\frac{\tau_2^\alpha}{\lambda_1^\alpha}\right)\left(\frac{T_{2j}^\alpha - \tau_2^\alpha}{\lambda_2^\alpha}\right)}{\left(1 + \frac{\tau_2^\alpha}{\lambda_1^\alpha} + \frac{T_{2j}^\alpha - \tau_2^\alpha}{\lambda_2^\alpha}\right)^2}\right] \\
&= (I_1 - I_2) x_1^2 + I_2 x_2^2 + I_3 (x_1 - x_2)^2, \\
F_{12} &= F_{21} = \left[\frac{-\partial^2 \log L}{\partial \beta_0 \partial \beta_1}\right] \\
&= \alpha^2 E[n_c] \left[\frac{x_1 \frac{\tau_2^\alpha}{\lambda_1^\alpha} + x_2 \frac{T^\alpha - \tau_2^\alpha}{\lambda_2^\alpha}}{\left(1 + \frac{T^\alpha - \tau_2^\alpha}{\lambda_2^\alpha} + \frac{\tau_2^\alpha}{\lambda_1^\alpha}\right)^2}\right] + 2\alpha^2 E[n_1] E\left[\frac{x_1 \frac{T_{1j}^\alpha}{\lambda_1^\alpha}}{\left(1 + \frac{T_{1j}^\alpha}{\lambda_1^\alpha}\right)^2}\right] \\
&\quad + 2\alpha^2 E[n_2] E\left[\frac{x_1 \frac{\tau_2^\alpha}{\lambda_1^\alpha} + x_2 \frac{T_{2j}^\alpha - \tau_2^\alpha}{\lambda_2^\alpha}}{\left(1 + \frac{\tau_2^\alpha}{\lambda_1^\alpha} + \frac{T_{2j}^\alpha - \tau_2^\alpha}{\lambda_2^\alpha}\right)^2}\right] \\
&= (I_1 - I_2) x_1 + I_2 x_2,
\end{aligned}
$$

and the resultant Fisher information matrix F is

$$F = \frac{1}{3} n\alpha^2 \begin{bmatrix} I_1 & (I_1 - I_2)x_1 + I_2 x_2 \\ (I_1 - I_2)x_1 + I_2 x_2 & (I_1 - I_2)x_1^2 + I_2 x_2^2 + I_3(x_1 - x_2)^2 \end{bmatrix},$$

where

$$I_1 = 1 - p_1'^3,$$

$$I_2 = (p_2'^3 - p_1'^3)\left(1 + \frac{\tau_2^\alpha}{\lambda_1^\alpha}\right),$$

$$I_3 = \frac{\tau_2^\alpha}{\lambda_1^\alpha}\left(I_2 - 3\frac{(T^\alpha - \tau_2^\alpha)}{\lambda_2^\alpha} p_1'^3\right),$$

$$p'_1 = \frac{1}{1 + \frac{T^\alpha - \tau_2^\alpha}{\lambda_2^\alpha} + \frac{\tau_2^\alpha}{\lambda_1^\alpha}},$$

and

$$p_2' = \frac{1}{\left(1 + \frac{\tau_2^\alpha}{\lambda_1^\alpha}\right)}.$$

2.3.6 Optimum Plan

The asymptotic variance of the log of median life at the design stress is given by

$$\begin{aligned} \text{n.AsVar}(\log \hat{\lambda}_0) &= \text{n.AsVar}(\hat{\beta}_0 + \hat{\beta}_1 x_0) \\ &= n[\text{var}[\hat{\beta}_0] + x_0^2 \text{var}(\hat{\beta}_1) + 2x_0 \text{cov}(\hat{\beta}_0, \hat{\beta}_1)] \\ &= \frac{3}{\alpha^2}\left[\frac{I_1\xi^2 + 2I_2\xi + (I_2 + I_3)}{I_1(I_2 + I_3) - I_2^2}\right], \end{aligned} \tag{2.29}$$

where ^ implies the MLE.

The optimal τ_2^* is found by minimizing (2.29) using NMinimize option of *Mathematica 5.0*. It is found that

(i) As $T \rightarrow \infty$, $\text{n.AsVar}(\log \hat{\lambda}_0)$ tends to $\text{n.AsVar}(\log \hat{\lambda}_0)$ for uncensored case given by

$$\text{n.AsVar}[\log \hat{\lambda}_0] = \frac{3}{\alpha^2}\left[\frac{\xi^2 + 2A_2\xi + (A_2 + A_3)}{(A_2 + A_3) - A_2^2}\right],$$

where

$$A_1 = \frac{\left(2 + \frac{\tau_2^\alpha}{\lambda_1^\alpha}\right)\frac{\tau_2^\alpha}{\lambda_1^\alpha}}{\left(1 + \frac{\tau_2^\alpha}{\lambda_1^\alpha}\right)^2},$$

$$A_2 = 1 - A_1,$$

and

$$A_3 = \frac{\frac{\tau_2^\alpha}{\lambda_1^\alpha}}{\left(1+\frac{\tau_2^\alpha}{\lambda_1^\alpha}\right)^2}.$$

(ii) As λ_1 increases, τ_2^* increases for given λ_2, ξ, n_c, and T as shown in Table 2.2

(iii) τ_2^* increases in T for given $\lambda_1, \lambda_2, n_c$, and ξ as given in Table 2.3.

2.4 TESTING OF HYPOTHESES

2.4.1 For Complete Data Set

Tests of hypotheses for parameters can be obtained either by using the likelihood ratio method or the asymptotic normality of MLEs in large samples. In the latter case,

$$(\hat{\beta}_0, \hat{\beta}_1) \sim N((\beta_0, \beta_1), F^{-1}).$$

Table 2.2: Optimal time changing stress $\tau_2^*(\xi, \lambda_1, T, n_c)$ ($T = 20, \beta_0 = 3, \beta_1 = -1, \xi = 0.3, n_c = 6, \lambda_2 = 4$)

	α				
λ_1	0.5	1	1.5	2	2.5
6	4.9915	7.8982	8.8907	8.0328	8.3802
8	5.8673	9.5027	11.1309	11.5284	11.1354
10	6.5948	10.7564	12.8658	13.8312	13.7897
12	7.2162	11.7490	14.1465	15.5323	16.0844
14	7.7578	12.5486	15.0775	16.6438	17.5691
16	8.2371	13.2046	15.7600	17.3459	18.3255

Table 2.3: Optimal time changing stress $\tau_2^*(\xi, n_c, \lambda_1, \lambda_2)$ ($\beta_0 = 3$, $\beta_1 = -1, \xi = 0.3, n_c = 6, \lambda_1 = 12.1825, \lambda_2 = 4.48169$)

	α				
T	**0.5**	**1**	**1.5**	**2**	**2.5**
20	7.1782	11.6459	13.9698	15.3555	16.0185
25	8.2446	13.5101	16.1794	17.2496	16.9340
30	9.2182	15.1832	17.8676	18.0986	17.0826
35	10.1215	16.6749	19.0644	18.4131	17.1156
40	10.9685	17.9979	19.8646	18.5360	17.1259
45	11.7685	19.1667	20.3834	18.5906	17.1300
60	12.5281	20.196	20.7186	18.6179	17.1319

An important inference problem concerning the regression coefficients (β_0, β_1) is the test of hypothesis $H_0 : \beta_1 = 0$ against $H_1 : \beta_1 < 0$. To test H_0 against H_1, one can use the likelihood ratio statistic

$$\Lambda = -2 \log \left[\frac{L(\tilde{\beta}_0, 0)}{L(\hat{\beta}_0, \hat{\beta}_1)} \right],$$

where $\tilde{\beta}_0$ is the MLE of β_0 under H_0 and $\hat{\beta}_0, \hat{\beta}_1$ are the unrestricted MLEs. Large values of Λ provide evidence against H_0, and approximate significance levels can be calculated by using the fact that in large samples Λ is approximately distributed as χ_1^2 under H_0. Similarly one can test for $H'_0 : \beta_0 = 0$ against $H'_1 : \beta_0 \neq 0$.

2.4.2 For Time-Censored Data Set

Confidence interval approach to hypothesis testing is used. Under this approach if the confidence interval, referred to as acceptance region, includes the null-hypothesized value, the null hypothesis is not rejected. On the other hand, if the null hypothesized value lies outside the confidence interval, i.e., in the region of rejection, one can reject the null hypothesis.

Remark 2.1. It is found that as T increases, n_c decreases and the confidence intervals for β_0 and β_1 reduce to their respective confidence intervals for uncensored case.

Table 2.4: Simple-step stress with two stress levels simulated data (complete data set) ($\beta_0 = 3$, $\beta_1 = -1$, $\alpha = 0.8$, and $\tau_1^* = 35.3702$)

Step Stress	Failure Times				
1. $x_1 = 0.5$	22.5218	31.0146	12.1952	28.2783	3.6311
	0.6417	3.6444	27.3618	0.4580	27.3504
	6.8547	6.3763	18.2242	0.6721	16.3055
	2.4938	1.0398	3.7583		
2. $x_2 = 1.5$	53.6776	75.5377	54.8459	67.0387	73.1331
	63.1528	110.6640	62.6622	51.1622	179.2730
	269.3810	35.6830			

2.5 NUMERICAL EXAMPLES

2.5.1 For Complete Data Set

Example 2.1. The data in Table 2.4 include 30 simulated observations based on $x_0 = 0.2$, $x_1 = 0.5$, $x_2 = 1.5$, $\alpha = 0.8$, $\beta_0 = 3$, $\beta_1 = -1$, $\lambda_1 = 12.1825$.

The optimal value of τ obtained using (2.17) is $\tau_1^* = 35.3702$. The following model is fitted:

$$\log(\lambda_i) = \beta_0 + \beta_1 x_i, \ i = 1, 2.$$

The MLEs of β_0 and β_1 are readily obtained using NMaximize option of *Mathematica 5.0.* The MLEs obtained are:

$$\hat{\beta}_0 = 3.6972 \quad \text{and} \quad \hat{\beta}_1 = -1.1703.$$

The estimate of the median life at design stress is

$$\hat{\lambda}_0 = 31.9177.$$

The inverse of Fisher observed information matrix $\hat{F}^{-1}$ is given as follows:

$$\hat{F}^{-1} = \begin{bmatrix} 0.3389 & -0.2734 \\ -0.2734 & 0.4091 \end{bmatrix}$$

The observed values of F^{-1}, that is $\hat{F}^{-1}$, were determined by substituting the estimated parameters $\hat{\beta}_0, \hat{\beta}_1$, for the true parameters in the asymptotic covariance matrix. To find the standard errors of $\hat{\beta}_0, \hat{\beta}_1$, the square root of the diagonal elements of $\hat{F}^{-1}$ is used. 95% confidence intervals for the parameters are

$$\hat{\beta}_k \pm z_{0.025}\text{S.E.}(\hat{\beta}_k), \quad k = 0, 1.$$

Thus,

$$2.9508 \leq \beta_0 \leq 4.4436$$

and

$$-1.9903 \leq \beta_1 \leq -0.3502.$$

Table 2.5 shows the likelihood ratio statistics Λ for tests of various submodels against the full model (β_0, β_1). The column d.f. indicates the degree of freedom of the Chi-square approximation.

The tests $H'_0 : \beta_0 = 0$ and $H_0 : \beta_1 = 0$ are rejected.

2.5.2 For Time-Censored Data Set

Example 2.2. Suppose an optimum simple step-stress plan is designed using the model

$$\log(\lambda_i) = \beta_0 + \beta_1 x_i, \quad i = 1, 2.$$

The data in Table 2.6 includes 35 simulated observations from cumulative log-logistic distribution (see (2.19)) based on data: $\alpha = 1.5$, $\beta_0 = 3$, $\beta_1 = -1$, $x_1 = 0.5$, $x_2 = 1.5$, $x_0 = 0.2$, $\xi = 0.3$, $T = 20$, $n = 35$ then the optimal value of τ_2 obtained by using NMinimize option of *Mathematica 5.0* is $\tau_2^* = 13.9698$.

Table 2.5: Tests for parameters in the model $\log(\lambda_i) = \beta_0 + \beta_1 x_i, i = 1, 2$

Model	Λ	d.f.	$\chi^2_{(0.05)}$
Full model (β_0, β_1)			
$\beta_0 = 0$	76.4807	1	3.84
$\beta_1 = 0$	8.3455	1	3.84

Table 2.6: Simple time-censored step-stress with two stress levels simulated data ($\beta_0 = 3, \beta_1 = -1, \alpha = 1.5$, and $\tau_2^* = 13.9698, n_c = 6$)

Step Stress	Failure Times
1. $x_1 = 0.5$	8.6730, 2.0152, 8.3673, 7.6020, 10.0332, 2.5650, 9.5276, 4.1290, 8.9773, 6.9223, 10.0849, 2.5480, 7.6780, 10.7510, 6.2867, 2.6169, 8.8257, 9.2201, 4.4801, 5.7062
2. $x_2 = 1.5$	16.9226, 14.4341, 16.9479, 15.0116, 19.5461, 14.5657, 15.4186, 15.9957, 19.364

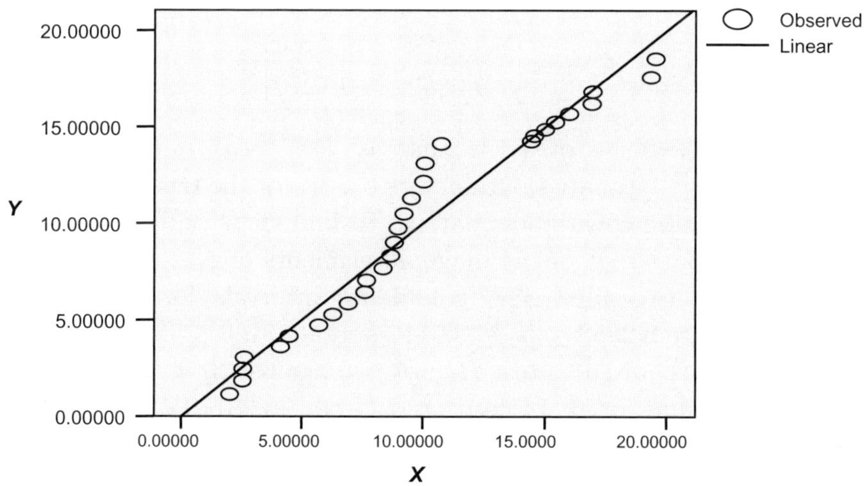

Figure 2.1: The cumulative exposure log-logistic probability plot for the simulated data

Figure 2.1 shows the associated log-logistic cumulative exposure probability plot. The failure times in Table 2.6 are arranged in increasing order and are ranked from $1, 2, \ldots, k, \ldots, n$. The plot is obtained by taking these ranked failure times on X-axis, and their corresponding ordinate values, viz., $\lambda_1(k/(n+1-k))^{1/\alpha}$ for failure times less than τ_2^* and $((k/(n+1-k) - (\tau_2^\alpha/\lambda_1^\alpha))\lambda_2^\alpha + \tau_2^\alpha)^{1/\alpha}$ for failure times greater than or equal to τ_2^*, on Y-axis (Grosh, 1989). These ordinate values are obtained using (2.19). The plotted points tend to follow straight line, which is substantiated by fitting straight line to these points resulting in high value of coefficient of determination $r^2 = 0.988$. The log-logistic cumulative exposure distribution therefore appears to describe the data adequately.

The MLEs of model parameters β_0 and β_1 obtained using NMaximize option of *Mathematica 5.0* is

$$\hat{\beta}_0 = 2.8981 \quad \text{and} \quad \hat{\beta}_1 = -0.8174.$$

The estimate of the median life at the design stress is

$$\hat{\lambda}_0 = 15.4035.$$

The inverse of observed Fisher information matrix $\hat{F}^{-1}$ is obtained as

$$\hat{F}^{-1} = \begin{bmatrix} 0.0978 & -0.0869 \\ -0.0869 & 0.1270 \end{bmatrix}.$$

The observed values of F^{-1}, that is $\hat{F}^{-1}$, was determined by substituting the estimated parameters $\hat{\beta}_0, \hat{\beta}_1$, for the true parameters in the asymptotic covariance matrix. To find the standard errors of $\hat{\beta}_0, \hat{\beta}_1$, the square root of the diagonal elements of $\hat{F}^{-1}$ is taken.

Thus, the two-sided 95% confidence intervals for β_0 and β_1, respectively, are $2.2852 \leq \beta_0 \leq 3.5110$ and $-1.5160 \leq \beta_1 \leq -0.1188$.

Since it is required to test $H_0 : \beta_1 = 0$ against $\beta_1 < 0$, lower 95% confidence bound for β_1 is used, and $(-\infty, -0.231195]$ is obtained. H_0 is rejected as it does not contain the null hypothesized value.

2.6 SENSITIVITY ANALYSIS

The sensitivity analysis identifies the sensitive parameters which need to be estimated with special care for the purpose of minimizing the risk of obtaining an erroneous optimal solution. The effects of $\pm 1\%$ to $\pm 5\%$ change in the pre-estimated parameters β_0, β_1 on τ_1^* and τ_2^* are presented for complete data set and Type-I censored data set in Tables 3.7 and 3.8, respectively.

$$\%\Delta\tau^* = (|(\tau^* - \tau^0)/\tau^*|)100,$$

where τ^* is the optimal stress change point for the plans obtained with the correctly specified values, and τ^0 is the stress change point

Table 2.7: Effects of incorrect pre-estimates of β_0, β_1 (Complete data set) ($\%\Delta\tau_1^*$ due to $\pm1\% - \pm5\%$ change in β_0 and β_1)

Percentage Deviation (%)	β_0	β_1	τ_1^*	$\%\Delta\tau_1^*$
+1	3.03	−1.01	36.265	2.531
−1	2.97	−0.99	34.496	2.469
+2	3.06	−1.02	37.183	5.127
−2	2.94	−0.98	33.645	4.876
+3	3.09	−1.03	38.125	7.788
−3	2.91	−0.97	32.814	7.225
+4	3.12	−1.04	39.090	10.517
−4	2.88	−0.96	32.004	9.516
+5	3.15	−1.05	40.079	13.314
−5	2.85	−0.95	31.214	11.750

Table 2.8: Effects of incorrect pre-estimates of β_0, β_1 (Type-I censored data set) ($\%\Delta\tau_2^*$ due to $\pm1\% - \pm5\%$ change in β_0 and β_1)

Percentage Deviation (%)	β_0	β_1	τ_2^*	$\%\Delta\tau_2^*$
+1	3.03	−1.01	14.095	0.901
−1	2.97	−0.99	13.841	0.921
+2	3.06	−1.02	14.219	1.785
−2	2.94	−0.98	13.709	1.862
+3	3.09	−1.03	14.340	2.650
−3	2.91	−0.97	13.575	2.823
+4	3.12	−1.04	14.458	3.496
−4	2.88	−0.96	13.438	3.804
+5	3.15	−1.05	14.574	4.326
−5	2.85	−0.95	13.298	4.806

for the plans obtained with misspecified values. Table 2.7 shows that if the misspecified values are not too far from the true values the increase is small. Table 2.8 indicates that because the misspecified values have a very small effect on the optimal values, τ_2^*, they are not sensitive.

2.7 SUMMARY

In this chapter, the design of optimum simple step-stress accelerated life test for complete data set as well as for Type-I censored data set has been explained. A log-logistic life distribution with median life that is log linear function of stress and a cumulative exposure model are assumed in both the cases. The model suggested is appropriate for highly reliable components such as insulation system. The optimum test plans are obtained by minimizing the asymptotic variance of the maximum likelihood estimator of the median life at a design stress for some selected values of parameter. As censoring time increases optimal stress change time also increases. Inferential problems have been also discussed. Sensitivity analyses show that initial estimates have a small effect on optimal values.

CHAPTER 3

OPTIMUM STEP-STRESS PARTIALLY ACCELERATED LIFE TEST PLANS WITH TYPE-I AND TYPE-II CENSORING

3.1 INTRODUCTION

In step-stress PALT, a test item is first run at normal use condition and if it does not fail then, it is run at accelerated condition until failure occurs or the observation is censored.

This chapter focuses on formulation of optimum step-stress PALT for the truncated logistic distribution with Type-I and Type-II censoring. Truncated distributions arise when sample selection is not possible in some sub-region of the sample space (see Appendix A.01). The failure rate of truncated logistic distribution truncated at point zero, is increasing and is more realistically bounded below and above by a nonzero finite quantity (see Section 1.4.2.8).

The work done in the literature on the design of step-stress PALT has been reviewed in 1.19.2.

3.1.1 Notation

q_l	proportion of units failed at use condition
nq_l or n_u	number of units failed at use condition
q_r	proportion of units failed before censoring
r	number of units failed before censoring at which the test is terminated under Type-II censoring (case I: number of failures before censoring pre-specified)

nq_r	number of units failed before censoring at which the test is terminated under Type-II censoring (case II: proportion of units failing before censoring pre-specified)
$q_r - q_l$	proportion of units failed at accelerated condition
$n(q_r - q_l)$ or n_a	number of units failed at accelerated condition
$1 - q_r$	proportion of censored units
$n(1 - q_r)$ or n_c	number of censored units
n	total number of test items in a step-stress PALT ($n = nq_r + n(1 - q_r)$ or $n = n_u + n_a + n_c$)
β	acceleration factor, $\beta > 1$
T	lifetime of an item at use condition
Y	total lifetime of an item under Type-I censoring and Type-II censoring (case I: number of failures before censoring pre-specified)
Y_i	observed value of the total lifetime Y_i of item i, $i = 1, 2, \ldots, n$ under Type-I censoring
$Y_{(i)}$	observed value of the total lifetime $Y_{(i)}$ of item i, $i = 1, 2, \ldots, n$ under Type-II censoring (case I: number of failures pre-specified)
U	total lifetime of an item under Type-II censoring (case II: proportion of units failing before censoring pre-specified)
$U_{(i)}$	observed value of the total lifetime $U_{(i)}$ of item i, $i = 1, 2, \ldots, n$ under Type-II censoring (case II: proportion of units failing before censoring pre-specified)
η	censoring time, $\eta > 0$
μ	location parameter, $-\infty < \mu < \infty$
σ	scale parameter, $\sigma > 0$
τ	stress change time, that is, $\tau < \eta$, under Type-I censoring, and $\tau < y_{(r)}$, under Type-II censoring
$F(t)$	cumulative distribution function
$f(t)$	probability density function
$R(t)$	reliability functions at time 't' at use condition
$h(t)$	hazard (failure) function at time 't' at use condition
$\wedge$	implies a MLE

δ_{1i}, δ_{2i} indicator functions
for Type-I censoring

$$\delta_{1i} = \begin{cases} 1, & y_i \le \tau, i = 1, 2, \ldots, n_u. \\ 0, & \text{otherwise.} \end{cases}$$

$$\delta_{2i} = \begin{cases} 1, & \tau < y_i \le \eta, i = n_u + 1, \ldots, n_u + n_a, \\ 0, & \text{otherwise.} \end{cases}$$

for Type-II censoring (case I: number of failures specified)

$$\delta_{1i} = \begin{cases} 1, & y_i \le \tau, i = 1, 2, \ldots, n_u \\ 0, & \text{otherwise} \end{cases}$$

$$\delta_{2i} = \begin{cases} 1, & \tau < y_i \le y_{(r)}, i = n_u + 1, \ldots, r, \\ 0, & \text{otherwise.} \end{cases}$$

for Type-II censoring (case II: proportion of units failing before censoring pre-specified)

$$\delta_{1i} = \begin{cases} 1, & u_{(i)} \le u_{(nq_l)}, \quad i = 1, 2, \ldots, nq_l \\ 0, & \text{otherwise} \end{cases}$$

$$\delta_{2i} = \begin{cases} 1, & u_{(nq_l)} < u_{(i)} \le u_{(nq_r)}, \quad i = (nq_l + 1), \ldots, nq_r \\ 0, & \text{otherwise} \end{cases}$$

$\bar{\delta}_{1i}$ $1 - \delta_{1i}$

$\bar{\delta}_{2i}$ $1 - \delta_{2i}$

A $\dfrac{1}{1 + \exp\left\{-\frac{\mu}{\sigma}\right\}}$

P_u probability of an item failing at use condition

P_u $\dfrac{1}{A}\left(\dfrac{1}{1 + \exp\left\{-\frac{\tau-\mu}{\sigma}\right\}} - \dfrac{1}{1 + \exp\left\{\frac{\mu}{\sigma}\right\}}\right),$

under Type-I censoring and Type-II censoring (case I: number of failures pre-specified)

P_u $\dfrac{1}{A}\left(\dfrac{1}{1 + \exp\left\{-\frac{u_{(nq_l)}-\mu}{\sigma}\right\}} - \dfrac{1}{1 + \exp\left\{\frac{\mu}{\sigma}\right\}}\right),$

under Type-II censoring (case II: proportion of units failing before censoring pre-specified)

P_a probability of an item failing at accelerated condition

P_a $\frac{1}{A}\left(\frac{1}{1+\exp\left\{-\frac{(\eta-\tau)\beta+\tau-\mu}{\sigma}\right\}}-\frac{1}{1+\exp\left\{-\frac{\tau-\mu}{\sigma}\right\}}\right)$,

under Type-I censoring

P_a $\frac{1}{A}\left(\frac{1}{1+\exp\left\{-\frac{(y_{(r)}-\tau)\beta+\tau-\mu}{\sigma}\right\}}-\frac{1}{1+\exp\left\{-\frac{\tau-\mu}{\sigma}\right\}}\right)$,

under Type-II censoring (case I: number of failures pre-specified), and

P_a $\frac{1}{A}\left(\frac{1}{1+\exp\left\{-\frac{u_{(nq_l)}+(u_{(nq_r)}-u_{(nq_l)})\beta-\mu}{\sigma}\right\}}-\frac{1}{1+\exp\left\{-\frac{u_{(nq_l)}-\mu}{\sigma}\right\}}\right)$,

under Type-II censoring (case II: proportion of units failing before censoring pre-specified)

P_c $1-(P_u+P_a)$

P_c $1-\frac{1}{A}\left(\frac{1}{1+\exp\left\{-\frac{(\eta-\tau)\beta+\tau-\mu}{\sigma}\right\}}-\frac{1}{1+\exp\left\{\frac{\mu}{\sigma}\right\}}\right)$,

under Type-I censoring

P_c $1-\frac{1}{A}\left(\frac{1}{1+\exp\left\{-\frac{(y_{(r)}-\tau)\beta+\tau-\mu}{\sigma}\right\}}-\frac{1}{1+\exp\left\{\frac{\mu}{\sigma}\right\}}\right)$,

under Type-II censoring (case I: number of failures pre-specified), and

P_c $1-\frac{1}{A}\left(\frac{1}{1+\exp\left\{-\frac{u_{(nq_l)}+(u_{(nq_r)}-u_{(nq_l)})\beta-\mu}{\sigma}\right\}}-\frac{1}{1+\exp\left\{\frac{\mu}{\sigma}\right\}}\right)$

under Type-II censoring (case II: proportion of units failing before censoring pre-specified)

$y_{(1)} < \cdots < y_{(n_u)} < \tau < y_{(n_u+1)} < \cdots < y_{(r)}$	ordered failure times under Type-II censoring (case I: number of failures pre-specified)
$u_{(1)} < \cdots < u_{(nq_l)} < u_{(nq_l+1)} < \cdots < u_{(nq_r)}$	ordered failure times under Type-II censoring (case II: proportion of units failing before censoring pre-specified)

3.1.2 Truncated Logistic Distribution under Step-Stress PALT

The cumulative distribution function (cdf), probability density function (pdf), reliability function, and hazard function, respectively, of the truncated logistic distribution are

$$F(t) = \frac{1}{A}\left(\frac{1}{1+\exp\{-\frac{t-\mu}{\sigma}\}} - \frac{1}{1+\exp\{\frac{\mu}{\sigma}\}}\right), \quad t > 0, \tag{3.1}$$

(Mood, Graybill, and Boes, 1974)

$$f(t) = \frac{\exp\{-\frac{t-\mu}{\sigma}\}}{A\sigma\left(1+\exp\{-\frac{t-\mu}{\sigma}\}\right)^2}, \quad t > 0, \tag{3.2}$$

$$R(t) = \frac{\exp\{-\frac{t-\mu}{\sigma}\}}{A\left(1+\exp\left\{-\frac{t-\mu}{\sigma}\right\}\right)}, \quad t > 0 \tag{3.3}$$

and

$$h(t) = \frac{1}{\sigma\left(1+\exp\{-\frac{t-\mu}{\sigma}\}\right)}, \quad t > 0. \tag{3.4}$$

The hazard function in (3.4) is an increasing function of t, and is bounded by

$$\frac{1}{\sigma\left(1+\exp\{\frac{\mu}{\sigma}\}\right)}, \quad \text{and} \quad \frac{1}{\sigma}.$$

Thus, the total lifetime Y of an item under Type-I censoring and Type-II censoring (case I: number of failures pre-specified) is

$$Y = \begin{cases} T, & T \le \tau \\ \tau + \beta^{-1}(T-\tau), & \text{otherwise.} \end{cases} \tag{3.5}$$

The cdf and pdf of total lifetime Y of an item under both Type-I censoring and Type-II censoring (case I: number of failures pre-specified), respectively, are

$$F(y) = \begin{cases} \dfrac{1}{A}\left(\dfrac{1}{1+\exp\{-\frac{y-\mu}{\sigma}\}} - \dfrac{1}{1+\exp\{\frac{\mu}{\sigma}\}}\right), & 0 < y \le \tau \\ \dfrac{1}{A}\left(\dfrac{1}{1+\exp\left\{-\frac{\tau+(y-\tau)\beta-\mu}{\sigma}\right\}} - \dfrac{1}{1+\exp\{\frac{\mu}{\sigma}\}}\right), & \tau < y \end{cases} \tag{3.6}$$

and

$$f(y) = \begin{cases} \dfrac{\exp\{-\frac{y-\mu}{\sigma}\}}{A\sigma\left(1+\exp\{-\frac{y-\mu}{\sigma}\}\right)^2}, & 0 < y \le \tau \\ \dfrac{\beta\exp\left\{-\frac{\tau+(y-\tau)\beta-\mu}{\sigma}\right\}}{A\sigma\left(1+\exp\left\{-\frac{\tau+(y-\tau)\beta-\mu}{\sigma}\right\}\right)^2}, & y > \tau. \end{cases} \tag{3.7}$$

Also, the total lifetime U of an item under Type-II censoring (case II: proportion of units failing before censoring pre-specified) is

$$U = \begin{cases} T, & T \le u_{(nq_l)} \\ u_{(nq_l)} + \beta^{-1}(T - u_{(nq_l)}), & \text{otherwise.} \end{cases} \tag{3.8}$$

The cdf and pdf of total lifetime U of an item under Type-II censoring (case II: proportion of units failing before censoring pre-specified), respectively, are

$$F(u) = \begin{cases} \dfrac{1}{A}\left(\dfrac{1}{1+\exp\{-\frac{u-\mu}{\sigma}\}} - \dfrac{1}{1+\exp\{\frac{\mu}{\sigma}\}}\right), & 0 < u \le u_{(nq_l)} \\ \dfrac{1}{A}\left(\dfrac{1}{1+\exp\left\{-\frac{u_{(nq_l)}+\beta(u-u_{(nq_l)})-\mu}{\sigma}\right\}} - \dfrac{1}{1+\exp\left\{\frac{\mu}{\sigma}\right\}}\right), & u_{(nq_l)} < u \end{cases} \tag{3.9}$$

and

$$f(u) = \begin{cases} \dfrac{\exp\{-\frac{u-\mu}{\sigma}\}}{A\sigma\left(1+\exp\{-\frac{u-\mu}{\sigma}\}\right)^2}, & 0 < u \leq u_{(nq_l)} \\ \dfrac{\beta\exp\left\{-\frac{u_{(nq_l)}+(u-u_{(nq_l)})\beta-\mu}{\sigma}\right\}}{A\sigma\left(1+\exp\left\{-\frac{u_{(nq_l)}+(u-u_{(nq_l)})\beta-\mu}{\sigma}\right\}\right)^2}, & u > u_{(nq_l)}. \end{cases} \tag{3.10}$$

3.2 OPTIMUM TIME-CENSORED STEP-STRESS PALT MODEL

In the present section, an optimum time-censored step-stress PALT plan for the truncated logistic distribution proposed by Srivastava and Mittal (2010) have been described.

3.2.1 Assumptions

1. The lifetime of an item tested at both use and accelerated condition follows the truncated logistic distribution.
2. The lifetimes of test items are independent and identically distributed random variables.

3.2.2 Test Procedure

1. All 'n' items are first tested at use condition.
2. If any item out of 'n' items does not fail at use condition by pre-specified time 'τ', then it is put on accelerated condition, and run until censoring time 'η'.
3. The test is continued until:

   (a) all test items fail, or
   (b) a prescribed censoring time η,

whichever occurs earlier, and the test conditions remain the same.

3.2.3 Log-Likelihood Function

The likelihood function L for n independent observations is

$$\begin{aligned}
L = L(\beta,\mu,\sigma) &= \prod_{i=1}^{n} L_i(\beta,\mu,\sigma;(y_i,\delta_{1i},\delta_{2i}))\\
&= \prod_{i=1}^{n}(f_1(y_i))^{\delta_{1i}}(f_2(y_i))^{\delta_{2i}}(R(\eta))^{\bar{\delta}_{1i}\bar{\delta}_{2i}}\\
&= \prod_{i=1}^{n}\left(\frac{\exp\{-\frac{y_i-\mu}{\sigma}\}}{A\sigma\left(1+\exp\{-\frac{y_i-\mu}{\sigma}\}\right)^2}\right)^{\delta_{1i}}\\
&\quad\times\left(\frac{\beta\exp\left\{-\frac{(y_i-\tau)\beta+\tau-\mu}{\sigma}\right\}}{A\sigma\left(1+\exp\left\{-\frac{(y_i-\tau)\beta+\tau-\mu}{\sigma}\right\}\right)^2}\right)^{\delta_{2i}}\\
&\quad\times\left(\frac{\exp\left\{-\frac{(\eta-\tau)\beta+\tau-\mu}{\sigma}\right\}}{A\left(1+\exp\left\{-\frac{(\eta-\tau)\beta+\tau-\mu}{\sigma}\right\}\right)}\right)^{\bar{\delta}_{1i}\bar{\delta}_{2i}}.
\end{aligned} \tag{3.11}$$

It is usually easier to maximize the natural logarithm of the likelihood function rather than the likelihood function itself.

Thus, the log-likelihood of (3.11) is

$$\begin{aligned}
\ln L = \ln L(\beta,\mu,\sigma) &= n_a\ln\beta-(n_a+n_u)\ln\sigma+n\ln\left(1+\exp\left\{-\frac{\mu}{\sigma}\right\}\right)\\
&-\sum_{i=1}^{n}2\delta_{1i}\ln\left(1+\exp\left\{-\frac{y_i-\mu}{\sigma}\right\}\right)\\
&-\sum_{i=1}^{n}\delta_{2i}\left(\frac{(y_i-\tau)\beta+\tau-\mu}{\sigma}\right)\\
&-\sum_{i=1}^{n}2\delta_{2i}\ln\left(1+\exp\left\{-\frac{(y_i-\tau)\beta+\tau-\mu}{\sigma}\right\}\right)
\end{aligned}$$

$$-\sum_{i=1}^{n} \bar{\delta}_{1i}\bar{\delta}_{2i}\left(\frac{(\eta-\tau)\beta+\tau-\mu}{\sigma}\right)$$

$$-\sum_{i=1}^{n} \bar{\delta}_{1i}\bar{\delta}_{2i}\ln\left(1+\exp\left\{-\frac{(\eta-\tau)\beta+\tau-\mu}{\sigma}\right\}\right)$$

$$-\sum_{i=1}^{n} \delta_{1i}\left(\frac{y_i-\mu}{\sigma}\right). \tag{3.12}$$

The model parameters are estimated using the method of maximum likelihood and are obtained by maximizing (3.12).

3.2.4 Fisher Information Matrix

The Fisher information matrix for n independent observations is

$$F = \begin{pmatrix} E\left[-\frac{\partial^2 \ln L}{\partial \beta^2}\right] & E\left[-\frac{\partial^2 \ln L}{\partial \beta \partial \mu}\right] & E\left[-\frac{\partial^2 \ln L}{\partial \beta \partial \sigma}\right] \\ E\left[-\frac{\partial^2 \ln L}{\partial \mu \partial \beta}\right] & E\left[-\frac{\partial^2 \ln L}{\partial \mu^2}\right] & E\left[-\frac{\partial^2 \ln L}{\partial \mu \partial \sigma}\right] \\ E\left[-\frac{\partial^2 \ln L}{\partial \sigma \partial \beta}\right] & E\left[-\frac{\partial^2 \ln L}{\partial \sigma \partial \mu}\right] & E\left[-\frac{\partial^2 \ln L}{\partial \sigma^2}\right] \end{pmatrix}. \tag{3.13}$$

The calculation of the elements of the Fisher information matrix is shown in Appendix A.3.

3.2.5 Optimal Test Plan

The optimal value of 'τ' is found by minimizing the asymptotic variance of the acceleration factor.

The NMinimize option of *Mathematica 6.0* has been used to formulate the optimal plan.

3.3 OPTIMUM FAILURE-CENSORED STEP-STRESS PALT MODEL

This section provides theory for the optimum failure-censored step-stress PALT plan for the truncated logistic distribution using two different methods. In the first method, it has been assumed that number of failures is known (Srivastava and Mittal (2010)). This method has been discussed in 3.3.1. In the second method, it has been assumed that proportion of units which will fail before censoring, is known (Srivastava and Mittal (2012c)). The latter assumption is more realistic than the former one. The second method has been discussed in 3.3.2.

3.3.1 Case I: Number of Failures Pre-Specified

3.3.1.1 *Assumptions*

1. The lifetime of an item tested at both use and at accelerated condition follows the truncated logistic distribution.
2. The lifetimes of test items are independent and identically distributed random variables.
3. The number of failures is pre-specified.

The failure time of the last unit is estimated using:

$$E[X_{(r)}] = F^{-1}\left(\frac{r}{n+1}\right), \quad \text{(David, 1981)}$$

where 'r' is total number of failures before censoring.

3.3.1.2 *Test Procedure*

1. All 'n' items are first tested at use condition.
2. If any item out of 'n' items do not fail at use condition by stress changing time 'τ', then it is put on accelerated condition and run until the pre-determined number of failures 'r' is reached.

3.3.1.3 *Log-Likelihood Function*

The likelihood function L for n independent observations is

$$\begin{aligned} L = L(\beta,\mu,\sigma) &= \prod_{i=1}^{n} L_i\left(\beta,\mu,\sigma;(y_i,\delta_{1i},\delta_{2i})\right) \\ &= \prod_{i=1}^{n} (f_1(y_i))^{\delta_{1i}} (f_2(y_i))^{\delta_{2i}} (R(y_{(r)}))^{\bar{\delta}_{1i}\bar{\delta}_{2i}} \end{aligned}$$

$$= \prod_{i=1}^{n} \left(\frac{\exp\{-\frac{y_i-\mu}{\sigma}\}}{A\sigma\left(1+\exp\{-\frac{y_i-\mu}{\sigma}\}\right)^2} \right)^{\delta_{1i}}$$

$$\times \left(\frac{\beta \exp\left\{-\frac{(y_i-\tau)\beta+\tau-\mu}{\sigma}\right\}}{A\sigma\left(1+\exp\left\{-\frac{(y_i-\tau)\beta+\tau-\mu}{\sigma}\right\}\right)^2} \right)^{\delta_{2i}}$$

$$\times \left(\frac{\exp\left\{-\frac{(y_{(r)}-\tau)\beta+\tau-\mu}{\sigma}\right\}}{A\left(1+\exp\left\{-\frac{(y_{(r)}-\tau)\beta+\tau-\mu}{\sigma}\right\}\right)} \right)^{\bar{\delta}_{1i}\bar{\delta}_{2i}}. \tag{3.14}$$

It is usually easier to maximize the natural logarithm of the likelihood function rather than the likelihood function itself.

Thus, the log-likelihood of (3.14) is

$$\begin{aligned}
\ln L &= \ln L(\beta,\mu,\sigma) \\
&= n_a \ln\beta - r\ln\sigma + n\ln\left(1+\exp\left\{-\frac{\mu}{\sigma}\right\}\right) - \sum_{i=1}^{n}\delta_{1i}\left(\frac{y_i-\mu}{\sigma}\right) \\
&\quad - \sum_{i=1}^{n} 2\delta_{1i}\ln\left(1+\exp\left\{-\frac{y_i-\mu}{\sigma}\right\}\right) \\
&\quad - \sum_{i=1}^{n}\delta_{2i}\left(\frac{(y_i-\tau)\beta+\tau-\mu}{\sigma}\right) \\
&\quad - \sum_{i=1}^{n} 2\delta_{2i}\ln\left(1+\exp\left\{-\frac{(y_i-\tau)\beta+\tau-\mu}{\sigma}\right\}\right) \\
&\quad - \sum_{i=1}^{n}\bar{\delta}_{1i}\bar{\delta}_{2i}\left(\frac{(y_{(r)}-\tau)\beta+\tau-\mu}{\sigma}\right) \\
&\quad - \sum_{i=1}^{n}\bar{\delta}_{1i}\bar{\delta}_{2i}\ln\left(1+\exp\left\{-\frac{(y_{(r)}-\tau)\beta+\tau-\mu}{\sigma}\right\}\right).
\end{aligned} \tag{3.15}$$

The method of maximum likelihood is used for estimating model parameters and are obtained by maximizing (3.15).

3.3.1.4 *Fisher Information Matrix*

The Fisher information matrix for n independent observations is

$$F = \begin{pmatrix} E\left[-\frac{\partial^2 \ln L}{\partial \beta^2}\right] & E\left[-\frac{\partial^2 \ln L}{\partial \beta \partial \mu}\right] & E\left[-\frac{\partial^2 \ln L}{\partial \beta \partial \sigma}\right] \\ E\left[-\frac{\partial^2 \ln L}{\partial \mu \partial \beta}\right] & E\left[-\frac{\partial^2 \ln L}{\partial \mu^2}\right] & E\left[-\frac{\partial^2 \ln L}{\partial \mu \partial \sigma}\right] \\ E\left[-\frac{\partial^2 \ln L}{\partial \sigma \partial \beta}\right] & E\left[-\frac{\partial^2 \ln L}{\partial \sigma \partial \mu}\right] & E\left[-\frac{\partial^2 \ln L}{\partial \sigma^2}\right] \end{pmatrix}. \tag{3.16}$$

The calculation of the elements of the Fisher information matrix is shown in Appendix A.4.

3.3.1.5 *Optimal Test Plan*

The optimal value of 'τ' is found by minimizing the asymptotic variance of the acceleration factor.

The NMinimize option of *Mathematica 6.0* has been used to formulate the optimal plan.

3.3.2 Case II: Proportion of Units Failing Before Censoring Pre-Specified

Here, optimal step-stress PALT using Type-II censoring and truncated logistic distribution due to Srivastava and Mittal (2012c) has been discussed.

3.3.2.1 *Assumptions*

1. The lifetime of an item tested at both use, and at accelerated condition follows the truncated logistic distribution.
2. The lifetimes of test items are independent and identically distributed random variables.
3. The proportion of items failing before censoring is pre-specified.

3.3.2.2 *Test Procedure*

1. All 'n' items should be first run simultaneously at use condition.
2. When 'nq_l' items have failed at use condition, the surviving '$n(1-q_l)$' items are put to test at accelerated condition.
3. The test is terminated when $n(q_r - q_l)$ units have failed at accelerated condition.

3.3.2.3 *Log-Likelihood Function*

The likelihood function L for n independent observations is

$$
\begin{aligned}
L = L(\beta,\mu,\sigma) &= \prod_{i=1}^{n} L_i(\beta,\mu,\sigma;(u_{(i)},\delta_{1i},\delta_{2i})) \\
&= \prod_{i=1}^{n} (f_1(u_{(i)}))^{\delta_{1i}} (f_2(u_{(i)}))^{\delta_{2i}} (R(u_{(nq_r)}))^{\bar{\delta}_{1i}\bar{\delta}_{2i}} \\
&= \prod_{i=1}^{n} \left(\frac{\exp\left\{-\frac{u_{(i)}-\mu}{\sigma}\right\}}{A\sigma\left(1+\exp\left\{-\frac{u_{(i)}-\mu}{\sigma}\right\}\right)^2} \right)^{\delta_{1i}} \\
&\quad \times \left(\frac{\beta\exp\left\{-\frac{u_{(nq_l)}+(u_{(i)}-u_{(nq_l)})\beta-\mu}{\sigma}\right\}}{A\sigma\left(1+\exp\left\{-\frac{u_{(nq_l)}+(u_{(i)}-u_{(nq_l)})\beta-\mu}{\sigma}\right\}\right)^2} \right)^{\delta_{2i}} \\
&\quad \times \left(\frac{\exp\left\{-\frac{u_{(nq_l)}+(u_{(nq_r)}-u_{(nq_l)})\beta-\mu}{\sigma}\right\}}{A\left(1+\exp\left\{-\frac{u_{(nq_l)}+(u_{(nq_r)}-u_{(nq_l)})\beta-\mu}{\sigma}\right\}\right)} \right)^{\bar{\delta}_{1i}\bar{\delta}_{2i}}. \qquad (3.17)
\end{aligned}
$$

It is usually easier to maximize the natural logarithm of the likelihood function rather than the likelihood function itself.

Thus, the log-likelihood of (3.17) is

$$
\begin{aligned}
\ln L &= \ln L(\beta,\mu,\sigma) \\
&= (nq_r - nq_l)\ln\beta - nq_r\ln\sigma + n\ln\left(1+\exp\left\{-\frac{\mu}{\sigma}\right\}\right)
\end{aligned}
$$

$$
\begin{aligned}
&-\sum_{i=1}^{n} 2\delta_{1i} \ln\left(1+\exp\left\{-\frac{u_{(i)}-\mu}{\sigma}\right\}\right) \\
&-\sum_{i=1}^{n} \delta_{2i}\left(\frac{u_{(nq_l)}+(u_{(i)}-u_{(nq_l)})\beta-\mu}{\sigma}\right) \\
&-\sum_{i=1}^{n} 2\delta_{2i} \ln\left(1+\exp\left\{-\frac{u_{(nq_l)}+(u_{(i)}-u_{(nq_l)})\beta-\mu}{\sigma}\right\}\right) \\
&-\sum_{i=1}^{n} \delta_{1i}\left(\frac{u_{(i)}-\mu}{\sigma}\right) \\
&-\sum_{i=1}^{n} \bar{\delta}_{1i}\bar{\delta}_{2i}\left(\frac{u_{(nq_l)}+(u_{(nq_r)}-u_{(nq_l)})\beta-\mu}{\sigma}\right) \\
&-\sum_{i=1}^{n} \bar{\delta}_{1i}\bar{\delta}_{2i} \ln\left(1+\exp\left\{-\frac{u_{(nq_l)}+(u_{(nq_r)}-u_{(nq_l)})\beta-\mu}{\sigma}\right\}\right).
\end{aligned}
\tag{3.18}
$$

The method of maximum likelihood is used for estimation of the model parameters, and are obtained by maximizing (3.18).

3.3.2.4 *Fisher Information Matrix*

The Fisher information matrix for n independent observations is

$$
F = \begin{pmatrix}
E\left[-\frac{\partial^2 \ln L}{\partial \beta^2}\right] & E\left[-\frac{\partial^2 \ln L}{\partial \beta \partial \mu}\right] & E\left[-\frac{\partial^2 \ln L}{\partial \beta \partial \sigma}\right] \\
E\left[-\frac{\partial^2 \ln L}{\partial \mu \partial \beta}\right] & E\left[-\frac{\partial^2 \ln L}{\partial \mu^2}\right] & E\left[-\frac{\partial^2 \ln L}{\partial \mu \partial \sigma}\right] \\
E\left[-\frac{\partial^2 \ln L}{\partial \sigma \partial \beta}\right] & E\left[-\frac{\partial^2 \ln L}{\partial \sigma \partial \mu}\right] & E\left[-\frac{\partial^2 \ln L}{\partial \sigma^2}\right]
\end{pmatrix}. \tag{3.19}
$$

The calculation of the elements of the Fisher information matrix is shown in Appendix A.5.

3.3.2.5 *Optimal Test Plan*

The optimal value of 'q_l' is found by using the D-optimality criterion which consists in minimizing the reciprocal of the determinant of the Fisher information matrix.

The NMinimize option of *Mathematica 6.0* has been used to formulate the optimal plan.

3.4 NUMERICAL EXAMPLES

3.4.1 Type-I Censoring

The following hypothetical data set has been considered.

$$n = 35, \quad \beta = 1.1, \quad \mu = 0.5, \quad \sigma = 3, \quad \eta = 6.8.$$

3.4.1.1 *Optimal Plan*

For $n = 35$, $\beta = 1.1$, $\mu = 0.5$, $\sigma = 3$, and $\eta = 6.8$, the optimal value of 'τ' is obtained as $\tau^* = 4.41263$ using NMinimize option of *Mathematica* 6.0.

3.4.1.2 *Simulated Data*

The data in Table 3.1 gives 35 simulated observations based on data $n = 35$, $\beta = 1.1$, $\mu = 0.5$, $\sigma = 3$, $\eta = 6.8$, and $\tau^* = 4.41263$. Thus,

Table 3.1: Simulated failure times ($n = 35$, $\beta = 1.1$, $\mu = 0.5$, $\sigma = 3$, $\eta = 6.8$, $\tau^* = 4.41263$, and $n_c = 9$).

Step-Stress	Failure Times
Use Condition	0.941951, 2.176240, 2.207960, 0.244373, 2.128270, 0.403601, 1.763830, 1.385200, 2.158640, 3.006270, 1.154010, 1.010620, 4.407310, 3.481320, 3.896940, 3.004100, 0.555658
Accelerated Condition	5.593110, 6.753510, 5.324340, 6.525530, 6.051710, 4.764450, 5.809210, 4.949270, 5.633080

the total number of units tested at use condition is $n_u = 17$, and at accelerated condition it is $n_a = 9$.

3.4.1.3 *MLEs of the Design Parameters*

The MLEs of the model parameters μ, σ, and the acceleration factor β obtained using simulated data given in Table 3.1 are $\hat{\mu} = -0.957196$, $\hat{\sigma} = 4.41033$, and $\hat{\beta} = 1.46292$.

These estimates are obtained using the NMaximize option of *Mathematica 6.0.*

3.4.1.4 *Confidence Intervals*

95% confidence intervals for the parameters are obtained using the inverse of the observed Fisher information matrix $\hat{F}^{-1}$, and it is given by

$$\hat{F}^{-1} = \begin{pmatrix} 0.898411 & 10.0328 & 3.06744 \\ 10.03280 & 181.509 & 49.9538 \\ 3.067440 & 49.9538 & 14.5182 \end{pmatrix}.$$

The observed value of F^{-1}, that is, $\hat{F}^{-1}$ is determined by substituting the estimated parameters $\hat{\beta}$, $\hat{\mu}$, and $\hat{\sigma}$ for the true parameters in the asymptotic variance–covariance matrix. The square root of the diagonal elements of $\hat{F}^{-1}$ gives the standard error of an estimator. Thus, the 95% approximate confidence intervals for β, μ, and σ, respectively, are

$$1.000000 < \beta \leq 5.20877,$$
$$-27.36330 \leq \mu \leq 25.4489,$$

and

$$0.811086 \leq \sigma \leq 23.9814.$$

Because, the range of parameter 'β' is greater than one, therefore lower limit of its confidence interval cannot be less than one. Thus, the lower limit is replaced by one whenever the lower limit comes out to be less than one.

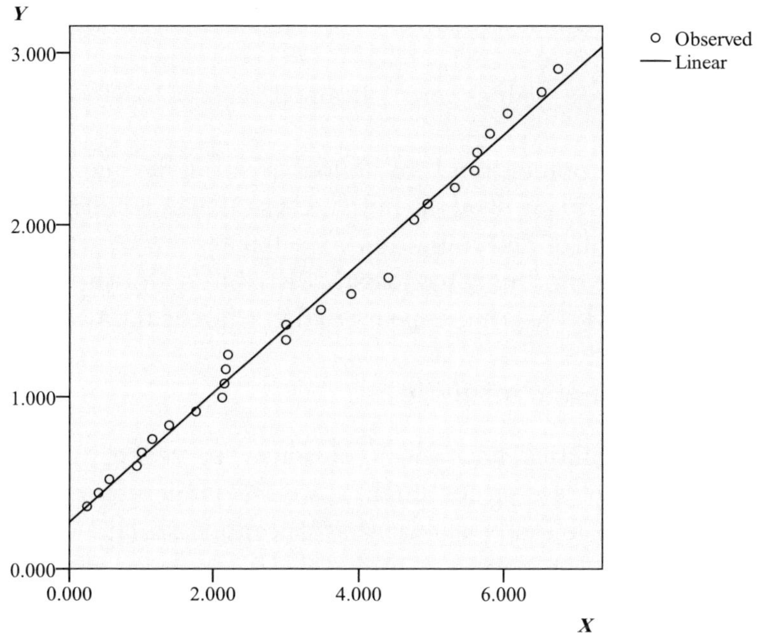

Figure 3.1: Truncated logistic distribution probability plot for the simulated data under Type-I censoring

3.4.1.5 *Graphical Goodness of Fit*

Figure 3.1 shows the associated step-stress truncated logistic probability plot at normal use and at accelerated condition with Type-I censoring.

The failure times at normal use and at accelerated condition given in Table 3.1 are both arranged in increasing order, and are ranked from $i = 1, 2, \ldots, k, \ldots, n$. The plot is obtained by taking these ranked failure times on X-axis, and their corresponding ordinate values, viz,

$$\mu - \sigma \ln\left(\left(\frac{Ai}{n+1} + \frac{1}{1+\exp\{\mu/\sigma\}}\right)^{-1} - 1\right)$$

at normal use, and

$$\tau^* - \frac{1}{\beta}\left(\tau^* - \mu + \sigma \ln\left(\left(\frac{Ai}{n+1} + \frac{1}{1+\exp\{\mu/\sigma\}}\right)^{-1} - 1\right)\right)$$

at accelerated condition, where $\tau^* = 4.41263$ on Y-axis (Grosh, 1989).

These ordinate values are obtained using (3.6) and (3.7), respectively.

The plotted points tend to follow a straight line, which is substantiated by fitting straight line to these points resulting in high value of coefficient of determination $r^2 = 0.990$.

Therefore, for the truncated logistic distribution for the simulated data under Type-I censoring appears to describe the data adequately.

3.4.1.6 *Coverage Probabilities*

Coverage probabilities have been obtained to see if the normal approximation holds even for small as well as moderate n values.

The simulation study shows that the approximate confidence intervals possess the coverage probabilities closer to the nominal levels for the parameters β and σ. However, this is not so for the parameter μ. This shows that normal approximation can be used even for small, and moderate samples for $\hat{\beta}$ and $\hat{\sigma}$ but not for $\hat{\mu}$. To compute confidence intervals for μ when n is small, other methods like bootstrap percentile method may be used (Efron, 1982).

Table 3.2 gives the estimated coverage probabilities of the two-sided 95% approximate confidence intervals for $\hat{\mu}$, $\hat{\sigma}$, and $\hat{\beta}$, for selected values of n, and given values of $\beta = 1.1$, $\mu = 0.5$, $\sigma = 3$,

Table 3.2: Estimated coverage probabilities (in %) of the approximate 95% confidence interval based on 1,000 simulations of Type-I censored samples of size n with ($\beta = 1.1, \mu = 0.5, \sigma = 3$, and $\eta = 6.8$)

n	β	μ	σ
10	84.8	63.5	91.3
15	91.0	55.6	82.9
20	96.0	68.8	97.3
25	92.8	67.7	91.6
30	96.6	70.7	97.1
35	95.6	73.2	99.2

and $\eta = 6.8$. These results are based on 1,000 pseudorandom Type-I censored samples of size n from a step-stress truncated logistic model. The simulation results show that even for small values of n the normal approximation for the pdfs of $\hat{\sigma}$, and $\hat{\beta}$ perform well.

3.4.2 Type-II Censoring: Case I — Number of Failures Pre-Specified

The following hypothetical data set has been considered.

$$n = 35, \quad \beta = 1.1, \quad \mu = 0.5, \quad \sigma = 3, \quad y_{(r)} = 7.$$

3.4.2.1 *Optimal Plan*

For the data set $n = 35$, $\beta = 1.1$, $\mu = 0.5$, $\sigma = 3$, $y_{(r)} = 7$, and the optimal value of 'τ' is as $\tau^* = 4.49458$, using NMinimize option of *Mathematica 6.0*.

3.4.2.2 *Simulated Data*

The data in Table 3.3 gives 35 simulated observations based on data $n = 35$, $\beta = 1.1$, $\mu = 0.5$, $\sigma = 3$, $y_{(r)} = 7$, and $\tau^* = 4.49458$. Thus, the total number of units tested at use condition is $n_u = 20$, and at accelerated condition it is $n_a = 8$.

Table 3.3: Simulated failure times ($n = 35$, $\beta = 1.1$, $\mu = 0.5$, $\sigma = 3$, $y_{(r)} = 7$, $\tau^* = 4.49458$, and $n_c = 7$)

Step-Stress	Failure Times
Use Condition	1.897660, 4.290860, 2.933900, 0.521004, 0.326628, 4.355820, 1.680090, 1.857180, 2.703160, 3.812230, 0.879353, 1.582120, 4.063950, 3.582280, 0.711965, 1.669980, 0.104429, 2.914920, 3.148060, 1.137920
Accelerated Condition	5.654130, 6.945540, 5.767420, 4.560850, 5.697860, 5.122740, 5.980470, 5.892920

3.4.2.3 MLEs of the Design Parameters

The MLEs of the model parameters μ, σ, and the acceleration factor β obtained using simulated data given in Table 3.6 are $\hat{\mu} = 1.57225$, $\hat{\sigma} = 2.97842$, and $\hat{\beta} = 1.10622$.

These estimates are obtained using the NMaximize option of *Mathematica 6.0.*

3.4.2.4 Confidence Intervals

95% confidence intervals for the parameters are obtained using the inverse of the observed Fisher information matrix $\hat{F}^{-1}$, and it is given by

$$\hat{F}^{-1} = \begin{pmatrix} 0.378404 & 1.57076 & 0.806715 \\ 1.570760 & 14.5333 & 5.858750 \\ 0.806715 & 5.85875 & 2.757020 \end{pmatrix}.$$

The observed value of F^{-1}, that is, $\hat{F}^{-1}$ is determined by substituting the estimated parameters $\hat{\beta}$, $\hat{\mu}$, and $\hat{\sigma}$ for the true parameters in the asymptotic variance–covariance matrix. The square root of the diagonal elements of $\hat{F}^{-1}$ gives the standard error of an estimator. Thus, the 95% approximate confidence intervals for β, μ, and σ, respectively, are

$$1.000000 < \beta \leq 3.28992,$$

$$-5.899770 \leq \mu \leq 9.04428$$

and

$$0.998718 \leq \sigma \leq 8.88235.$$

Since the range of parameter 'β' is greater than one, therefore the lower limit of its confidence interval cannot be less than one, and so is replaced by one whenever it comes out to be so.

3.4.2.5 Graphical Goodness of Fit

Figure 3.2 shows the associated step-stress truncated logistic probability plot at normal use and at accelerated condition with Type-II

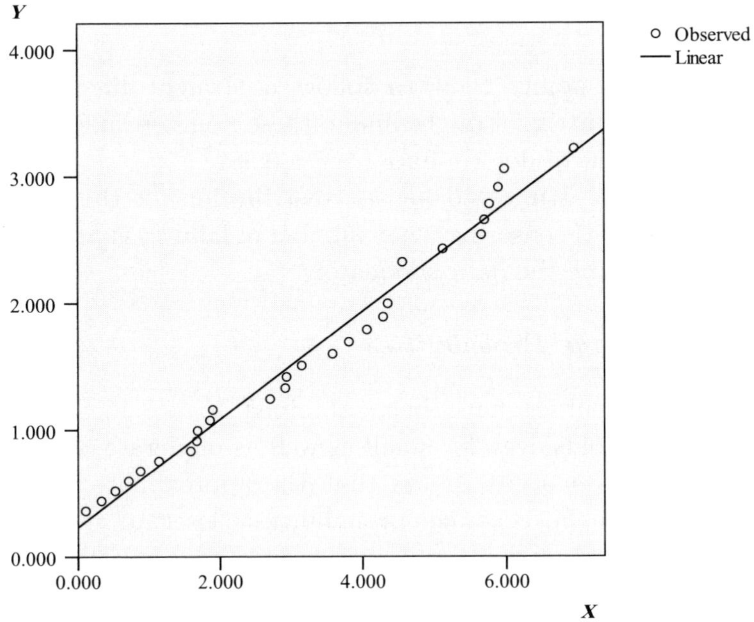

Figure 3.2: Truncated logistic distribution probability plot for the simulated data under Type-II censoring (case I)

censoring. This figure is being added to verify the results obtained, through graphical approach also.

The failure times at normal use and at accelerated condition given in Table 3.3 are both arranged in increasing order, and are ranked from $i = 1, 2, \ldots, k, \ldots, n$. The plot is obtained by taking these ranked failure times on X-axis, and their corresponding ordinate values, viz,

$$\mu - \sigma \ln\left(\left(\frac{Ai}{n+1} + \frac{1}{1+\exp\{\mu/\sigma\}}\right)^{-1} - 1\right)$$

at normal use, and

$$\tau^* - \frac{1}{\beta}\left(\tau^* - \mu + \sigma \ln\left(\left(\frac{Ai}{n+1} + \frac{1}{1+\exp\{\mu/\sigma\}}\right)^{-1} - 1\right)\right)$$

at accelerated condition, where $\tau^* = 4.49458$ on Y-axis (Grosh, 1989).

These ordinate values are obtained using (3.6) and (3.7), respectively.

The plotted points tend to follow a straight line, which is substantiated by fitting straight line to these points resulting in high value of coefficient of determination $r^2 = 0.982$.

Therefore, the truncated logistic distribution for the simulated data under Type-II censoring when number of failures is pre-specified appears to describe the data adequately.

3.4.2.6 *Coverage Probabilities*

Coverage probabilities have been obtained to see if the normal approximation holds even for small as well as moderate n values.

The simulation study shows that the approximate confidence intervals possess the coverage probabilities closer to the nominal levels for the parameters β and σ. However, this is not so for the parameter μ. This shows that normal approximation can be used even for small, and moderate samples for $\hat{\beta}$ and $\hat{\sigma}$ but not for $\hat{\mu}$. To compute confidence intervals for μ when n is small, other methods like bootstrap percentile method may be used (Efron, 1982).

Table 3.4 gives the estimated coverage probabilities of the two-sided 95% approximate confidence intervals for $\hat{\mu}$, $\hat{\sigma}$, and $\hat{\beta}$, for selected values of n, $\beta = 1.1$, $\mu = 0.5$, $\sigma = 3$, and $y_{(r)} = 7$. These

Table 3.4: Estimated coverage probabilities (in %) of the approximate 95% confidence interval based on 1,000 simulations of Type-II censored samples case I of size n with ($\beta = 1.1$, $\mu = 0.5$, $\sigma = 3$, and $y_{(r)} = 7$)

n	β	μ	σ
10	85.1	61.8	91.5
15	91.8	58.8	84.4
20	89.8	63.1	89.4
25	94.5	70.6	92.7
30	93.5	66.5	90.0
35	88.2	69.3	89.4

results are based on 1,000 pseudorandom Type-II censored samples of size n from a step-stress truncated logistic model. The simulation results show that even for small values of n the normal approximation for the pdfs of $\hat{\sigma}$, and $\hat{\beta}$ perform well.

3.4.3 Type-II Censoring: Case II — Proportion of Censoring Pre-Specified

The following hypothetical data set has been considered.

$$n = 36, \quad \beta = 3.5, \quad \mu = 3, \quad \sigma = 2, \quad q_r = 0.8.$$

3.4.3.1 *Optimal Plan*

Assuming $n = 36$, $\beta = 3.5$, $\mu = 3$, $\sigma = 2$, and $q_r = 0.8$, the optimal value of 'q_l' is obtained by D-optimality criterion using NMinimize option of *Mathematica 6.0.* It is obtained as $q_l^* = 0.537308$.

Table 3.5 gives the optimal value of 'q_l' for various sets of parametric values.

Table 3.5: Optimum failure-censored step-stress PALT for $n = 36$

q_r	β	μ	σ	q_l^*
0.3	2.5	3.0	2.0	0.186342
	3.0	3.0	2.0	0.186342
	3.5	3.0	2.0	0.186342
	4.0	3.0	2.0	0.186342
	3.5	2.0	2.0	0.190166
	3.5	2.5	2.0	0.188581
	3.5	3.5	2.0	0.183423
	3.5	3.0	1.0	0.158032
	3.5	3.0	1.5	0.179808
	3.5	3.0	2.5	0.188949
0.4	2.5	3.0	2.0	0.259427
	3.0	3.0	2.0	0.259427
	3.5	3.0	2.0	0.259427
	4.0	3.0	2.0	0.259427

(*Continued*)

Table 3.5: (*Continued*)

q_r	β	μ	σ	q_l^*
	3.5	2.0	2.0	0.266285
	3.5	2.5	2.0	0.263185
	3.5	3.5	2.0	0.255025
	3.5	3.0	1.0	0.225038
	3.5	3.0	1.5	0.250013
	3.5	3.0	2.5	0.263857
0.5	2.5	3.0	2.0	0.330905
	3.0	3.0	2.0	0.330905
	3.5	3.0	2.0	0.330905
	4.0	3.0	2.0	0.330905
	3.5	2.0	2.0	0.340668
	3.5	2.5	2.0	0.336119
	3.5	3.5	2.0	0.325077
	3.5	3.0	1.0	0.289845
	3.5	3.0	1.5	0.318713
	3.5	3.0	2.5	0.337084
0.6	2.5	3.0	2.0	0.401011
	3.0	3.0	2.0	0.401012
	3.5	3.0	2.0	0.401012
	4.0	3.0	2.0	0.401012
	3.5	2.0	2.0	0.413640
	3.5	2.5	2.0	0.407660
	3.5	3.5	2.0	0.393791
	3.5	3.0	1.0	0.353060
	3.5	3.0	1.5	0.386092
	3.5	3.0	2.5	0.408886
0.7	2.5	3.0	2.0	0.470086
	3.0	3.0	2.0	0.470090
	3.5	3.0	2.0	0.470076
	4.0	3.0	2.0	0.470065
	3.5	2.0	2.0	0.485497
	3.5	2.5	2.0	0.478096
	3.5	3.5	2.0	0.461425
	3.5	3.0	1.0	0.415176
	3.5	3.0	1.5	0.452428
	3.5	3.0	2.5	0.479954

(*Continued*)

Table 3.5: (*Continued*)

q_r	β	μ	σ	q_l^*
0.8	2.5	3.0	2.0	0.537878
	3.0	3.0	2.0	0.537155
	3.5	3.0	2.0	0.537308
	4.0	3.0	2.0	0.537544
	3.5	2.0	2.0	0.556477
	3.5	2.5	2.0	0.547546
	3.5	3.5	2.0	0.527894
	3.5	3.0	1.0	0.476486
	3.5	3.0	1.5	0.519722
	3.5	3.0	2.5	0.548966
0.9	2.5	3.0	2.0	0.593930
	3.0	3.0	2.0	0.609393
	3.5	3.0	2.0	0.616109
	4.0	3.0	2.0	0.617383
	3.5	2.0	2.0	0.627229
	3.5	2.5	2.0	0.621397
	3.5	3.5	2.0	0.591588
	3.5	3.0	1.0	0.537065
	3.5	3.0	1.5	0.583533
	3.5	3.0	2.5	0.562226

3.4.3.2 *Simulated Data*

The data in Table 3.6 gives 36 simulated observations based on data $n = 36$, $\beta = 3.5$, $\mu = 3$, $\sigma = 2$, $q_r = 0.8$, and $q_l^* = 0.537308$. Thus, the total number of units tested at use condition is $nq_l = 19$, and at accelerated condition it is $nq_r - nq_l = 10$.

3.4.3.3 *MLEs of the Design Parameters*

The MLEs of the model parameters μ, σ, and the acceleration factor β obtained using simulated data given Table 3.6 are $\hat{\mu} = 2.32716$, $\hat{\sigma} = 2.37193$, and $\hat{\beta} = 2.28489$.

These estimates are obtained using the NMaximize option of *Mathematica 6.0.*

Table 3.6: Simulated failure times ($n = 36$, $\beta = 3.5$, $\mu = 3$, $\sigma = 2$, $q_r = 0.8$, $q_l^* = 0.537308$, and $n_c = 7$)

Step-Stress	Failure Times
Use Condition	0.825647, 1.274270, 3.522210, 1.689260, 3.634120, 2.678840, 0.509556, 0.876301, 0.928494, 3.127680, 2.878130, 0.384488, 1.666690, 1.443160, 3.433290, 1.752420, 3.139340, 1.500500, 3.893300
Accelerated Condition	3.929780, 4.483510, 3.982710, 4.721040, 5.092100, 4.265770, 4.561690, 4.841640, 4.271310, 4.049440

3.4.3.4 *Confidence Intervals*

95% confidence intervals for the parameters are obtained using the inverse of the observed Fisher information matrix $\hat{F}^{-1}$, and it is given by

$$\hat{F}^{-1} = \begin{pmatrix} 1.34435 & -1.34934 & 1.09024 \\ -1.34934 & 3.93506 & -2.04478 \\ 1.09024 & -2.04478 & 1.42118 \end{pmatrix}.$$

The observed value of F^{-1}, that is, $\hat{F}^{-1}$ is determined by substituting the estimated parameters $\hat{\beta}$, $\hat{\mu}$, and $\hat{\sigma}$ for the true parameters in the asymptotic variance–covariance matrix. The square root of the diagonal elements of $\hat{F}^{-1}$ gives the standard error of an estimator. Thus, the 95% approximate confidence intervals for β, μ, and σ, respectively, are:

$$1.000000 < \beta \leq 6.17751,$$

$$-1.560890 \leq \mu \leq 6.21521,$$

and

$$0.885686 \leq \sigma \leq 6.35219.$$

Since the range of parameter 'β' is greater than one, therefore the lower limit of its confidence interval cannot be less than one, and so is replaced by one whenever it comes out to be so.

3.5 SENSITIVITY ANALYSIS

3.5.1 Type-I Censoring and Type-II Censoring (Case: I — Number of Failures Pre-Specified)

To use an optimum test plan, one must have information about acceleration factor, β, and parameters of the model, μ and σ, which are usually unknown. Therefore they have to be approximated from experience, similar data, or preliminary tests. Incorrect choice of these gives a nonoptimal test plan. With the true values of parameters, the effects of incorrect pre-estimates of β, μ, and σ in terms of the relative increase of asymptotic variance for truncated logistic distribution under Type-I censoring and Type-II censoring (case I: number of failures pre-specified) have been studied and presented in Tables 3.7 and 3.8, respectively.

$$\mathrm{RV} = \left|\frac{\mathrm{AsVar}^* - \mathrm{AsVar}^0}{\mathrm{AsVar}^*}\right| \times 100,$$

where AsVar* is the asymptotic variance for the plan obtained with the correctly specified values, and AsVar^0 is the asymptotic variance for the plan obtained with mis-specified values.

Table 3.7: Sensitivity analysis for step-stress PALT plan with Type-I censoring ($n = 35$, $\eta = 6.8$, $\tau^* = 4.41263$, and AsVar$^* = 0.388484$)

Percentage Change (%)	β	μ	σ	RV	Incorrect τ^*	AsVar^0
−1	1.089	0.495	2.97	2.38054	4.38924	0.379236
+1	1.111	0.505	3.03	2.42030	4.43563	0.397887
−2	1.078	0.49	2.94	4.72143	4.36546	0.370142
+2	1.122	0.510	3.06	4.88102	4.45826	0.407446
−3	1.067	0.485	2.91	7.02320	4.34128	0.361200
+3	1.133	0.515	3.09	7.38524	4.48052	0.417164
−4	1.056	0.480	2.88	9.28661	4.31670	0.352407
+4	1.144	0.520	3.12	9.92525	4.50242	0.427042
−5	1.045	0.475	2.85	11.5117	4.29172	0.343763
+5	1.155	0.525	3.15	12.5097	4.52395	0.437082

Table 3.8: Sensitivity analysis for step-stress PALT plan with Type-II censoring (case I) (n = 35, $y_{(r)}$ = 7, τ^* = 4.49458, and AsVar* = 0.381725)

Percentage Change (%)	β	μ	σ	RV	Incorrect τ^*	AsVar^0
−1	1.089	0.495	2.97	2.36217	4.46998	0.372708
+1	1.111	0.505	3.03	2.40094	4.51878	0.390890
−2	1.078	0.490	2.94	4.68531	4.44498	0.363840
+2	1.122	0.510	3.06	4.84171	4.54260	0.400207
−3	1.067	0.485	2.91	6.97020	4.41958	0.355118
+3	1.133	0.515	3.09	7.32229	4.56603	0.409676
−4	1.056	0.480	2.88	9.21737	4.39376	0.346540
+4	1.144	0.520	3.12	9.84347	4.58909	0.419300
−5	1.045	0.475	2.85	11.4268	4.36753	0.338106
+5	1.155	0.525	3.15	12.4055	4.61178	0.429080

It has been found that if they are not too far from the true values the increase is small, as shown in Tables 3.7 and 3.8. The results of sensitivity analysis show that the plan described is robust.

3.5.2 Type-II Censoring (Case II: Proportion of Items Failing before Censoring Pre-Specified)

To use an optimum test plan, one needs estimates of the design parameters β, μ, and σ. These estimates sometimes may significantly affect the values of the resulting decision variables; therefore, their incorrect choice may give a poor estimate of the quantile at a design constant stress. Hence, it is important to conduct sensitivity analysis to evaluate the robustness of the resulting PALT plan.

The percentage deviations (PDs) of the optimal settings are measured by

$$\mathrm{PD} = \frac{|Z^{**} - Z^*|}{Z^*} \times 100,$$

where Z^* is the optimal asymptotic variance obtained with the given design parameters, and Z^{**} is the one obtained when a parameter is mis-specified.

Table 3.9: Sensitivity analysis for step-stress PALT plan with Type-II censoring (case II) ($n = 36$, $q_r = 0.8$, and $q_l^* = 0.537308$)

Parameter	Percentage Change (%)	q_l	PD
β	+5	0.537394	0.0159323
β	−5	0.537228	0.0149052
μ	+5	0.534043	0.607694
μ	−5	0.540533	0.600216
σ	+5	0.540506	0.595246
σ	−5	0.534825	0.462128

Table 3.10: Comparative study of step-stress PALT models with Type-II censoring (case II)

PALT Model	Log-Likelihood Function
Srivastava and Mittal (2012c) Model	−62.3714
Abdel-Ghaly, Attia, and Abdel-Ghani (2002) Model	−63.1796

Table 3.9 shows the optimal test plan for various deviations from the design parameter estimates. The results show that the optimal setting of Z is robust to the deviations of those baseline parameter estimates.

3.6 COMPARATIVE STUDY FOR TYPE-II CENSORING (CASE II — PROPORTION OF ITEMS FAILING BEFORE CENSORING PRE-SPECIFIED)

In this section, the Srivastava and Mittal (2012c) step-stress PALT model has been compared with the one designed by Abdel-Ghaly, Attia, and Abdel-Ghani (2002) in terms of log-likelihood functions using the hypothetical failure time data set given in Table 3.6.

Table 3.10 shows that the Srivastava and Mittal (2012c) model performs better for the given data set.

3.7 SUMMARY

In this chapter, optimum step-stress PALT for the truncated logistic distribution with Type-I and Type-II censoring has been dealt with. The truncated logistic life distribution truncated at point zero, has the failure rate that is increasing, and is more realistically bounded below, and above by a nonzero finite quantity. The procedures developed have been explained using the examples, and sensitivity analysis carried out. The results of sensitivity analysis show that the optimum plans described are robust.

CHAPTER 4

OPTIMUM RAMP-STRESS FULLY ACCELERATED LIFE TEST PLANS UNDER TYPE-I CENSORING

4.1 INTRODUCTION

Life testing under fully accelerated environmental condition consists of testing all the items at accelerated conditions using some life–stress relationship. Unlike in time varying ALT models with step-stress loading, in a progressive-stress test, the stress applied to a test unit continuously increases with time. In ramp-test, the stress is increased linearly. In particular, the ramp-test with two different linearly increasing stresses is a simple ramp-test. Ramp-tests are used for example in fatigue testing (Prot, 1948), capacitors (Endicott, Hatch, and Schmer, 1965; Starr and Endicott, 1961), insulation (Goba, 1969; Solomon, Klein, and Albert, 1976), and integrated circuits (Chan, 1990).

Progressive-stress accelerated life test (ALT) models have been studied by several authors (see Section 1.18.3). Bai, Cha, and Chung (1992) have considered the design of simple ramp tests for Weibull distribution under Type-I censoring, and determined low ramp rate by minimizing the asymptotic variance of the MLE of a specified quantile at design stress. Srivastava and Shukla (2009) have dealt with design of optimum simple ramp-test for the log-logistic distribution with Type-I censoring. There are situations where the stress cannot be increased indefinitely in a ramp-test because too high stress can cause failure modes other than that under consideration, and/or the test equipment might not be able to provide such a high stress. This has led Bai, Chun, and Cha (1997) to formulate optimum time-censored ramp-test for items with Weibull distribution when there is a stress upper bound.

In practicing ALT, one is interested in the lower tails of the life distribution. However, the single objective optimal plans separately minimize the asymptotic variances of the MLEs of the log of quantiles 1%, 10%, and 50% of the distribution at a design constant-stress. It is impractical to obtain the optimal plans using a single objective after conducting such costly tests. This has necessitated formulation of multi-objective ramp-stress ALT plans and has been brought forth in this chapter. The optimum plans have been obtained with multiple linearly increasing stresses as well as with stress upper bound using Burr Type-XII life distribution (see Section 1.4.2.9) and time-censored data.

4.1.1 Notation

M	total number of stress levels
I	index, implies ith stress level, $i = 1, 2, \ldots, m$
β	ramp or stress rate, $t_2(=(s_2 - s_0)/\beta)$
β^*	optimal stress rate
β_i	stress rate at level i, $\beta_i > 0$
ξ_i	stress rate factor; $\xi_1 < \xi_2 < \cdots < \xi_m = 1$; $\xi_i = \dfrac{\beta_i}{\beta_m}$
ξ_i^*	optimal stress rate factor
π_i	proportion of units tested at ξ_i
π_i^*	proportion of units tested at ξ_i^*
N	total number of test units
n_i	total number of test units allocated to stress level i, $n_i = n\,\pi_i; n = n_1 + n_2 + \cdots + n_m$
n_{c_i}	number of items censored at stress level i
n_{f_i}	number of items failed at stress level i

n_c	number of items censored under the stress upper bound test plan
η	censoring time, $\eta > 0$
$\text{burr}(c, k, \alpha)$	Burr Type-XII distribution with parameters c, k, α
c, k	shape parameters of the Burr Type-XII distribution, $c > 0,\ k > 0$
α	scale parameter of the Burr Type-XII distribution, $\alpha > 0$
γ_0, γ_1	parameters of the inverse power law, $\gamma_0 > 0, -\infty < \gamma_1 < \infty$
s_0 or s_d	stress level under normal operating conditions or design stress
s_h	upper bound of test stress
c_1	$c(1 + \gamma_1)$
α_1	$\left(e^{\gamma_0}\left(\frac{s_0}{\beta}\right)^{\gamma_1}(1 + \gamma_1)\right)^{\frac{1}{1+\gamma_1}}$
$T_1(\xi_i)$	$1 - \gamma_0 + \ln\left(\frac{s_0}{\beta_m}\right) - \ln(\xi_i) - \ln(1 + \gamma_1)$
τ	$\frac{s_h}{\beta}$
τ^*	$\frac{s_h}{\beta^*}$
$s(t)$	stress at time t
$c_1(t)$	$\frac{\tau^{-\gamma_1} t^{\gamma_1+1}}{\gamma_1 + 1}$
$c_2(t)$	$t - \frac{\tau\gamma_1}{\gamma_1 + 1}$
D_1	$\{t{:}0 \le t < \infty\}$
D_2	$\{t{:}\tau \le t < \eta\}$
μ, σ	location and scale parameters of the transformed distribution, $-\infty < \mu < \infty, \sigma > 0$

z	$\dfrac{y-\mu(\xi)}{\sigma}$; Standardized log failure time
η'	$\dfrac{\ln(\eta)-\mu(\xi)}{\sigma}$; Standardized log censoring time
p_i	Probability that a unit will fail by time η at ξ_i, and is given by: $1-\left\{1+\left[\dfrac{\eta}{\left(e^{\gamma_0}\left(\frac{s_0}{\beta_m \xi_i}\right)^{\gamma_1}(1+\gamma_1)\right)^{\frac{1}{1+\gamma_1}}}\right]^{c_1}\right\}^{-k}$
$I(z)$	Indicator function under the multi-level test plan: $I(z) \equiv I(z \leq \eta')$
$\delta(t)$	Indicator function under the stress upper bound test plan: $\delta(\text{True}) = 1, \delta(\text{False}) = 0$
$u_i(t)$	$\delta(t \in D_i), i = 1, 2$
$u_3(t)$	$1 - u_1(t) - u_2(t)$
$\wedge$	Implies a MLE
q	Quantile
AsVar*	Asymptotic variance under multi-objective plan
AsVar_q	Asymptotic variance under single-objective plan at quantile q

4.1.2 Burr Type-XII Life Distribution

The pdf, cdf, reliability function, and hazard function of the Burr Type-XII distribution are

$$g(t) = \left(\frac{kc}{\alpha}\right)\left(\frac{t}{\alpha}\right)^{c-1}\left(1+\left(\frac{t}{\alpha}\right)^{c}\right)^{-(k+1)}, \quad t \geq 0, \tag{4.1}$$

$$G(t) = 1-\left(1+\left(\frac{t}{\alpha}\right)^{c}\right)^{-k}, \quad t \geq 0, \tag{4.2}$$

$$R(t) = \left(1+\left(\frac{t}{\alpha}\right)^{c}\right)^{-k}, \quad t \geq 0, \tag{4.3}$$

and

$$h(t) = \left(\frac{kc}{\alpha}\right)\left(\frac{t}{\alpha}\right)^{c-1}\left(1+\left(\frac{t}{\alpha}\right)^{c}\right)^{-1}, \quad t \geq 0, \tag{4.4}$$

respectively. See Section 1.4.2.9 for details.

4.2 OPTIMUM MULTI-LEVEL RAMP-STRESS ALT PLAN WITHOUT STRESS UPPER BOUND

In the present section, design of an optimum time-censored multi-level ramp-stress ALT plan with multi-objectives has been explained (Srivastava and Mittal, 2012a) for the Burr Type-XII life distribution.

4.2.1 Assumptions

1. The lifetimes are independently and identically distributed.
2. The stress level (s_0) under the product's normal operating condition is constant, and specified.
3. For the effect of changing stress, the linear cumulative exposure model holds (Nelson, 1980, 1990; Nilsson, 1985; Yin and Sheng, 1987).
4. The test uses 'm' stress levels.
5. 'm' stress rates $\beta_1, \beta_2, \ldots, \beta_m (\beta_1 < \beta_2 < \cdots < \beta_m)$ are used in the ramp-test.
6. The shape parameter 'k' does not depend on the stress level and is assumed to be known.
7. At any constant-stress s, the lifetime of a unit follows the Burr Type-XII model with scale parameter $\alpha(s)$, and shape parameters c and k; and the inverse power law:

$$\alpha(s) = e^{\gamma_0}\left(\frac{s_0}{s}\right)^{\gamma_1} \tag{4.5}$$

holds.

4.2.2 Test Procedure

1. The stress applied to test units continuously increases with constant rate β from zero.

   The stress is an increasing function of time or it is directly proportional to time. This is quite reasonable as some specimens fail while voltage is being brought to the selected test stress while others last too long. These are called zeros and run-outs, respectively. By applying a uniformly increasing stress started from zero voltage, no zeros or run-outs will occur. Also, there is no problem in choosing test voltage stress.

2. Test units are subject to Type-I censoring with the censoring time η.
3. The test is continued until:
   (a) all test items fail, or
   (b) a prescribed censoring time η

   whichever occurs earlier, and the test conditions remain the same.
4. 'n' test units are tested under 'm' linearly increasing stress conditions
5. '$n_i\,\pi_i$' units randomly chosen among n are allocated to stress rate β_i, $i = 1, 2, \ldots, m$.

4.2.3 Life Distribution under Multi-Level Ramp-Test without Stress Upper Bound

The stress at time t is $s(t) = \beta t$ (from Test Procedure 1).

From the linear cumulative exposure model, and the inverse power law (see Assumption 7), the cdf of the lifetime T of a unit tested under stress rate 'β' is

$$F(t) = G(\varepsilon(t)), \tag{4.6}$$

where $G(\cdot)$ is the assumed cdf (4.2) with the scale parameter α set equal to one (see Appendix A.6), and

$$\varepsilon(t) = \int_0^t \frac{1}{\alpha(s(u))} du \tag{4.7}$$

is the cumulative exposure (damage) model.

Hence, the cdf is

$$F(t) = 1 - \left\{1 + \left[\int_0^t \frac{1}{\alpha(s(u))} du\right]^c\right\}^{-k} \quad \text{(using (4.2), (4.6), and (4.7))}$$

$$= 1 - \left\{1 + \left[\frac{t}{\alpha_1}\right]^{c_1}\right\}^{-k} \tag{4.8}$$

(4.8) is the Burr Type-XII distribution with scale parameter α_1, and shape parameters k and c_1.

4.2.4 Log-Likelihood Function

Let for convenience, $Y = \ln T$, then, pdf and cdf for Y, respectively, are

$$h(y) = \frac{k \exp\left\{\frac{y-\mu}{\sigma}\right\}}{\sigma\left(1 + \exp\left\{\frac{y-\mu}{\sigma}\right\}\right)^{(k+1)}}, \quad -\infty < y < \infty \tag{4.9}$$

and

$$H(y) = 1 - \left(1 + \exp\left\{\frac{y-\mu}{\sigma}\right\}\right)^{-k}, \quad -\infty < y < \infty \tag{4.10}$$

with location parameter

$$\mu = \ln(\alpha_1) = \left(\frac{1}{1+\gamma_1}\right)\left(\gamma_0 + \gamma_1 \ln\left(\frac{s_0}{\beta}\right) + \ln(1+\gamma_1)\right) \tag{4.11}$$

and scale parameter

$$\sigma = \frac{1}{c_1}. \tag{4.12}$$

The log-likelihood of an observation y at stress rate, $\beta_i, i = 1, 2, \ldots, m$, under Type-I censoring is derived. The stress rate factor is

$$\xi_i = \frac{\beta_i}{\beta_m}, \quad i = 1, 2, \ldots, m. \tag{4.13}$$

For the high stress rate, $\beta_m, \xi_m = 1$;

The lower stress rate factors are $\xi_i = \frac{\beta_i}{\beta_m}$, $i = 1, 2, \ldots, (m-1)$. Therefore, the location parameter of lifetime distribution of a unit tested under stress rate, β_i, $i = 1, 2, \ldots, m$ is

$$\mu(\xi_i) = \left(\frac{1}{1+\gamma_1}\right)\left(\gamma_0 + \gamma_1 \ln\left(\frac{s_0}{\beta_i}\right) + \ln(1+\gamma_1)\right) \quad \text{(using (4.11))}$$
$$= \left(\frac{1}{1+\gamma_1}\right)\left(\gamma_0 + \gamma_1 \ln\left(\frac{s_0}{\beta_m}\right) - \gamma_1 \ln(\xi_i) + \ln(1+\gamma_1)\right)$$
$$\text{(using (4.13))}.$$

The log-likelihood of a single observation y is

$$\begin{aligned} L &= L(\gamma_0, \gamma_1, c) \\ &= I(z)\left[\ln k - zk - \ln\sigma - (k+1)\ln(1+e^{-z})\right] \\ &\quad - \bar{I}(z)\left[k \ln(1+e^{\eta'})\right]. \end{aligned} \tag{4.14}$$

Let the log-likelihood of unit j be L_j. The log-likelihood L_0 for n independent observations is

$$L_0 = L_1 + \cdots + L_n. \tag{4.15}$$

4.2.5 Parameter Estimation

For a single observation, the first order partial derivatives of (4.14) are

$$\frac{\partial L}{\partial \gamma_0} = c\left[I(z)\left\{k - (k+1)\left(\frac{e^{-z}}{1+e^{-z}}\right)\right\} + \bar{I}(z)\left\{\frac{k}{1+e^{-\eta'}}\right\}\right], \tag{4.16}$$

$$\begin{aligned} \frac{\partial L}{\partial \gamma_1} &= \left(\frac{1}{1+\gamma_1}\right)\left[I(z)\left\{1 - \left(k - \frac{(k+1)e^{-z}}{1+e^{-z}}\right)(z - T_1 c)\right\}\right. \\ &\quad \left. - \bar{I}(z)\left\{\frac{k(\eta' - T_1 c)}{1+e^{-\eta'}}\right\}\right], \end{aligned} \tag{4.17}$$

and

$$\frac{\partial L}{\partial c} = \frac{1}{c}\left[I(z)\left\{1 - \left(k - \frac{(k+1)e^{-z}}{1+e^{-z}}\right)z\right\} - \bar{I}(z)\left\{\frac{k\,\eta'}{1+e^{-\eta'}}\right\}\right]. \tag{4.18}$$

From (4.16)–(4.18), one obtains

$$\frac{\partial L}{\partial \gamma_1} = \left(\frac{1}{1+\gamma_1}\right)\left[c\left(\frac{\partial L}{\partial c}\right) + T_1\left(\frac{\partial L}{\partial \gamma_0}\right)\right]. \tag{4.19}$$

The likelihood equations are obtained by setting (4.16)–(4.18) equal to zero.

The parameter values that solve "these equations summed over all test units" are the MLEs $\hat{\gamma}_0, \hat{\gamma}_1$, and $\hat{c}$. As the system of likelihood equations has no closed form solution in γ_0, γ_1, and c, the MLEs $\hat{\gamma}_0, \hat{\gamma}_1$, and $\hat{c}$ are obtained by maximizing (4.15) using the NMaximize option of *Mathematica 6.0.*

4.2.6 Fisher Information Matrix

The Fisher information matrix is obtained by taking the expectations of the negative of the second-order partial derivatives of the log-likelihood function with respect to γ_0, γ_1, and c. For an observation at ξ_i it is given by

$$F_{\xi_i} = \begin{bmatrix} E\left[-\frac{\partial^2 L}{\partial \gamma_0^2}\right] & E\left[-\frac{\partial^2 L}{\partial \gamma_0 \partial \gamma_1}\right] & E\left[-\frac{\partial^2 L}{\partial \gamma_0 \partial c}\right] \\ E\left[-\frac{\partial^2 L}{\partial \gamma_0 \partial \gamma_1}\right] & E\left[-\frac{\partial^2 L}{\partial \gamma_1^2}\right] & E\left[-\frac{\partial^2 L}{\partial \gamma_1 \partial c}\right] \\ E\left[-\frac{\partial^2 L}{\partial \gamma_0 \partial c}\right] & E\left[-\frac{\partial^2 L}{\partial \gamma_1 \partial c}\right] & E\left[-\frac{\partial^2 L}{\partial c^2}\right] \end{bmatrix}. \tag{4.20}$$

The calculation of the elements of the Fisher information matrix is shown in Appendix A.8.

Since $n\,\pi_i$ units are tested under stress rate factor ξ_i, the Fisher information matrix, F, for the plan with a sample of n independent units at 'm' stress levels is

$$F = \sum_{i=1}^{m} n\pi_i F_{\xi_i}. \tag{4.21}$$

Because, for some set of values of $\gamma_0, \gamma_1, s_0, \eta, \beta_m, \xi_i, k$, and c; $|F|$ or the variance function may be negative. Therefore, only that set of values for which $|F| > 0$ or the variance function is positive, is chosen. A similar observation has been made by Gouno, Sen, and Balakrishnan (2004).

4.2.7 Asymptotic Variance of the MLE of Logarithm of Quantile at Design Stress

For any plan, the asymptotic variance–covariance matrix of the MLEs $\hat{\gamma}_0, \hat{\gamma}_1$, and $\hat{c}$ is the inverse of the corresponding Fisher information matrix,

$$\sum = F^{-1} = \begin{bmatrix} \mathrm{Var}(\hat{\gamma}_0) & \mathrm{Cov}(\hat{\gamma}_0, \hat{\gamma}_1) & \mathrm{Cov}(\hat{\gamma}_0, \hat{c}) \\ \mathrm{Cov}(\hat{\gamma}_0, \hat{\gamma}_1) & \mathrm{Var}(\hat{\gamma}_1) & \mathrm{Cov}(\hat{\gamma}_1, \hat{c}) \\ \mathrm{Cov}(\hat{\gamma}_0, \hat{c}) & \mathrm{Cov}(\hat{\gamma}_1, \hat{c}) & \mathrm{Var}(\hat{c}) \end{bmatrix}.$$

The asymptotic variance of the MLE of logarithm of quantile q of the distribution at a design constant stress s_0 is

$$\mathrm{AsVar}\left[\log \hat{y}_q\right] = \mathrm{AsVar}\left[\hat{h}(\hat{\gamma}_0, \hat{\gamma}_1, \hat{c})\right],$$

where

$$\mathrm{AsVar}\left[\hat{h}(\hat{\gamma}_0, \hat{\gamma}_1, \hat{c})\right] = \underset{\sim}{\hat{H}'}\, F^{-1}\, \underset{\sim}{\hat{H}},$$

vector $\underset{\sim}{\hat{H}'}$ is a transpose of vector $\underset{\sim}{\hat{H}}$,

$$\begin{aligned} \hat{h} &= \hat{\gamma}_0 + \left(\frac{1}{\hat{c}}\right) \ln\left(-1 + (1-q)^{-\frac{1}{k}}\right) \\ &= \hat{\gamma}_0 + \left(\frac{1+\hat{\gamma}_1}{\hat{c}_1}\right) \ln\left(-1 + (1-q)^{-\frac{1}{k}}\right) \end{aligned}$$

and

$$\underset{\sim}{\hat{H}} = \left[\frac{\partial \hat{h}}{\partial \hat{\gamma}_0} \quad \frac{\partial \hat{h}}{\partial \hat{\gamma}_1} \quad \frac{\partial \hat{h}}{\partial \hat{c}}\right]' \text{ with } \hat{h} \equiv \hat{h}(\hat{\gamma}_0, \hat{\gamma}_1, \hat{c}).$$

Thus,

$$\text{AsVar}\left[\log \hat{y}_q\right] = \text{AsVar}\left[\hat{h}(\hat{\gamma}_0, \hat{\gamma}_1, \hat{c})\right]. \tag{4.22}$$

4.3 OPTIMUM MULTI-OBJECTIVE RAMP-STRESS ALT PLAN WITH STRESS UPPER BOUND

In this section, formulation of an optimum multi-objective ramp-stress ALT plan with stress upper bound for the Burr Type-XII distribution under Type-I censoring designed by Srivastava and Mittal (2012b) has been explained.

4.3.1 Assumptions

(1) The test times are independent and identically distributed.
(2) The highest (s_h) and design (s_d) test stresses are specified.
(3) For changing stress in the ramp test, a cumulative exposure model holds (Nelson, 1980).
(4) The shape parameter 'k' does not depend on the stress level.
(5) At any constant-stress s, the lifetime of a unit follows the Burr Type-XII model with scale parameter $\alpha(s)$, and shape parameters c and k; and the inverse power law holds for

$$\alpha(s) = e^{\gamma_0}\left(\frac{s_h}{s}\right)^{\gamma_1}, \tag{4.23}$$

assuming k is known for the sake of mathematical convenience.

4.3.2 Test Procedure

1. 'n' test units are tested under a linearly increasing stress condition.
2. If the stress reaches s_h, it remains constant at that level.

3. The test is continued until:

   (a) all test items fail, or
   (b) a prescribed censoring time η

   whichever occurs earlier, and the test conditions remain the same.

If the stress does not reach s_h before η, the stress upper bound does not affect the test, and this test situation is equivalent to the case where there is no stress upper bound (Bai, Chun, and Cha, 1997).

4.3.3 Life Distribution under Ramp-Stress with Stress Upper Bound

From the linear cumulative exposure model, and the inverse power law (see Assumption 5), the cdf of the lifetime T of a unit tested under stress rate 'β' is

$$F(t) = G(\varepsilon(t)), \tag{4.24}$$

where $G(\cdot)$ is the assumed cdf (4.2) with the scale parameter α set equal to one (see Appendix A.7), and

$$\varepsilon(t) = \int_0^t \frac{1}{\alpha(s(u))} du \tag{4.25}$$

is the cumulative exposure (damage) model.

Hence, the cdf is

$$F(t) = 1 - \left\{1 + \left[\int_0^t \frac{1}{\alpha(s(u))} du\right]^c\right\}^{-k} \quad \text{(using (4.2), (4.24), and (4.25))}$$

$$= 1 - \left\{1 + \left[\int_0^t \frac{1}{e^{\gamma_0}(s_h/s(u))^{\gamma_1}} du\right]^c\right\}^{-k} \quad \text{(using (4.23)).} \tag{4.26}$$

Let, $s(t) = \min(\beta t, s_h)$, $0 < t < \eta$.

Then,

$$F(t) = \begin{cases} 1 - \left\{1 + [e^{-\gamma_0} c_1(t)]^c\right\}^{-k}, & 0 \le t < \tau \\ 1 - \left\{1 + [e^{-\gamma_0} c_2(t)]^c\right\}^{-k}, & \tau \le t < \eta. \end{cases} \tag{4.27}$$

The pdf is given by

$$f(t) = \begin{cases} f_1(t), & 0 \le t < \tau \\ f_2(t), & \tau \le t < \eta, \end{cases} \tag{4.28}$$

where

$$f_1(t) = kc\tau^{-\gamma_1} t^{\gamma_1} e^{-\gamma_0 c} (c_1(t))^{c-1} \left(1 + (e^{-\gamma_0} c_1(t))^c\right)^{-(k+1)} \tag{4.29}$$

and

$$f_2(t) = kce^{-\gamma_0 c} (c_2(t))^{c-1} \left(1 + (e^{-\gamma_0} c_2(t))^c\right)^{-(k+1)}. \tag{4.30}$$

4.3.4 Log-Likelihood Function

The log-likelihood of an observation at time t at stress rate β under Type-I censoring is derived.

The log-likelihood of a single observation at time t at stress rate β is

$$\begin{aligned} L &= L(\gamma_0, \gamma_1, c) \\ &= (\ln k + \ln c - \gamma_0 c) \sum_{i=1}^{2} u_i(t) - k u_3(t) \ln\left(1 + (e^{-\gamma_0} c_2(\eta))^c\right) \\ &\quad + (c-1) \sum_{i=1}^{2} u_i(t) \ln c_i(t) - (k+1) \sum_{i=1}^{2} u_i(t) \ln\left(1 + (e^{-\gamma_0} c_i(t))^c\right) \\ &\quad + u_1(t) \gamma_1 \ln\left(\frac{t}{\tau}\right). \end{aligned} \tag{4.31}$$

Let the log-likelihood of unit j be L_j. The log-likelihood L_0 for n independent observations is

$$L_0 = L_1 + \cdots + L_n. \tag{4.32}$$

4.3.5 Parameter Estimation

As the system of likelihood equations obtained using (4.31) has no closed form solution in γ_0, γ_1, and c, therefore, the MLEs $\hat{\gamma}_0, \hat{\gamma}_1$, and $\hat{c}$ are obtained by maximizing (4.32) using the NMaximize option of *Mathematica 6.0*.

4.3.6 Fisher Information Matrix

The Fisher information matrix for an observation at τ is

$$F_\tau = \begin{bmatrix} E\left[-\frac{\partial^2 L}{\partial \gamma_0^2}\right] & E\left[-\frac{\partial^2 L}{\partial \gamma_0 \partial \gamma_1}\right] & E\left[-\frac{\partial^2 L}{\partial \gamma_0 \partial c}\right] \\ E\left[-\frac{\partial^2 L}{\partial \gamma_0 \partial \gamma_1}\right] & E\left[-\frac{\partial^2 L}{\partial \gamma_1^2}\right] & E\left[-\frac{\partial^2 L}{\partial \gamma_1 \partial c}\right] \\ E\left[-\frac{\partial^2 L}{\partial \gamma_0 \partial c}\right] & E\left[-\frac{\partial^2 L}{\partial \gamma_1 \partial c}\right] & E\left[-\frac{\partial^2 L}{\partial c^2}\right] \end{bmatrix}. \tag{4.33}$$

The calculation of the elements of the Fisher information matrix is shown in Appendix A.9.

Because n units are tested at τ, the Fisher information matrix, F, for the plan with a sample of n independent units at stress rate β is

$$F = nF_\tau. \tag{4.34}$$

For some set of values of $\gamma_0, \gamma_1, s_0, \eta, \beta, k$, and c; $|F|$ or the variance function may be negative, therefore, only that set of values for which $|F| > 0$ or the variance function is positive, is chosen. A similar observation has been made by Gouno, Sen and Balakrishnan (2004).

4.3.7 Asymptotic Variance of the MLE of Logarithm of Quantile at Design Stress

For any plan, the asymptotic variance–covariance matrix of the model parameters is given by the inverse of the corresponding Fisher information matrix,

$$F^{-1} = \begin{bmatrix} \mathrm{Var}(\hat{\gamma}_0) & \mathrm{Cov}(\hat{\gamma}_0, \hat{\gamma}_1) & \mathrm{Cov}(\hat{\gamma}_0, \hat{c}) \\ \mathrm{Cov}(\hat{\gamma}_0, \hat{\gamma}_1) & \mathrm{Var}(\hat{\gamma}_1) & \mathrm{Cov}(\hat{\gamma}_1, \hat{c}) \\ \mathrm{Cov}(\hat{\gamma}_0, \hat{c}) & \mathrm{Cov}(\hat{\gamma}_1, \hat{c}) & \mathrm{Var}(\hat{c}) \end{bmatrix}.$$

The asymptotic variance of the MLEs of the log of quantile q of the distribution at a design constant-stress s_d is

$$\mathrm{AsVar}\left[\log \hat{\theta}_q\right] = \mathrm{AsVar}\left[\hat{h}(\hat{\gamma}_0, \hat{\gamma}_1, \hat{c})\right],$$

where $\mathrm{AsVar}[\hat{h}(\hat{\gamma}_0, \hat{\gamma}_1, \hat{c})] = \underset{\sim}{\hat{H}}' F^{-1} \underset{\sim}{\hat{H}}$, vector $\underset{\sim}{\hat{H}}'$ is a transpose of vector $\underset{\sim}{\hat{H}}$.

$$\hat{h} = \hat{\gamma}_0 + \hat{\gamma}_1 \ln\left(\frac{s_h}{s_d}\right) + \frac{1}{\hat{c}} \ln\left(-1 + (1-q)^{-\frac{1}{k}}\right)$$

and

$$\underset{\sim}{\hat{H}} = \left[\frac{\partial \hat{h}}{\partial \hat{\gamma}_0} \quad \frac{\partial \hat{h}}{\partial \hat{\gamma}_1} \quad \frac{\partial \hat{h}}{\partial \hat{c}}\right]' \text{ with } \hat{h} \equiv \hat{h}(\hat{\gamma}_0, \hat{\gamma}_1, \hat{c}).$$

Thus,

$$\mathrm{AsVar}\left[\log \hat{\theta}_q\right] = \mathrm{AsVar}\left[\hat{h}(\hat{\gamma}_0, \hat{\gamma}_1, \hat{c})\right]. \tag{4.35}$$

4.4 FORMULATION OF A MULTI-OBJECTIVE OPTIMIZATION PROBLEM

A variety of approaches can be used to solve the multi-objective optimization problem. One popular approach is to combine these objectives into one single composite objective (Liao and Li, 2008). This approach has been used in the present section.

4.4.1 Optimization Problem For Multi-Level Ramp-Stress ALT Plan without Stress Upper Bound

The ramp test is specified by 'm' stress rates, $\beta_1, \beta_2, \ldots, \beta_m$, and proportion of units, π_i allocated to stress rate β_i. The high stress rate β_m is known from technical considerations. The stress rate factor, ξ_i, and proportion of units, π_i allocated to ξ_i are determined by minimizing the weighted sum of the asymptotic variances of the ML estimators of the quantile lifetimes at a design constant stress.

The asymptotic variance of the ML estimators is a function of $\pi_1, \pi_2, \ldots, \pi_m, \xi_1, \xi_2, \ldots, \xi_{(m-1)}, s_0, \eta, \gamma_0, \gamma_1, c$, and k.

Thus, the optimal design problem can be formulated as a nonlinear optimization problem:

$$
\begin{aligned}
\text{Minimize} \quad & \sum_{j=1}^{b} w_j V_j \\
\text{Subject to} \quad & 0 \leq \pi_i \leq 1, \quad i = 1, 2, \ldots, m \\
& \sum_{i=1}^{m} \pi_i = 1 \\
& \sum_{j=1}^{b} w_j = 1 \\
& 0 < \xi_{i'} < 1, \quad i' = 1, 2, \ldots, (m-1) \\
& \xi_1 < \xi_2 < \cdots < \xi_{(m-1)} \\
& n\,\pi_i\, p_i \geq \text{MMNF}, \quad i = 1, 2, \ldots, m,
\end{aligned} \tag{4.36}
$$

where 'w_j' is the weight, $j = 1, 2, \ldots, b$, 'b' is the number of quantile lifetimes,

$$
\begin{aligned}
V_j &= \text{AsVar}(\hat{y}_{q_j}) \\
&= \left(1 \quad \frac{\ln\left((1-q_j)^{-\frac{1}{k}} - 1\right)}{c_1} \quad - \frac{\ln\left((1-q_j)^{-\frac{1}{k}} - 1\right)}{c^2}\right)
\end{aligned}
$$

$$\times F^{-1}\left(1 \quad \frac{\ln\left((1-q_j)^{-\frac{1}{k}}-1\right)}{c_1} \quad - \frac{\ln\left((1-q_j)^{-\frac{1}{k}}-1\right)}{c^2}\right)'$$

(using (4.22)),

and MMNF is the minimum mean number of failures taken to be 5.

Another suitable number for MMNF can be specified according to cost, time, and precision implications (Meeker and Hahn, 1985; Yang, 1994).

Because, an optimum ramp test depends on the values of $s_0, \eta, \gamma_0, \gamma_1, c$, and k; one must obtain their values from experience, similar data, or a preliminary test. The choice of the sample size, and stress levels depend upon the plan designer, and cost constraints. Cost factor has not been considered. The optimum values of $\pi_1, \pi_2, \ldots, \pi_m, \xi_1, \xi_2, \ldots, \xi_{(m-2)}$, and $\xi_{(m-1)}$ has been found by using the NMinimize option of *Mathematica 6.0*.

4.4.2 Optimization Problem For Ramp-Stress ALT Plan with Stress Upper Bound

The ramp test is specified by stress rate β, and 'τ' is determined by minimizing the weighted sum of the asymptotic variances of the ML estimators of the quantile lifetimes at a design constant stress.

The asymptotic variance is a function of $s_d, \eta, \gamma_0, \gamma_1, c, k, s_h$, and τ.

Thus, the optimal design problem can be formulated as a nonlinear optimization problem:

$$\begin{aligned} &\text{Minimize } \sum_{j=1}^{b} w_j V_j \\ &\text{subject to } 0 < \tau < \eta \\ &\sum_{j=1}^{b} w_j = 1, \end{aligned} \tag{4.37}$$

where 'w_j' is the weight, $j = 1, 2, \ldots, b$,

'b' is the number of quantile lifetimes, and

$$\begin{aligned}
V_j &= \text{AsVar}(\log \hat{\theta}_{q_j}) \\
&= \left(1 \quad \ln\left(\frac{s_h}{s_d}\right) \quad -\frac{\ln\left((1-q_j)^{-\frac{1}{k}}-1\right)}{c^2}\right) \\
&\quad \times F^{-1}\left(1 \quad \ln\left(\frac{s_h}{s_d}\right) \quad -\frac{\ln\left((1-q_j)^{-\frac{1}{k}}-1\right)}{c^2}\right)' \quad \text{(using (4.35))}.
\end{aligned}$$

It is necessary to impose the constraint $0 < \tau < \eta$, as $\eta > \tau$ implies the ramp test without stress upper bound (Bai, Chun, and Cha, 1997).

Because, an optimum ramp test depends on the values of $s_d, \eta, \gamma_0, \gamma_1, c, k$, and s_h; one must obtain their values from experience, similar data, or a preliminary test. The choice of the sample size depends upon the plan designer, and cost constraints. Cost factor has not been considered. The optimum value of τ has been found by using the NMinimize option of *Mathematica 6.0.*

In practicing ALT, one is interested in the lower tails of the life distribution, i.e., early failures, so 1%, 10%, and 50% quantiles need to be considered.

The weights are chosen according to the needs of the plan designed.

4.5 NUMERICAL EXAMPLES

4.5.1 Multi-Level Ramp-Stress ALT Plan without Stress Upper Bound

The following hypothetical data set has been considered:

$$n = 35, \quad \gamma_0 = 4.5, \quad \gamma_1 = 6, \quad s_0 = 30, \quad \eta = 100, \quad \beta_4 = 9.9,$$

$$k = 1, \quad c = 2, \quad b = 3, \quad w_1 = w_2 = w_3 = \frac{1}{3}.$$

4.5.1.1 *Optimal Plan*

The optimal allocations, and the optimal stress rate factors are obtained by minimizing the weighted sum of the asymptotic variances of the MLEs of the log of quantiles 1%, 10%, and 50% of the distribution at a design constant-stress, (4.22), using the NMinimize option of *Mathematica 6.0.* These are obtained as

$$\xi_1^* = 0.0481666, \quad \xi_2^* = 0.40404, \quad \xi_3^* = 0.808081, \quad \text{and}$$
$$\pi_1^* = 0.571429, \quad \pi_2^* = 0.142857, \quad \pi_3^* = 0.142857, \quad \pi_4^* = 0.142857.$$

Thus, the optimum lower stress rates are

$$\begin{aligned} \beta_1^* &= \beta_4\, \xi_1^* = 0.4768493, \\ \beta_2^* &= \beta_4\, \xi_2^* = 3.999996, \quad \text{and} \\ \beta_3^* &= \beta_4\, \xi_3^* = 8.0000019. \end{aligned}$$

The optimal number of units to be tested at lower stress rates β_1, β_2, and β_3 are 20, 5, and 5, respectively, and at high stress rate, β_4 it is 5.

4.5.1.2 *Simulated Data*

Type-I censored sample is simulated under the ramp-test using (4.8) and is given in Table 4.1.

Table 4.1: Simulated failure times ($n = 35$, $\gamma_0 = 4.5$, $\gamma_1 = 6$, $s_0 = 30$, $\eta = 100$, $\beta_4 = 9.9$, $k = 1$, $c = 2$, $b = 3$, and $w_1 = w_2 = w_3 = 1/3$)

Stress Rate						n_{c_i}
$\xi_1^* = \mathbf{0.0481666}$	4.41532,	4.41756,	4.29717,	4.33470,	4.48297,	**3**
	4.57291,	4.22077,	4.48894,	4.25115,	4.44575,	
	4.55067,	4.47870,	4.47940,	4.41341,	4.54703,	
	4.42514,	4.45133				
$\xi_2^* = \mathbf{0.4040400}$	2.86518,	2.66943,	2.49939,	2.35022,	2.61562	**0**
$\xi_3^* = \mathbf{0.8080810}$	2.17541,	2.08735,	1.72644,	2.00088,	1.99236	**0**
$\xi_4 = \mathbf{1}$	1.90145,	1.86930,	1.91432,	1.90792,	1.71252	**0**

4.5.1.3 MLEs of the Design Parameters

The MLEs of the design parameters obtained using simulated data given in Table 4.1 are $\hat{\gamma}_0 = 4.48803$, $\hat{\gamma}_1 = 6.28148$, and $\hat{c} = 1.81917$.

These estimates are obtained using the NMaximize option of *Mathematica 6.0*.

4.5.1.4 Confidence Intervals

95% confidence intervals for the parameters are obtained using the inverse of the observed Fisher information matrix $\hat{F}^{-1}$, and it is given by

$$\hat{F}^{-1} = \begin{pmatrix} 0.1146630 & 0.2606830 & -0.0732092 \\ 0.2606830 & 0.7665200 & -0.2060410 \\ -0.0732092 & -0.2060410 & 0.1872330 \end{pmatrix}.$$

The observed value of F^{-1}, that is, $\hat{F}^{-1}$ is determined by substituting the estimated parameters $\hat{\gamma}_0, \hat{\gamma}_1$, and $\hat{c}$ for the true parameters in the asymptotic variance–covariance matrix. The square root of the diagonal elements of $\hat{F}^{-1}$ gives the standard error of an estimator. Thus, the 95% approximate confidence intervals for γ_0, γ_1, and c, respectively, are

$$\begin{aligned} &3.82433 \leq \gamma_0 \leq 5.15172, \\ &4.77991 \leq \gamma_1 \leq 8.25476, \quad \text{and} \\ &1.14131 \leq c \leq 2.89963. \end{aligned}$$

4.5.2 Ramp-Stress ALT Plan with Stress Upper Bound

Consider the following hypothetical data set.

$$n = 35, \quad \gamma_0 = 0.5, \quad \gamma_1 = 5, \quad s_d = 20, \quad s_h = 30,$$

$$\eta = 19, \quad k = 2, \quad c = 1, \quad b = 3, \quad w_1 = w_2 = w_3 = \frac{1}{3}.$$

4.5.2.1 *Optimal Plan*

The optimal τ is obtained by minimizing the weighted sum of the asymptotic variances of the MLEs of the log of quantiles 1%, 10%, and 50% of the distribution at a design constant-stress, (4.35), using the NMinimize option of *Mathematica 6.0*. It is obtained as $\tau^* = 7.06961$.

Thus, the optimum stress rate is

$$\beta^* = \frac{s_h}{\tau^*} = 4.2435156.$$

4.5.2.2 *Simulated Data*

Type-I censored sample is simulated under the ramp-test using (4.27), and is displayed in Table 4.2.

4.5.2.3 *MLEs of the Design Parameters*

The MLEs of the design parameters obtained using simulated data given in Table 4.2 are $\hat{\gamma}_0 = 0.861125$, $\hat{\gamma}_1 = 5.13551$, and $\hat{c} = 0.963443$.

These estimates are obtained using the NMaximize option of *Mathematica 6.0*.

Table 4.2: Simulated failure times ($n = 35$, $\gamma_0 = 0.5$, $\gamma_1 = 5$, $s_d = 20$, $s_h = 30$, $\eta = 19$, $k = 2$, $c = 1$, $n_c = 1$, $b = 3$, and $w_1 = w_2 = w_3 = 1/3$)

Intervals	**Failure Times**					
$0 \le t < \tau$	5.87369,	5.12560,	4.37980,	6.81937,	5.11455,	5.12643,
	5.68369,	6.43288,	5.60970,	3.39309,	6.32190,	5.31403,
	5.82308,	6.90719,	4.94967,	6.36280,	6.08122,	5.98342
$\tau \le t < \eta$	10.2812,	12.1832,	7.40265,	7.26162,	7.42935,	11.0032,
	8.23416,	8.12027,	9.56197,	7.08203,	13.7645,	9.75099,
	10.7121,	7.43437,	7.23713,	7.57809		

4.5.2.4 *Confidence Intervals*

95% confidence intervals for the parameters are obtained using the inverse of the observed Fisher information matrix $\hat{F}^{-1}$, and is given by

$$\hat{F}^{-1} = \begin{pmatrix} 0.2710840 & -1.3057700 & 0.0139789 \\ -1.3057700 & 8.3525100 & -0.1516730 \\ 0.0139789 & -0.1516730 & 0.0206541 \end{pmatrix}.$$

The observed value of F^{-1}, that is, $\hat{F}^{-1}$ is determined by substituting the estimated parameters $\hat{\gamma}_0, \hat{\gamma}_1$, and $\hat{c}$ for the true parameters in the asymptotic variance–covariance matrix. The square root of the diagonal elements of $\hat{F}^{-1}$ gives the standard error of an estimator. Thus, the 95% approximate confidence intervals for γ_0, γ_1, and c, respectively, are

$$\begin{aligned} &-0.159362 \le \gamma_0 \le 1.88161, \\ &1.704320 \le \gamma_1 \le 15.4745, \quad \text{and} \\ &0.719203 \le c \le 1.29063. \end{aligned}$$

4.6 SENSITIVITY ANALYSIS

To use an optimum test plan, one needs estimates of the design parameters γ_0, γ_1, and c. These estimates sometimes may significantly affect the values of the resulting decision variables; therefore, their incorrect choice may give a poor estimate of the quantile at a design constant-stress. Hence, it is important to conduct sensitivity analysis to evaluate the robustness of the resulting fully ALT plan.

The PDs of the optimal settings are measured by

$$\text{PD} = \frac{|Z^{**} - Z^*|}{Z^*} \times 100,$$

where Z^* is the optimum variance obtained with the given design parameters, and Z^{**} is the one obtained when a parameter is misspecified.

Table 4.3: Sensitivity analysis for the multi-level ramp-stress ALT plan without stress upper bound ($n = 35$, $s_0 = 30$, $\eta = 100$, $\beta_4 = 9.9$, $k = 1$, $b = 3$, and $w_1 = w_2 = w_3 = 1/3$)

Parameter	% Change	PD
γ_0	+5	1.680280
γ_0	−5	8.213110
γ_1	+5	0.269643
γ_1	−5	8.042450
c	+5	11.68650
c	−5	14.44950

Table 4.4: Sensitivity analysis for the ramp-stress ALT plan with stress upper bound $n = 35$, $s_d = 20$, $s_h = 30$, $\eta = 19$, $b = 3$, $w_1 = w_2 = w_3 = 1/3$, and $k = 2$)

Parameter	% Change	τ	PD
γ_0	+10	7.45491	0.748054
γ_0	−10	6.69959	0.827434
γ_1	+10	8.01436	13.18640
γ_1	−10	6.30171	12.61680
c	+10	7.27117	1.218340
c	−10	7.08173	4.217200

Tables 4.3 and 4.4 show the optimal test plans for various deviations from the design parameter estimates. The results show that the optimal setting of Z is robust to the deviations from baseline parameters.

4.7 COMPARATIVE STUDY

4.7.1 Multi-Level Ramp-Stress ALT Plan without Stress Upper Bound

In this section, Srivastava and Mittal (2012a) multilevel ramp-stress ALT model has been compared with the ones designed by

Table 4.5: Hypothetical Failure time data set for 2-level multi-objective ramp-stress ALT plan ($n = 35$, $\gamma_0 = 4.5$, $\gamma_1 = 6$, $s_0 = 30$, $\eta = 100$, $\beta_4 = 9.9$, $k = 1$, $c = 2$, $b = 3$, and $w_1 = w_2 = w_3 = 1/3$)

Stress Rate	Failure Times					n_{c_i}
$\xi_1^* = 0.0481818$	4.30685,	4.37833,	4.44029,	4.44406,	4.52003,	6
	4.21793,	4.15934,	4.40244,	4.44000,	4.50176,	
	4.55287,	4.53443,	4.50120,	4.45627,	4.44219,	
	4.57741,	4.48570,	4.18338,	4.39215		
$\xi_2 = 1$	1.79279,	1.87561,	1.80571,	1.97435,	1.88249,	0
	1.85218,	1.82612,	1.93664,	1.88418,	1.87163	

Table 4.6: Comparative study of ramp-stress ALT models

ALT Model	Log-Likelihood Function
Srivastava and Mittal (2012a) Model	11.7506
Bai, Cha and Chung (1992) Model	11.5252
Srivastava and Shukla (2009) Model	−111.120

Bai, Cha, and Chung (1992) and Srivastava and Shukla (2009) in terms of log-likelihood functions using the hypothetical failure time data set under ramp-stress ALT with Type-I censoring given in Table 4.5.

Table 4.6 shows that the Srivastava and Mittal (2012a) model performs better for the given data set.

Also Srivastava and Mittal (2012a) ALT model has been compared with respect to number of stress levels on the basis of changes in parametric values in Table 4.7, and on the basis of equal and unequal weights in Table 4.8.

Tables 4.7 and 4.8 show that it is better to choose two stress levels or three stress levels for testing as higher number of stress levels would result in increased value of the weighted sum of the asymptotic variances of the ML estimators of the quantile lifetimes at a design constant stress.

Table 4.7: Comparative study of ramp-stress model with respect to number of stress levels on the basis of changes in parametric values ($n = 35$, $s_0 = 30$, $\eta = 100$, $\beta_4 = 9.9$, $k = 1$, $b = 3$, and $w_1 = w_2 = w_3 = 1/3$)

γ_0	γ_1	c	2 Levels AsVar	3 Levels AsVar	4 Levels AsVar
5.5	6.0	2.0	0.145753	0.145753	0.174758
3.5	6.0	2.0	0.139979	0.142992	0.155838
4.5	6.2	2.0	0.129769	0.132810	0.151734
4.5	5.8	2.0	0.130528	0.133605	0.161742
4.5	6.0	2.2	0.100694	0.103099	0.118036
4.5	6.0	1.8	0.174826	0.179073	0.206597

Table 4.8: Comparative study of ramp-stress model with respect to number of stress levels on the basis of equal and unequal weights ($n = 35$, $\gamma_0 = 4.5$, $\gamma_1 = 6$, $s_0 = 30$, $\eta = 100$, $\beta_4 = 9.9$, $k = 1$, $c = 2$, and $b = 3$)

w_1	w_2	w_3	2 Levels AsVar	3 Levels AsVar	4 Levels AsVar
0.33	0.33	0.33	0.1301350	0.1331930	0.149947
0.60	0.30	0.10	0.1791770	0.1830790	0.203261
0.60	0.10	0.30	0.1829980	0.1872850	0.208669
0.10	0.30	0.60	0.0873967	0.0933234	0.104638
0.30	0.10	0.60	0.1284030	0.1284030	0.149548
0.30	0.60	0.10	0.1189160	0.1213460	0.136212
0.10	0.60	0.30	0.0817869	0.0919010	0.968694

4.7.2 Ramp-Stress ALT Plan with Stress Upper Bound

In this section, comparative study has been carried out to highlight the merits of the plan. Table 4.9 depicts comparison of a multi-objective plan, and the single-objective plans with respect to changes in parametric values. Table 4.10 shows the comparison of a multi-objective plan with equal and unequal weights. Table 4.11 shows the

Table 4.9: Comparison of a multi-objective plan, and the single-objective plans with respect to changes in parametric values ($n = 35$, $s_d = 20$, $s_h = 30$, $\eta = 19$, $k = 2$, and $n_c = 1$)

γ_0	γ_1	c	AsVar*	$\text{AsVar}_{0.01}$	$\text{AsVar}_{0.1}$	$\text{AsVar}_{0.5}$
0.3	5.0	1.0	0.419535	0.587926	0.364997	0.1304960
0.5	5.0	1.0	0.436116	0.684533	0.367840	0.1412690
0.7	5.0	1.0	0.447748	0.687784	0.374100	0.1607740
0.5	4.5	1.0	0.381092	0.623777	0.306567	0.0944853
0.5	5.5	1.0	0.493624	0.773821	0.440446	0.2006960
0.5	5.0	0.5	1.012860	2.084100	0.696693	0.2397690
0.5	5.0	1.5	0.604746	0.896441	0.548685	0.3047940

Table 4.10: Comparison of a multi- objective plan with equal and unequal weights ($n = 35$, $\gamma_0 = 0.5$, $\gamma_1 = 5$, $s_d = 20$, $s_h = 30$, $\eta = 19$, $k = 2$, $c = 1$, $b = 3$, and $n_c = 1$)

w_1	w_2	w_3	AsVar*
0.33	0.33	0.33	0.436116
0.60	0.30	0.10	0.544612
0.60	0.10	0.30	0.527115
0.10	0.30	0.60	0.308344
0.30	0.10	0.60	0.387668
0.30	0.60	0.10	0.409113
0.10	0.60	0.30	0.289845

Table 4.11: Comparison of a multi-objective plan, and the single-objective plans with respect to changes in stress spread ($n = 35$, $\gamma_0 = 0.5$, $\gamma_1 = 5$, $\eta = 19$, $k = 2$, and $c = 1$)

s_d	s_h	AsVar*	$\text{AsVar}_{0.01}$	$\text{AsVar}_{0.1}$	$\text{AsVar}_{0.5}$
20	50	2.647240	3.189010	2.56383	2.225030
20	40	1.382210	1.791500	1.30280	0.956553
20	30	0.436116	0.684533	0.36784	0.141269
15	30	1.382210	1.791500	1.30280	0.956553

comparison of a multi-objective plan, and the single-objective plans with respect to changes in stress spread.

Tables 4.9–4.11 show that the multi-objective plan yields a smaller variance compared to the single-objective plan at 1% quantile.

4.8 SUMMARY

Optimum design of time-censored multi-objective ramp-stress plan with multiple linearly increasing stresses, and that of ramp-stress plan with stress upper bound under fully accelerated environmental condition has been explained using the Burr Type-XII life distribution. The procedures developed have been explained using numerical examples, and sensitivity analyses carried out. The results of sensitivity analyses show that the optimum plans discussed are robust. Comparative study has also been carried out to highlight the merits of the plans described.

CHAPTER 5

OPTIMUM FULLY ACCELERATED LIFE TEST PLANS WITH MODIFIED STRESS LOADING SCHEMES UNDER TYPE-I CENSORING

5.1 INTRODUCTION

The traditional constant-stress and step-stress fully accelerated life tests (ALTs) assume instantaneous changes in the stress levels. The use of modified stress loading schemes, viz., modified constant-stress loading and modified step-stress loading ensure gradual changes in stress levels thereby preventing a stress (thermal) shock or undesirable failure modes not observed under normal operating conditions. **Modified constant-stress** is also referred to as **ramp-constant stress** and **modified step-stress** as **ramp-step stress** (see Sections 1.2.6 and 1.2.7).

Several authors have studied life testing models under fully accelerated condition using combination of stress loadings (see Section 1.18.4). Park and Yum (1998) were the first to propose the single-objective optimum ALT plans for modified stress loadings that minimize the asymptotic variance of the MLE of the log mean lifetime for the use condition.

The present chapter deals with the optimum design of time-censored multi-objective test plans using the Burr Type-XII distribution with modified stress schemes under fully accelerated conditions.

5.1.1 Notation

i	index for the modified constant-stress fully ALT plan, $i = 1, 2$
β	stress rate, $\beta > 0$
s_0	design stress
s_1	low stress
s_1^*	optimal low stress
s_2	high stress
π	proportion of units allocated to the low stress level
π^*	optimal proportion of units allocated to the low stress level
$1 - \pi$	proportion of units allocated to the high stress level
n	total number of test units
n_c	number of items censored under modified step-stress fully ALT plan
n_{c1}	number of items censored at low stress level under modified constant-stress fully ALT plan
n_{c2}	number of items censored at high stress level under modified constant-stress fully ALT plan
η	censoring time, $\eta > 0$
Burr (c, k, α)	Burr Type-XII distribution with parameters c, k, α
c, k	shape parameters of the Burr Type-XII distribution, $c > 0$, $k > 0$
α	scale parameter of the Burr Type-XII distribution, $\alpha > 0$
γ_0, γ_1	parameters of the inverse power law, $\gamma_1 > 0$, $-\infty < \gamma_0 < \infty$
T	lifetime of an item under modified step-stress fully ALT plan
X	lifetime of an item under modified constant-stress fully ALT plan
t_1	time at which stress changes from s_0 to s_1
t_2	time at which stress starts increasing at the rate β from s_1

t_2^*	optimal time at which stress starts increasing at the rate β from s_1
t_3	time at which stress changes from s_1 to s_2
x_1	time at which stress reaches s_1
x_2	time at which stress reaches s_2
$s(t)$	stress at time t
$\varepsilon(t)$	cumulative exposure at time t
$G_1(t)$	$\dfrac{e^{-\gamma_0} s_0^{-\gamma_1}\left((\beta t + s_0)^{(1+\gamma_1)} - s_0^{(1+\gamma_1)}\right)}{\beta(1+\gamma_1)}$
$G_2(t)$	$\dfrac{e^{-\gamma_0} s_0^{-\gamma_1}(t - t_1)}{s_1^{-\gamma_1}}$
$G_3(t)$	$\dfrac{e^{-\gamma_0} s_0^{-\gamma_1}\left((\beta t - \beta t_2 + s_1)^{(1+\gamma_1)} - s_1^{(1+\gamma_1)}\right)}{\beta(1+\gamma_1)}$
$G_4(t)$	$\dfrac{e^{-\gamma_0} s_0^{-\gamma_1}(t - t_3)}{s_2^{-\gamma_1}}$
Q_1	$kce^{-\gamma_0} s_0^{-\gamma_1}$
$R_1(t)$	$\dfrac{s_0^{(1+\gamma_1)} \ln s_0 - (\beta t + s_0)^{(1+\gamma_1)} \ln(\beta t + s_0)}{(\beta t + s_0)^{(1+\gamma_1)} - s_0^{(1+\gamma_1)}}$
$R_2(t)$	$\dfrac{s_1^{(1+\gamma_1)} \ln s_1 - (\beta t - \beta t_2 + s_1)^{(1+\gamma_1)} \ln(\beta t - \beta t_2 + s_1)}{(\beta t - \beta t_2 + s_1)^{(1+\gamma_1)} - s_1^{(1+\gamma_1)}}$
$A(t)$	$\dfrac{1}{(1+\gamma_1)^2} - \dfrac{s_0^{(1+\gamma_1)}(\beta t + s_0)^{(1+\gamma_1)}(\ln s_0 - \ln(\beta t + s_0))^2}{((\beta t + s_0)^{(1+\gamma_1)} - s_0^{(1+\gamma_1)})^2}$
$B(t)$	$G_1(t_1)A(t_1) + \left(\ln\left(\dfrac{s_0}{s_1}\right)\right)^2 G_2(t) + G_1(t_1)\left(\ln s_0 + \dfrac{1}{1+\gamma_1} + R_1(t_1)\right)^2$

$C(t)$ $\quad \frac{1}{(1+\gamma_1)^2} - \frac{s_1^{(1+\gamma_1)}(\beta t - \beta t_2 + s_1)^{(1+\gamma_1)}\left(\ln s_1 - \ln(\beta t - \beta t_2 + s_1)\right)^2}{\left((\beta t - \beta t_2 + s_1)^{(1+\gamma_1)} - s_1^{(1+\gamma_1)}\right)^2}$

$D(t)$ $\quad B(t_2) + G_3(t)C(t) + G_3(t)\left(\ln s_0 + \frac{1}{1+\gamma_1} + R_2(t)\right)^2$

$E(t)$ $\quad D(t_3) + G_4(t)\left(\ln\left(\frac{s_0}{s_2}\right)\right)^2$

$G_1(x)$ $\quad \frac{e^{-\gamma_0} s_0^{-\gamma_1}\left((\beta x + s_0)^{(1+\gamma_1)} - s_0^{(1+\gamma_1)}\right)}{\beta(1+\gamma_1)}$

$G_{i2}(x)$ $\quad \frac{e^{-\gamma_0} s_0^{-\gamma_1}(x - x_i)}{s_i^{-\gamma_1}}$

$A(x)$ $\quad \frac{1}{(1+\gamma_1)^2} - \frac{s_0^{(1+\gamma_1)}(\beta x + s_0)^{(1+\gamma_1)}(\ln s_0 - \ln(\beta x + s_0))^2}{((\beta x + s_0)^{(1+\gamma_1)} - s_0^{(1+\gamma_1)})^2}$

$R_1(x)$ $\quad \frac{s_0^{(1+\gamma_1)} \ln s_0 - (\beta x + s_0)^{(1+\gamma_1)} \ln(\beta x + s_0)}{(\beta x + s_0)^{(1+\gamma_1)} - s_0^{(1+\gamma_1)}}$

$K(x)$ $\quad G_1(x_i)A(x_i) + \left(\ln\left(\frac{s_0}{s_i}\right)\right)^2 G_{i2}(x)$

$\quad + G_1(x_i)\left(\ln s_0 + \frac{1}{1+\gamma_1} + R_1(x_i)\right)^2$

B_m $\quad \{t | t_m < t \leq t_{m+1}\}$, $m = 0, 1, 2, 3, 4,$ where $t_0 = 0$, $t_4 = \eta$, $t_5 = \infty$

$\delta_m = \delta_m(t)$ $\quad$ indicator function under modified step-stress fully ALT plan

$$\delta_m(t) = \begin{cases} 1, & \text{if } t \in B_m,\ m = 0, 1, 2, 3, 4 \\ 0, & \text{otherwise} \end{cases}$$

A_0 $\quad \{x | 0 < x \leq x_i\}$

A_1 $\quad \{x | x_i < x \leq \eta\}$

A_2	$\{x \mid x > \eta\}$
$I_j = I_j(x)$	indicator function under modified constant-stress fully ALT plan $$I_j(x) = \begin{cases} 1, & \text{if } x \in A_j,\ j = 0, 1, 2 \\ 0, & \text{otherwise} \end{cases}$$
$\wedge$	implies a MLE
q	quantile
AsVar*	Asymptotic variance under multi-objective plan
AsVar_q	Asymptotic variance under single-objective plan at quantile q.

5.1.2 Burr Type-XII Lifetime Distribution

The pdf, cdf, reliability function, and hazard function of the Burr Type-XII distribution are

$$g(t) = \left(\frac{kc}{\alpha}\right)\left(\frac{t}{\alpha}\right)^{c-1}\left(1 + \left(\frac{t}{\alpha}\right)^{c}\right)^{-(k+1)}, \quad t \geq 0, \tag{5.1}$$

$$G(t) = 1 - \left(1 + \left(\frac{t}{\alpha}\right)^{c}\right)^{-k}, \quad t \geq 0, \tag{5.2}$$

$$R(t) = \left(1 + \left(\frac{t}{\alpha}\right)^{c}\right)^{-k}, \quad t \geq 0 \tag{5.3}$$

and

$$h(t) = \left(\frac{kc}{\alpha}\right)\left(\frac{t}{\alpha}\right)^{c-1}\left(1 + \left(\frac{t}{\alpha}\right)^{c}\right)^{-1}, \quad t \geq 0, \tag{5.4}$$

respectively (see Section 1.4.2.9).

5.2 OPTIMUM MULTI-OBJECTIVE MODIFIED STEP-STRESS ALT PLAN

This section deals with formulation of optimum multi-objective modified step-stress test plan due to Srivastava and Mittal (2014) for the Burr Type-XII distribution under fully accelerated environmental condition and Type-I censoring. The optimum plan gives the value of optimum stress change time, and optimum lower level stress.

5.2.1 Assumptions

1. The lifetimes of test units are independent and identically distributed.
2. Failed items are not replaced during the test.
3. Two stress levels higher than design stress (s_0), viz.; the low stress level (s_1), and the high stress level (s_2) are employed, with s_0, and s_2 assumed to be pre-specified.
4. The censoring time η is pre-specified.
5. For the effect of changing stress levels, the cumulative exposure model holds (Nelson, 1980).
6. Between any two stress levels, the stress is increased at the rate β.
7. The shape parameter 'k' does not depend on the stress level and is assumed to be known for the sake of mathematical convenience.
8. At any constant-stress s, the lifetime of a unit follows a Burr Type-XII model with scale parameter $\alpha(s)$, and shape parameters c and k; and the inverse power law holds for

$$\alpha(s) = e^{\gamma_0} \left(\frac{s_0}{s}\right)^{\gamma_1}. \tag{5.5}$$

5.2.2 Test Procedure

The modified step-stress fully ALT proceeds as follows:

1. 'n' test units are put on test at time t_0.
2. The stress is increased at the rate β until it reaches s_1, which occurs at time $t_1 = \frac{s_1 - s_0}{\beta}$.
3. From t_1 to t_2, the stress is maintained at the level s_1.
4. At time t_2, the stress is increased again at the rate β until it reaches s_2, which occurs at time $t_3 = t_2 + \left(\frac{s_2 - s_1}{\beta}\right)$.
5. After t_3, the stress is maintained at the level s_2 until the pre-specified censoring time η is reached.

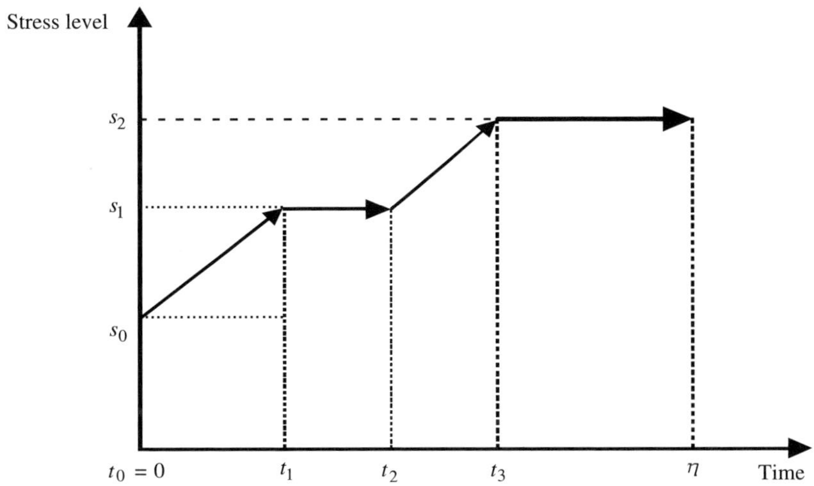

Figure 5.1: Modified step-stress loading

6. The test is continued until:
   (a) all test items fail, or
   (b) a prescribed censoring time η

whichever occurs earlier, and the test conditions remain the same.

The modified step-stress loading is depicted in Figure 5.1.

5.2.3 Life Distribution Under Modified Step-Stress

Based on the inverse power law (see Assumption 8), the cumulative exposure function $\varepsilon(t)$ at time t under stress level s is:

For $0 < t \leq t_1$,

$$\varepsilon(t) = \int_0^t \frac{1}{\alpha(s(y))} dy = \int_0^t \frac{1}{e^{\gamma_0}\left(\frac{s_0}{s(y)}\right)^{\gamma_1}} dy = \int_0^t \frac{1}{e^{\gamma_0}\left(\frac{s_0}{y\beta+s_0}\right)^{\gamma_1}} dy$$

$$= \frac{e^{-\gamma_0} s_0^{-\gamma_1}\left((\beta t + s_0)^{1+\gamma_1} - s_0^{1+\gamma_1}\right)}{\beta(1+\gamma_1)} = G_1(t). \tag{5.6}$$

For $t_1 < t \leq t_2$,

$$\varepsilon(t) = \int_0^t \frac{1}{\alpha(s(y))} dy = \varepsilon(t_1) + \int_{t_1}^t \frac{1}{e^{\gamma_0}\left(\frac{s_0}{s_1}\right)^{\gamma_1}} dy$$

$$= \varepsilon(t_1) + e^{-\gamma_0}(t - t_1)\left(\frac{s_0}{s_1}\right)^{-\gamma_1} = \varepsilon(t_1) + G_2(t). \quad (5.7)$$

For $t_2 < t \leq t_3$,

$$\varepsilon(t) = \int_0^t \frac{1}{\alpha(s(y))} dy = \varepsilon(t_2) + \int_{t_2}^t \frac{e^{-\gamma_0}}{\left(\frac{s_0}{s(y)}\right)^{\gamma_1}} dy$$

$$= \varepsilon(t_2) + \int_{t_2}^t \frac{1}{e^{\gamma_0}\left(\frac{s_0}{\beta y - \beta t_2 + s_1}\right)^{\gamma_1}} dy$$

$$= \varepsilon(t_2) + \frac{e^{-\gamma_0} s_0^{-\gamma_1}((\beta t - \beta t_2 + s_1)^{1+\gamma_1} - s_1^{1+\gamma_1})}{\beta(1+\gamma_1)}$$

$$= \varepsilon(t_2) + G_3(t) \quad (5.8)$$

For $t_3 < t \leq \eta$,

$$\varepsilon(t) = \int_0^t \frac{1}{\alpha(s(y))} dy = \varepsilon(t_3) + \int_{t_3}^t \frac{1}{e^{\gamma_0}\left(\frac{s_0}{s_2}\right)^{\gamma_1}} dy$$

$$= \varepsilon(t_3) + e^{-\gamma_0}(t - t_3)\left(\frac{s_0}{s_2}\right)^{-\gamma_1} = \varepsilon(t_3) + G_4(t). \quad (5.9)$$

Then, cdf of the lifetime T of a unit tested under modified step-stress is

$$F(t) = G(\varepsilon(t)), \quad (5.10)$$

where $G(\cdot)$ is the assumed cdf (5.2) with the scale parameter α set equal to one (see Appendix A.10), and $\varepsilon(t)$ is the cumulative exposure (damage) model.

Thus, the cdf is

$$F(t) = 1 - (1 + \varepsilon(t)^c)^{-k} \text{ (using (5.2) and (5.10))}.$$

Therefore,

$$F(t) = \begin{cases} F_1(t), & 0 < t \leq t_1 \\ F_2(t), & t_1 < t \leq t_2 \\ F_3(t), & t_2 < t \leq t_3 \\ F_4(t), & t_3 < t \leq \eta, \end{cases} \tag{5.11}$$

where

$$F_1(t) = 1 - (1 + (G_1(t))^c)^{-k},$$
$$F_2(t) = 1 - (1 + (G_1(t_1) + G_2(t))^c)^{-k},$$
$$F_3(t) = 1 - (1 + (G_1(t_1) + G_2(t_2) + G_3(t))^c)^{-k},$$

and

$$F_4(t) = 1 - (1 + (G_1(t_1) + G_2(t_2) + G_3(t_3) + G_4(t))^c)^{-k}.$$

The pdf is given by

$$f(t) = \begin{cases} f_1(t), & 0 < t \leq t_1 \\ f_2(t), & t_1 < t \leq t_2 \\ f_3(t), & t_2 < t \leq t_3 \\ f_4(t), & t_3 < t \leq \eta, \end{cases} \tag{5.12}$$

where

$$f_1(t) = Q_1(\beta t + s_0)^{\gamma_1}(G_1(t))^{c-1}(1 + (G_1(t))^c)^{-k-1},$$
$$f_2(t) = Q_1(s_1)^{\gamma_1}(G_1(t_1) + G_2(t))^{c-1}(1 + (G_1(t_1) + G_2(t))^c)^{-k-1},$$
$$f_3(t) = Q_1(\beta t - \beta t_2 + s_1)^{\gamma_1}(G_1(t_1) + G_2(t_2)$$
$$+ G_3(t))^{c-1}(1 + (G_1(t_1) + G_2(t_2) + G_3(t))^c)^{-k-1}$$

and

$$f_4(t) = Q_1(s_2)^{\gamma_1}(G_1(t_1) + G_2(t_2) + G_3(t_3) \\ + G_4(t))^{c-1}(1 + (G_1(t_1) + G_2(t_2) + G_3(t_3) + G_4(t))^c)^{-k-1}.$$

5.2.4 Log-Likelihood Function

The log-likelihood of an observation at time t under Type-I censoring is derived.

The log-likelihood of a single observation at time t is

$$\begin{aligned} L = L(\gamma_0, \gamma_1, c) &= \sum_{j=0}^{3} \delta_j \ln f_{j+1}(t) + \delta_4 \ln(1 - F_4(\eta)) \\ &= \delta_0[\ln Q_1 + \gamma_1 \ln(\beta t + s_0) + (c-1)\ln G_1(t) \\ &\quad - (k+1)\ln(1 + (G_1(t))^c)] + \delta_1[\ln Q_1 + \gamma_1 \ln(s_1) \\ &\quad + (c-1)\ln(G_1(t_1) + G_2(t)) - (k+1)\ln(1 + (G_1(t_1) + G_2(t))^c)] \\ &\quad + \delta_2[\ln Q_1 + \gamma_1 \ln(\beta t - \beta t_2 + s_1) + (c-1)\ln(G_1(t_1) + G_2(t_2) \\ &\quad + G_3(t)) - (k+1)\ln(1 + (G_1(t_1) + G_2(t_2) + G_3(t))^c)] \\ &\quad + \delta_3[\ln Q_1 + \gamma_1 \ln(s_2) + (c-1)\ln(G_1(t_1) + G_2(t_2) + G_3(t_3) \\ &\quad + G_4(t)) - (k+1)\ln(1 + (G_1(t_1) + G_2(t_2) + G_3(t_3) + G_4(t))^c)] \\ &\quad + \delta_4[-k\ln(1 + (G_1(t_1) + G_2(t_2) + G_3(t_3) + G_4(\eta))^c)]. \end{aligned} \tag{5.13}$$

Let the log-likelihood of unit j be L_j. The log-likelihood L_0 for n independent observations is

$$L_0 = L_1 + \cdots + L_n. \tag{5.14}$$

The model parameters are estimated using method of maximum likelihood, and are obtained by maximizing (5.14).

5.2.5 Fisher Information Matrix

The Fisher information matrix is obtained by taking the expectations of the negative of the second-order partial derivatives of the log-likelihood function with respect to γ_0, γ_1, and c.

The Fisher information matrix for an observation is

$$F_{\text{single}} = \begin{bmatrix} E\left[-\frac{\partial^2 L}{\partial \gamma_0^2}\right] & E\left[-\frac{\partial^2 L}{\partial \gamma_0 \partial \gamma_1}\right] & E\left[-\frac{\partial^2 L}{\partial \gamma_0 \partial c}\right] \\ E\left[-\frac{\partial^2 L}{\partial \gamma_0 \partial \gamma_1}\right] & E\left[-\frac{\partial^2 L}{\partial \gamma_1^2}\right] & E\left[-\frac{\partial^2 L}{\partial \gamma_1 \partial c}\right] \\ E\left[-\frac{\partial^2 L}{\partial \gamma_0 \partial c}\right] & E\left[-\frac{\partial^2 L}{\partial \gamma_1 \partial c}\right] & E\left[-\frac{\partial^2 L}{\partial c^2}\right] \end{bmatrix}. \quad (5.15)$$

The calculation of the elements of the Fisher information matrix is shown in the Appendix A.11.

Because, n units are tested, the Fisher information matrix, F, for the plan with a sample of n independent units is

$$F = nF_{\text{single}}. \quad (5.16)$$

Because for some set of values of γ_0, γ_1, s_0, s_1, s_2, η, β, k, c, t_1, t_2, and t_3; $|F|$ or the variance function may be negative. Therefore, only that set of values for which $|F| > 0$ or the variance function is positive, is chosen. A similar observation has been made by Gouno, Sen, and Balakrishnan (2004).

5.2.6 Asymptotic Variance of the MLEs of Logarithm of Quantile q at Design Stress

For any plan, the asymptotic variance–covariance matrix of the model parameters is given by the inverse of the corresponding Fisher information matrix

$$F^{-1} = \begin{bmatrix} \text{Var}(\hat{\gamma}_0) & \text{Cov}(\hat{\gamma}_0, \hat{\gamma}_1) & \text{Cov}(\hat{\gamma}_0, \hat{c}) \\ \text{Cov}(\hat{\gamma}_0, \hat{\gamma}_1) & \text{Var}(\hat{\gamma}_1) & \text{Cov}(\hat{\gamma}_1, \hat{c}) \\ \text{Cov}(\hat{\gamma}_0, \hat{c}) & \text{Cov}(\hat{\gamma}_1, \hat{c}) & \text{Var}(\hat{c}) \end{bmatrix}.$$

The asymptotic variance of the MLE of the log of quantile q of the distribution at a design constant stress s_0 is

$$\mathrm{AsVar}[\log \hat{\theta}_q] = \mathrm{AsVar}[\hat{h}(\hat{\gamma}_0, \hat{\gamma}_1, \hat{c})],$$

where $\mathrm{AsVar}[\hat{h}(\hat{\gamma}_0, \hat{\gamma}_1, \hat{c})] = \underset{\sim}{\hat{H}}' F^{-1} \underset{\sim}{\hat{H}}$ vector $\underset{\sim}{\hat{H}}'$ is a transpose of vector $\underset{\sim}{\hat{H}}$.

$$\hat{h} = \hat{\gamma}_0 + \frac{1}{\hat{c}} \ln\left(-1 + (1-q)^{-\frac{1}{k}}\right)$$

and

$$\underset{\sim}{\hat{H}} = \left[\frac{\partial \hat{h}}{\partial \hat{\gamma}_0} \quad \frac{\partial \hat{h}}{\partial \hat{\gamma}_1} \quad \frac{\partial \hat{h}}{\partial \hat{c}} \right]' \text{ with } \hat{h} \equiv \hat{h}(\hat{\gamma}_0, \hat{\gamma}_1, \hat{c}).$$

Thus,

$$\mathrm{AsVar}[\log \hat{\theta}_q] = \mathrm{AsVar}[\hat{h}(\hat{\gamma}_0, \hat{\gamma}_1, \hat{c})]. \tag{5.17}$$

5.3 OPTIMUM MULTI-OBJECTIVE MODIFIED CONSTANT-STRESS ALT PLAN

This section describes design of an optimum multi-objective modified constant-stress ALT plan using the Burr Type-XII distribution under Type-I censoring proposed by Srivastava and Mittal (2013a). The optimum plan gives the value of optimum proportion of units allocated to different stress levels, and optimum low stress level.

5.3.1 Assumptions

1. The lifetimes of test units are independent and identically distributed.
2. Failed items are not replaced during the test.
3. Two stress levels higher than design stress (s_0) are employed. The low stress level (s_1), and the high stress level (s_2), with s_0 and s_2 assumed to be pre-specified are used.

4. The censoring time η is pre-specified.
5. For the effect of changing stress levels, the cumulative exposure model holds (Nelson, 1980).
6. Between any two stress levels, the stress is increased at the rate β.
7. The shape parameter 'k' does not depend on the stress level.
8. At any constant-stress s, the lifetime of a unit follows a Burr Type-XII model with scale parameter $\alpha(s)$, and shape parameters c and k; and the inverse power law holds for $\alpha(s)$:

$$\alpha(s) = e^{\gamma_0}\left(\frac{s_0}{s}\right)^{\gamma_1}, \tag{5.18}$$

k is assumed to be known for the sake of mathematical convenience.

5.3.2 Test Procedure

The modified constant-stress fully ALT proceeds as follows:

1. '$n\pi$', and '$n(1-\pi)$' test units are put on test at time x_0.
2. For '$n\pi$' test units, the stress is increased at the rate β until it reaches s_1, which occurs at time $x_1 = \frac{s_1 - s_0}{\beta}$.
3. For '$n(1-\pi)$' test units, the stress is increased at the rate β until it reaches s_2, which occurs at time $x_2 = \frac{s_2 - s_0}{\beta}$.
4. After time x_1, the stress is maintained at the level s_1 until the pre-specified censoring time η is reached.
5. After time x_2, the stress is maintained at the level s_2 until the pre-specified censoring time η is reached.
6. The test is continued until:

   (a) all test items fail, or
   (b) a prescribed censoring time η

whichever occurs earlier, and the test conditions should remain the same.

Figure 5.2 depicts modified constant-stress loading.

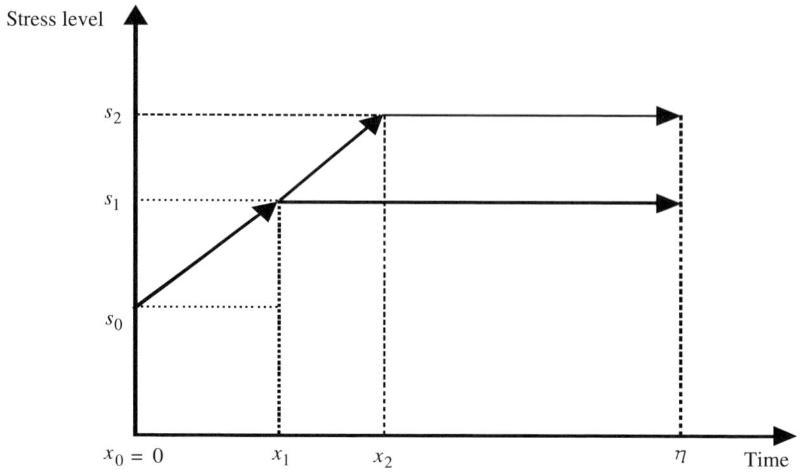

Figure 5.2: Modified constant-stress ALT

5.3.3 Life Distribution under Modified Constant-Stress

Based on the inverse power law (from Assumption 8), the cumulative exposure function $\varepsilon(x)$ at time x under stress level s is:

For $0 < x \le x_i$, $i = 1, 2$,

$$\varepsilon(x) = \int_0^x \frac{1}{\alpha(s(y))}dy = \int_0^x \frac{1}{e^{\gamma_0}\left(\frac{s_0}{s(y)}\right)^{\gamma_1}}dy = \int_0^x \frac{1}{e^{\gamma_0}\left(\frac{s_0}{\beta y + s_0}\right)^{\gamma_1}}dy$$

$$= \frac{e^{-\gamma_0}s_0^{-\gamma_1}((\beta x + s_0)^{1+\gamma_1} - s_0^{1+\gamma_1})}{\beta(1+\gamma_1)} = G_1(x). \tag{5.19}$$

For $x_i < x \le \eta$

$$\varepsilon(x) = \int_0^x \frac{1}{\alpha(s(y))}dy = \varepsilon(x_i) + \int_{x_i}^x \frac{1}{e^{\gamma_0}\left(\frac{s_0}{s_i}\right)^{\gamma_1}}dy$$

$$= \varepsilon(x_i) + e^{-\gamma_0}(x - x_i)\left(\frac{s_0}{s_i}\right)^{-\gamma_1} = \varepsilon(x_i) + G_{i2}(x). \tag{5.20}$$

Then, cdf of the lifetime X of a unit tested under modified constant-stress is

$$F(x) = G(\varepsilon(x)), \tag{5.21}$$

where $G(\cdot)$ is the assumed cdf (5.2) with the scale parameter α set equal to one (see Appendix A.12), and $\varepsilon(x)$ is the cumulative exposure (damage) model.

Hence, the cdf is

$$F(t) = 1 - (1 + \varepsilon(t)^c)^{-k} \quad \text{(using (5.2) and (5.21))}.$$

Therefore,

$$F(x) = \begin{cases} F_1(x), & 0 < x \le x_i \\ F_{i2}(x), & x_i < x \le \eta, \end{cases} \tag{5.22}$$

where $F_1(x) = 1 - (1 + (G_1(x))^c)^{-k}$, and $F_{i2}(x) = 1 - (1 + (G_1(x_i) + G_{i2}(x))^c)^{-k}$.

The pdf is given by

$$f(x) = \begin{cases} f_1(x), & 0 < x \le x_i \\ f_{i2}(x), & x_i < x \le \eta, \end{cases} \tag{5.23}$$

where $f_1(x) = Q_1(\beta x + s_0)^{\gamma_1}(G_1(x))^{c-1}(1 + (G_1(x))^c)^{-k-1}$ and $f_{i2}(x) = Q_1(s_i^{\gamma_1})(G_1(x_i) + G_{i2}(x))^{c-1}(1 + (G_1(x_i) + G_{i2}(x))^c)^{-k-1}$, $i = 1, 2$.

5.3.4 Log-Likelihood Function

The log-likelihood of an observation at time x under Type-I censoring is derived.

The log-likelihood of a single observation at time x is

$$\begin{aligned} L = L(\gamma_0, \gamma_1, c) &= I_0 \ln f_1(x) + I_1 \ln f_{i2}(x) + I_2 \ln(1 - F_{i2}(\eta)) \\ &= I_0[\ln Q_1 + \gamma_1 \ln(\beta x + s_0) + (c-1)\ln G_1(x) \\ &\quad - (k+1)\ln(1 + (G_1(x))^c)] + I_1[\ln Q_1 + \gamma_1 \ln(s_i) \\ &\quad + (c-1)\ln(G_1(x_i) + G_{i2}(x)) - (k+1)\ln(1 + (G_1(x_i) \\ &\quad + G_{i2}(x))^c)] + I_2[-k \ln(1 + (G_1(x_i) + G_{i2}(\eta))^c)]. \end{aligned} \tag{5.24}$$

Let the log-likelihood of unit j be L_j. The log-likelihood L_0 for n independent observations is

$$L_0 = L_1 + \cdots + L_n. \tag{5.25}$$

The model parameters are estimated by using method of maximum likelihood and are obtained by maximizing (5.25).

5.3.5 Fisher Information Matrix

The Fisher information matrix is obtained by taking the expectations of the negative of the second-order partial derivatives of the log-likelihood function with respect to γ_0, γ_1, and c. The Fisher information matrix for an observation is

$$F^{(i)} = \begin{bmatrix} E\left[-\frac{\partial^2 L}{\partial \gamma_0^2}\right] & E\left[-\frac{\partial^2 L}{\partial \gamma_0 \partial \gamma_1}\right] & E\left[-\frac{\partial^2 L}{\partial \gamma_0 \partial c}\right] \\ E\left[-\frac{\partial^2 L}{\partial \gamma_0 \partial \gamma_1}\right] & E\left[-\frac{\partial^2 L}{\partial \gamma_1^2}\right] & E\left[-\frac{\partial^2 L}{\partial \gamma_1 \partial c}\right] \\ E\left[-\frac{\partial^2 L}{\partial \gamma_0 \partial c}\right] & E\left[-\frac{\partial^2 L}{\partial \gamma_1 \partial c}\right] & E\left[-\frac{\partial^2 L}{\partial c^2}\right] \end{bmatrix}. \tag{5.26}$$

The calculation of the elements of the Fisher information matrix is shown in Appendix A.13.

Because, $n\pi$ and $n(1-\pi)$ test units are allocated to the low, and high stress levels, respectively. The Fisher information matrix, F, for the plan with a sample of n independent units is

$$F = n\pi F^{(1)} + n(1-\pi)F^{(2)}. \tag{5.27}$$

Because, for some set of values of γ_0, γ_1, s_0, s_1, s_2, η, β, k, c, x_1, and x_2; $|F|$ or the variance function may be negative. Therefore, only that set of values for which $|F| > 0$ or the variance function is positive, is chosen. A similar observation has been made by Gouno, Sen, and Balakrishnan (2004).

5.3.6 Asymptotic Variance of the MLEs of the Logarithm of Quantile q at Design Stress

For any plan, the asymptotic variance–covariance matrix of the model parameters is given by the inverse of the corresponding Fisher information matrix

$$F^{-1} = \begin{bmatrix} \mathrm{Var}(\hat{\gamma}_0) & \mathrm{Cov}(\hat{\gamma}_0, \hat{\gamma}_1) & \mathrm{Cov}(\hat{\gamma}_0, \hat{c}) \\ \mathrm{Cov}(\hat{\gamma}_0, \hat{\gamma}_1) & \mathrm{Var}(\hat{\gamma}_1) & \mathrm{Cov}(\hat{\gamma}_1, \hat{c}) \\ \mathrm{Cov}(\hat{\gamma}_0, \hat{c}) & \mathrm{Cov}(\hat{\gamma}_1, \hat{c}) & \mathrm{Var}(\hat{c}) \end{bmatrix}.$$

The asymptotic variance of the MLEs of the log of quantile q of the distribution at a design constant-stress s_0 is

$$\mathrm{AsVar}[\log \hat{\theta}_q] = \mathrm{AsVar}[\hat{h}(\hat{\gamma}_0, \hat{\gamma}_1, \hat{c})],$$

where $\mathrm{AsVar}[\hat{h}(\hat{\gamma}_0, \hat{\gamma}_1, \hat{c})] = \underset{\sim}{\hat{H}}' F^{-1} \underset{\sim}{\hat{H}}$, vector $\underset{\sim}{\hat{H}}'$ is a transpose of vector $\underset{\sim}{\hat{H}}$.

$$\hat{h} = \hat{\gamma}_0 + \frac{1}{\hat{c}} \ln(-1 + (1-q)^{-\frac{1}{k}})$$

and

$$\underset{\sim}{\hat{H}} = \begin{bmatrix} \frac{\partial \hat{h}}{\partial \hat{\gamma}_0} & \frac{\partial \hat{h}}{\partial \hat{\gamma}_1} & \frac{\partial \hat{h}}{\partial \hat{c}} \end{bmatrix}' \text{ with } \hat{h} \equiv \hat{h}(\hat{\gamma}_0, \hat{\gamma}_1, \hat{c}).$$

Thus,

$$\mathrm{AsVar}[\log \hat{\theta}_q] = \mathrm{AsVar}[\hat{h}(\hat{\gamma}_0, \hat{\gamma}_1, \hat{c})]. \tag{5.28}$$

5.4 FORMULATION OF A MULTI-OBJECTIVE OPTIMIZATION PROBLEM

A variety of approaches can be used to solve the multi-objective optimization problem. One popular approach is to combine these objectives into one single composite objective (Liao and Li, 2008). This approach has been used in the present section.

5.4.1 Modified Step-Stress ALT Plan

A modified step-stress test is specified by stress rate β. s_1 and t_2 are determined by minimizing the weighted sum of asymptotic variances of the ML estimators of the quantile lifetimes at a design constant-stress.

The asymptotic variance is a function of s_0, s_1, s_2, η, β, γ_0, γ_1, c, k, t_1, t_2, and t_3.

Thus, the optimal design problem can be formulated as a nonlinear optimization problem:

$$\begin{aligned} \text{Minimize} \quad & \sum_{j=1}^{b} w_j V_j \\ \text{subject to} \quad & s_0 < s_1 < s_2 \\ & \sum_{j=1}^{b} w_j = 1 \\ & t_1 < t_2 < t_3 < \eta, \end{aligned} \tag{5.29}$$

where 'w_j' is the weight, $j = 1, 2, \ldots, b$, 'b' is the number of quantile lifetimes, and

$$V_j = \text{AsVar}(\log \hat{\theta}_{q_j}) = \begin{pmatrix} 1 & 0 & -\dfrac{\ln\left((1-q_j)^{-\frac{1}{k}} - 1\right)}{c^2} \end{pmatrix} F^{-1} \times \begin{pmatrix} 1 & 0 & -\dfrac{\ln\left((1-q_j)^{-\frac{1}{k}} - 1\right)}{c^2} \end{pmatrix}' \quad \text{(using (5.17))}.$$

Because, the optimum modified step-stress test depends on the values of s_0, η, β, γ_0, γ_1, c, k, and s_2; one must obtain their values from experience, similar data, or a preliminary test. The choice of the sample size depends upon the plan designer, and cost constraints. Cost factor has not been considered. The optimum values of s_1, and t_2 has been found by using the NMinimize option of *Mathematica 6.0*.

In practicing ALT, one is interested in the lower tails of the life distribution, i.e., early failures, so 1%, 10%, and 50% quantiles need to be considered.

The weights are chosen according to the needs of the plan designed.

5.4.2 Modified Constant-Stress ALT Plan

A modified constant-stress test is specified by stress rate β. s_1 and π are determined by minimizing the weighted sum of asymptotic variances of the ML estimators of the quantile lifetimes at a design constant-stress.

The asymptotic variance is a function of s_0, s_1, s_2, η, β, γ_0, γ_1, c, k, x_1, and x_2.

Thus, the optimal design problem can be formulated as a nonlinear optimization problem:

$$\begin{aligned} \text{Minimize} \quad & \sum_{j=1}^{b} w_j V_j \\ \text{subject to} \quad & s_0 < s_1 < s_2 \\ & \sum_{j=1}^{b} w_j = 1 \\ & 0 < \pi < 1 \\ & t_1 < t_2 < \eta, \end{aligned} \tag{5.30}$$

where 'w_j' is the weight, $j = 1, 2, \ldots, b$, 'b' is the number of quantile lifetimes, and

$$V_j = \text{AsVar}(\log \hat{\theta}_{q_j}) = \begin{pmatrix} 1 & 0 & -\dfrac{\ln\left((1-q_j)^{-\frac{1}{k}} - 1\right)}{c^2} \end{pmatrix} F^{-1} \times \begin{pmatrix} 1 & 0 & -\dfrac{\ln\left((1-q_j)^{-\frac{1}{k}} - 1\right)}{c^2} \end{pmatrix}' \quad \text{(using (5.28))}.$$

Because, the optimum modified constant-stress test depends on the values of s_0, η, β, γ_0, γ_1, c, k, and s_2; one must obtain their values from experience, similar data, or a preliminary test. The choice of the sample size depends upon the plan designer, and cost constraints. Cost factor has not been considered. The optimum values of s_1, and π has been found by using the NMinimize option of *Mathematica 6.0*.

In practicing ALT, one is interested in the lower tails of the life distribution, i.e., early failures, so 1%, 10%, and 50% quantiles need to be considered.

The weights are chosen according to the needs of the plan designed.

5.5 NUMERICAL EXAMPLES

5.5.1 Modified Step-Stress ALT Plan

Considered the following hypothetical data set to illustrate the procedure developed.

$$n = 35, \quad \beta = 9.8, \quad \gamma_0 = 4.7, \quad \gamma_1 = 5.4, \quad s_0 = 30, \quad s_2 = 60,$$
$$\eta = 10, \quad k = 3, \quad c = 1, \quad b = 3, \quad w_1 = w_2 = w_3 = 1/3$$

5.5.1.1 *Optimal Plan*

The optimal s_1 and t_2 are obtained by minimizing the weighted sum of the asymptotic variances of the MLEs of the log of quantiles 1%, 10%, and 50% of the distribution at a design constant-stress, (5.17), using the NMinimize option of *Mathematica 6.0*. These are obtained as $s_1^* = 38.2825$, and $t_2^* = 7.24574$.

Thus, the optimum t_1^*, and t_3^* are

$$t_1^* = \frac{s_1^* - s_0}{\beta} = 0.845156 \quad \text{and} \quad t_3^* = t_2^* + \left(\frac{s_2 - s_1^*}{\beta}\right) = 9.46181.$$

Table 5.1 shows that AsVar* are not substantially affected by β when $\beta \geq 50$.

5.5.1.2 *Simulated Data*

Type-I censored sample is simulated under modified step-stress test using (5.11), and is displayed in Table 5.2.

Table 5.1: Optimal modified step-stress ALT plan ($n = 35, \gamma_0 = 4.7$, $\gamma_1 = 5.4$, $s_0 = 30$, $s_2 = 60$, $\eta = 10$, $k = 3$, $c = 1$, $b = 3$, and $w_1 = w_2 = w_3 = 1/3$)

β	s_1^*	t_2^*	AsVar*
5	37.3838	5.73818	0.540229
7	37.9534	6.70461	0.448281
9	38.2147	7.13900	0.415200
50	38.5591	7.71454	0.350359
100	38.5141	7.69727	0.344256
200	38.4807	7.68049	0.341119
500	38.4572	7.66649	0.339143
1,000	38.4491	7.66063	0.338448
10,000	38.4418	7.65440	0.337796
100,000	38.4411	7.65371	0.337728
1,000,000	38.4411	7.65364	0.337722
10,000,000	38.4411	7.65363	0.337721
100,000,000	38.4411	7.65363	0.337721
1,000,000,000	38.4411	7.65363	0.337721
∞	38.4411	7.65363	0.337721

Table 5.2: Simulated failure times ($n = 35$, $\beta = 9.8$, $\gamma_0 = 4.7$, $\gamma_1 = 5.4$, $s_0 = 30$, $s_2 = 60$, $\eta = 10$, $k = 3$, $c = 1$, $b = 3$, $w_1 = w_2 = w_3 = 1/3$, and $n_c = 7$)

Intervals	Failure Times
$0 < t \leq t_1$	0.110754
$t_1 < t \leq t_2$	1.091550, 3.025400, 7.185920, 5.862560, 5.844150, 2.219030, 4.401020, 1.955970, 2.447920, 6.015890
$t_2 < t \leq t_3$	9.259510, 8.584220, 9.458620, 8.258580, 8.378940, 8.457210, 8.581040, 9.322800, 8.698340, 9.304890, 8.025800, 8.828960, 8.297140
$t_3 < t \leq \eta$	9.701050, 9.825260, 9.485010, 9.543130

5.5.1.3 *MLEs of the Design Parameters*

The MLEs of the design parameters obtained using simulated data in Table 5.2 are $\hat{\gamma}_0 = 6.42675$, $\hat{\gamma}_1 = 8.2459$, and $\hat{c} = 0.817719$. These estimates are obtained using the NMaximize option of *Mathematica 6.0.*

5.5.1.4 *Confidence Intervals*

95% confidence intervals for the parameters are obtained using the inverse of the observed Fisher information matrix $\hat{F}^{-1}$, and it is given by

$$\hat{F}^{-1} = \begin{pmatrix} 2.065100 & 3.189140 & -0.3364550 \\ 3.189140 & 5.224230 & -0.5074580 \\ -0.336455 & -0.507458 & 0.0662519 \end{pmatrix}.$$

The observed value of F^{-1}, that is, $\hat{F}^{-1}$ is determined by substituting the estimated parameters $\hat{\gamma}_0$, $\hat{\gamma}_1$, and $\hat{c}$ for the true parameters in the asymptotic variance–covariance matrix. The square root of the diagonal elements of $\hat{F}^{-1}$ gives the standard error of an estimator. Thus, the 95% approximate confidence intervals for γ_0, γ_1, and c, respectively, are

$$3.61015 \leq \gamma_0 \leq 9.24336,$$

$$4.78952 \leq \gamma_1 \leq 14.1966,$$

and

$$0.44123 \leq c \leq 1.51545.$$

5.5.2 Modified Constant-Stress ALT Plan

The following hypothetical data set is considered:

$$n = 35, \quad \beta = 3.8, \quad \gamma_0 = 3.36, \quad \gamma_1 = 1.25, \quad s_0 = 30, \quad s_2 = 60,$$
$$\eta = 10, \quad k = 4, \quad c = 1, \quad b = 3, \quad w_1 = w_2 = w_3 = 1/3.$$

5.5.2.1 *Optimal Plan*

The optimal s_1 and π are obtained by minimizing the weighted sum of the asymptotic variances of the MLEs of the log of quantiles 1%,

Table 5.3: Optimal modified constant-stress ALT plan ($n = 35$, $\gamma_0 = 3.36$, $\gamma_1 = 1.25$, $s_0 = 30$, $s_2 = 60$, $\eta = 10$, $k = 4$, $c = 5$, $b = 3$, and $w_1 = w_2 = w_3 = 1/3$)

β	s_1^*	π^*	AsVar*
3	35.868	0.794534	0.184857
5	36.9755	0.608446	0.0343371
8	39.172	0.559391	0.0285545
10	40.104	0.490695	0.0265973
200	39.126	0.559737	0.0390489
500	39.8531	0.661459	0.0490663
1,000	40.2214	0.70013	0.0541278
10,000	40.5973	0.736958	0.0596919
100,000	40.6359	0.740721	0.0603055
1,000,000	40.6397	0.741098	0.0603674
10,000,000	40.6401	0.741136	0.0603736
100,000,000	40.6401	0.74114	0.0603743
10,00,000,000	40.6401	0.74114	0.0603743
∞	40.6401	0.74114	0.0603743

10%, and 50% of the distribution at a design constant-stress, (5.28), using the NMinimize option of *Mathematica 6.0*. They are obtained as $s_1^* = 31.0574$ and $\pi^* = 0.543488$.

Thus, the optimum x_1^* and x_2^* are

$$x_1^* = \frac{s_1^* - s_0}{\beta} = 0.277433 \quad \text{and} \quad x_2^* = \frac{s_2^* - s_0}{\beta} = 7.89474.$$

Table 5.3 shows that AsVar* has not been substantially affected by β when $\beta \geq 5$, for the modified constant-stress fully ALT plan.

5.5.2.2 *Simulated Data*

Type-I censored sample is simulated under modified constant-stress test using (5.22), and is displayed in Table 5.4.

5.5.2.3 *MLEs of the Design Parameters*

The MLEs of the design parameters obtained using simulated data given in Table 5.4 are $\hat{\gamma}_0 = 3.16493$, $\hat{\gamma}_1 = 2.57624$, and $\hat{c} = 1.02677$.

Table 5.4: Simulated failure times ($n = 35$, $\beta = 3.8$, $\gamma_0 = 3.36$, $\gamma_1 = 1.25$, $s_0 = 30$, $s_2 = 60$, $\eta = 10$, $k = 4$, $c = 1$, $b = 3$, $w_1 = w_2 = w_3 = 1/3$, $n_{c1} = 4$, and $n_{c2} = 1$)

Intervals	Failure Times
$0 < t \leq t_1$	0.153023
$t_1 < t \leq \eta$	0.852466, 2.194370, 4.691590, 0.484253, 1.429790, 7.412240, 5.005090, 1.236860, 1.225350, 0.677436, 7.515770, 5.964910, 5.079300, 8.059270
$0 < t \leq t_2$	0.497629, 4.070570, 1.909200, 6.260480, 6.785590, 1.606590, 0.814416, 0.636909, 4.850150, 1.617640, 5.195060, 3.160790, 3.943240, 4.234180
$t_2 < t \leq \eta$	7.962640

These estimates are obtained using the NMaximize option of *Mathematica 6.0.*

5.5.2.4 *Confidence Intervals*

95% confidence intervals for the parameters are obtained using the inverse of the observed Fisher information matrix $\hat{F}^{-1}$, and it is given by

$$\hat{F}^{-1} = \begin{pmatrix} 0.2259670 & 0.654954 & -0.0754897 \\ 0.6549540 & 2.857740 & -0.2086450 \\ -0.0754897 & -0.208645 & 0.0387004 \end{pmatrix}.$$

The observed value of F^{-1}, that is, $\hat{F}^{-1}$ is determined by substituting the estimated parameters $\hat{\gamma}_0$, $\hat{\gamma}_1$, and $\hat{c}$ for the true parameters in the asymptotic variance–covariance matrix. The square root of the diagonal elements of $\hat{F}^{-1}$ gives the standard error of an estimator. Thus, the 95% approximate confidence intervals for γ_0, γ_1, and c, respectively, are

$$2.233220 \leq \gamma_0 \leq 4.09663,$$

$$0.711919 \leq \gamma_1 \leq 9.32268,$$

and

$$0.705316 \leq c \leq 1.49473.$$

5.6 SENSITIVITY ANALYSIS

It is important to conduct sensitivity analysis to evaluate the robustness of the plans described as incorrect choice of design parameters, γ_0, γ_1, and c, may give poor estimate of the quantile at design constant-stress.

The percentage deviations (PDs) of the optimal settings are measured by

$$\text{PD} = \frac{|Z^{**} - Z^{*}|}{Z^{*}} \times 100,$$

where Z^{*} is the optimum asymptotic variance obtained with the given design parameters, and Z^{**} is the one obtained when a parameter is mis-specified.

Tables 5.5 and 5.6 show the optimal test plans for various deviations from the design parameters. The results show that the optimal setting of Z is robust to the small deviations from baseline parameters.

Table 5.5: Sensitivity analysis for modified step-stress ALT plan ($n = 35$, $\beta = 9.8$, $s_0 = 30$, $s_2 = 60$, $\eta = 10$, $b = 3$, $w_1 = w_2 = w_3 = 1/3$, and $k = 3$)

Parameters	% Change	s_1	t_2	PD
γ_0	+0.1	38.2914	7.24647	0.20608100
γ_0	−0.1	38.2736	7.24502	0.20496000
γ_1	+0.1	38.2824	7.24593	0.06737200
γ_1	−0.1	38.2826	7.24555	0.06752160
c	+0.1	38.2838	7.24585	0.14026100
c	−0.1	38.2813	7.24563	0.14076500

Table 5.6: Sensitivity analysis for modified constant-stress ALT plan ($n = 35$, $\beta = 3.8$, $s_0 = 30$, $s_2 = 60$, $\eta = 10$, $b = 3$, $w_1 = w_2 = w_3 = 1/3$, and $k = 4$)

Parameter	% Change	s_1	π	PD
γ_0	+2	30.4206	0.352336	1.2219000
γ_0	−2	31.1916	0.670041	1.3806600
γ_1	+2	31.0748	0.539974	0.0822561
γ_1	−2	31.0410	0.547831	0.0847198
c	+2	30.7164	0.418836	3.8576700
c	−2	31.1761	0.639154	3.9656000

Table 5.7: Comparative study of modified step-stress ALT models

ALT Model	Log-Likelihood Function
Srivastava and Mittal (2014) Model	−72.6653
Park and Yum (1998) Model	−72.9099

5.7 COMPARATIVE STUDY

In this section comparative study has been presented to highlight the merits of the plans discussed.

5.7.1 Modified Step-Stress ALT Plan

In Table 5.7, the Srivastava and Mittal (2014) modified step-stress ALT model has been compared with the one designed by Park and Yum (1998) in terms of log-likelihood functions using the hypothetical failure time data set with Type-I censoring given in Table 5.2. Table 5.8 depicts comparison of the multi-objective plan, and the single-objective plan with respect to the changes in parametric values. Table 5.9 shows the comparison of a multi-objective plan on the basis of equal and unequal weights.

Table 5.8: Comparison of a multi-objective plan, and the single-objective plans with respect to changes in parametric values ($n = 35$, $\beta = 9.8$, $s_0 = 30$, $s_2 = 60$, $\eta = 10$, and $k = 3$)

γ_0	γ_1	c	AsVar*	AsVar$_{0.01}$	AsVar$_{0.1}$	AsVar$_{0.5}$
14.2	15.4	1.0	48.8589	28.8940	48.7402	59.9847
14.7	15.4	1.0	81.9250	61.4429	81.8324	99.3590
15.2	15.4	1.0	138.773	106.390	138.745	166.752
14.7	15.9	1.0	60.9606	45.3858	60.8740	74.0811
14.7	16.4	1.0	45.6476	26.5458	45.6476	55.6625
14.7	15.4	1.5	486.695	326.862	488.407	544.094
14.7	15.4	2.0	2685.71	1658.36	2685.71	2903.62

Table 5.9: Comparison of a multi-objective plan with equal and unequal weights ($n = 35$, $\beta = 9.8$, $\gamma_0 = 14.7$, $\gamma_1 = 15.4$, $s_0 = 30$, $s_2 = 60$, $\eta = 10$, $k = 3$, $c = 1$, and $b = 3$)

w_1	w_2	w_3	AsVar*
0.33	0.33	0.33	81.9250
0.50	0.40	0.10	74.2890
0.50	0.10	0.40	80.1030
0.10	0.40	0.50	89.1038
0.40	0.10	0.50	83.7835
0.40	0.50	0.10	76.2293
0.10	0.50	0.40	87.3294

Table 5.7 shows that the Srivastava and Mittal (2014) model performs better for the given data set.

Table 5.8 shows that the multi-objective plan yields a smaller variance as compared to the single-objective plan at a 50% quantile.

Table 5.9 shows that a multi-objective plan with equal weights yields smaller variance as compared to a multi-objective plan with unequal weights when more weight is given to asymptotic variance corresponding to 50% quantile, and when more weight is given to asymptotic variance corresponding to 10% quantile with next higher weight to asymptotic variance corresponding to 50% quantile.

5.7.2 Modified Constant-Stress ALT Plan

In Table 5.10, the Srivastava and Mittal (2013a) modified constant-stress fully ALT model has been compared with the one designed by Park and Yum (1998) in terms of log-likelihood functions using the hypothetical failure time data set given in Table 5.4. Table 5.11 depicts comparison of the multi-objective plan and the single-objective plan with respect to the changes in parametric values. Table 5.12 shows the comparison of a multi-objective plan on the basis of equal and unequal weights.

Table 5.10 shows that the Srivastava and Mittal (2013a) model performs better for the given data set.

Table 5.10: Comparative study of modified constant-stress ALT models

ALT Model	Log-Likelihood Function
Srivastava and Mittal (2013a) Model	−75.0354
Park and Yum (1998) Model	−78.1304

Table 5.11: Comparison of a multi-objective plan, and the single-objective plans with respect to changes in parametric values ($n = 35$, $\beta = 4$, $s_0 = 30$, $s_2 = 60$, $k = 4$, and $\eta = 10$)

γ_0	γ_1	c	AsVar*	AsVar$_{0.01}$	AsVar$_{0.1}$	AsVar$_{0.5}$
5.66	1.95	2.9	125.605	61.3997	122.580	190.130
6.10	1.95	2.9	709.973	380.881	700.399	1041.86
6.60	1.95	2.9	4248.70	2515.38	4223.79	5983.78
5.66	1.45	2.9	244.438	110.929	237.776	380.572
5.66	1.75	2.9	170.487	80.3951	166.135	261.866
5.66	1.95	2.7	66.9156	30.3763	64.6965	103.391
5.66	1.95	3.1	228.240	119.304	224.292	337.722

Table 5.12: Comparison of a multi-objective plan with equal and unequal weights ($n = 35$, $\beta = 4$, $\gamma_0 = 5.66$, $\gamma_1 = 1.95$, $s_0 = 30$, $s_2 = 60$, $\eta = 10$, $k = 4$, $c = 2.9$, and $b = 3$)

w_1	w_2	w_3	AsVar*
0.33	0.33	0.33	125.6050
0.50	0.40	0.10	99.52850
0.50	0.10	0.40	120.2680
0.10	0.40	0.50	150.7090
0.40	0.10	0.50	133.0240
0.40	0.50	0.10	105.5540
0.10	0.50	0.40	143.9410

Table 5.11 shows that the multi-objective plan yields smaller variance as compared to the single-objective plan at a 50% quantile.

Table 5.12 shows that a multi-objective plan with equal weights yields smaller variance as compared to a multi-objective plan with unequal weights when more weight is given to asymptotic variance corresponding to 50% quantile, and when more weight is given to asymptotic variance corresponding to 10% quantile with next higher weight to asymptotic variance corresponding to 50% quantile.

5.8 SUMMARY

In this chapter, optimum time-censored multi-objective test plans using modified stress loading methods have been obtained under the fully accelerated environmental condition using the Burr Type-XII life distribution. The procedures developed have been explained using examples. The results of sensitivity analyses show that the optimum plans are robust for small deviations in the true values of the model parameters. Comparative study has also been carried out to bring forth the merits of the plans discussed.

CHAPTER 6

OPTIMUM TIME-CENSORED STEP-STRESS FULLY ALT WITH COMPETING RISKS FOR FAILURE

6.1 INTRODUCTION

Many products have more than one cause of failure. Examples include:

1. Ball bearing assemblies, in which a ball or the race can fail.
2. A semiconductor device can fail at a junction or at a lead.
3. Any solder joint on a circuit board can fail.
4. In motors, the Turn, Phase, or Ground insulation can fail, etc.

Accelerated tests that yield a mix of failure modes have for long attracted the attention of experimenters. They have endeavored to find a valid way to extrapolate such data to estimate product life distribution at a design stress as it is clearly wrong to use data on one failure mode to estimate the life distribution of another failure mode. This chapter presents a valid model and analysis of such a data.

Several papers have been written on analyzing ALT when failures can occur from any one of p statistically independent causes (see Section 1.18.5).

In this chapter, optimum simple time-censored step-stress life test under fully accelerated environment, with competing causes of failure due to Srivastava, Shukla, and Sen (2014) has been discussed. Here, sample data consist of a failure time and the cause of failure for each item. In simple time-censored step-stress ALT, the stress is changed to high level after the stress is run for specified time at low stress level. The Khamis–Higgins (KH) model (1999) used for the purpose is an alternative to the Weibull cumulative exposure model for the step-stress accelerated life testing and is based on a time transformation of the exponential cumulative exposure model. The life distribution of each failure cause, which is independent of the others, follows Weibull distribution with a characteristic life (see Definition 1.4) that is a log-linear function of the stress, and a cumulative exposure model is assumed. The sum of asymptotic variances of the MLEs of the log characteristic life at design stress due to all the failure causes is minimized, and the optimal stress change time determined. Inferential procedures involving design parameters have been also discussed, and sensitivity analysis carried out.

6.1.1 Notation

x_0, x_1, x_2	stress (design, low, high)
n	number of test units
n_i	number of failed units at stress x_i, $i = 1, 2$
n_c	number of units censored $n_c = n - n_1 - n_2$
T	censoring time
τ	stress change point
τ^*	optimal time of changing stress
δ	shape parameter (known)
$\theta_{ij}^{1/\delta}$	characteristic life for failure cause j at stress x_i, $i = 0, 1, 2$; $j = 1, 2$
$G_j(\cdot)$	cdf of the lifetime distribution under step-stress ALT for failure cause $j, j = 1, 2$

6.2 ASSUMPTIONS

1. Each unit has two statistically independent potential failure times corresponding to two causes of failure.

2. Failure time of a unit is the smallest of its two potential failure times.
3. Two stress levels x_1 and x_2 ($x_1 < x_2$) are used.
4. For any level of stress, the lifetime under each failure cause follows a Weibull distribution with shape parameter δ (known) and scale parameter θ_{ij}, $i, j = 1, 2$, i.e.,

$$G_j(w) = 1 - \exp\left\{\frac{-w^\delta}{\theta_{ij}}\right\}, \quad 0 \leq w < \infty. \tag{6.1}$$

5. The characteristic life (see Definition 1.4) of two potential failure times are log-linear functions of stress:

$$\log \theta_{ij}^{\frac{1}{\delta}} = \alpha_j + \beta_j x_i; \quad i = 0, 1, 2; \ j = 1, 2. \tag{6.2}$$

   α_j, β_j(<0) are unknown parameters depending on the nature of the product and the test method and δ is known. It can be showed as follows (using 1.5) that $\theta_{ij}^{\frac{1}{\delta}}$ is the characteristic life of (6.1):

$$G_j(\xi_p) = p \Rightarrow \xi_P = (-\theta_{ij} \log(1-p))^{1/\delta},$$

   $\Rightarrow \theta_{ij}^{1/\delta} \approx \xi_{0.632}$, $i, j = 1, 2$.
6. For each failure cause, a cumulative exposure model is assumed. That is, the remaining life of a test unit under each failure cause depends only on the exposure it has seen and does not remember how the exposure was accumulated (Nelson, 1990; Miller and Nelson, 1983) (see Section 1.9.1).

6.3 TEST PROCEDURE

1. Initially n test units are placed on low stress x_1 and run until time $\tau (0 < \tau < T)$, when the stress is changed to x_2 and the test is continued until a pre-assigned censoring time T.
2. Failure times and failure causes of test units are jointly observed continuously.

6.4 LIFE DISTRIBUTION UNDER STEP-STRESS WITH COMPETING FAILURE MODES

From the cumulative exposure model and Weibull distributed life assumptions, the cdf of failure cause $j = 1, 2$ under simple time step-stress test is the KH model:

$$
\begin{aligned}
G_j(w) &= G_j(w; \theta_{1j}, \theta_{2j}) \\
&= \begin{cases} 1 - \exp\left\{ \dfrac{-w^\delta}{\theta_{1j}} \right\}, & \text{if } 0 < w < \tau \\[2ex] 1 - \exp\left\{ -\dfrac{\tau^\delta}{\theta_{1j}} - \dfrac{(w^\delta - \tau^\delta)}{\theta_{2j}} \right\}, & \text{if } \tau \le w < \infty \end{cases}
\end{aligned}
\tag{6.3}
$$

for $j = 1, 2$, and the corresponding pdf of W_j is given by

$$
\begin{aligned}
g_j(w) &= g_j(w; \theta_{1j}, \theta_{2j}) \\
&= \begin{cases} \dfrac{\delta w^{\delta-1}}{\theta_{1j}} \exp\left\{ -\dfrac{w^\delta}{\theta_{1j}} \right\}, & \text{if } 0 < w < \tau \\[2ex] \dfrac{\delta w^{\delta-1}}{\theta_{2j}} \exp\left\{ -\dfrac{\tau^\delta}{\theta_{1j}} - \dfrac{(w^\delta - \tau^\delta)}{\theta_{2j}} \right\}, & \text{if } \tau \le w < \infty \end{cases}
\end{aligned}
\tag{6.4}
$$

for $j = 1, 2$. Since only the smaller of W_1 and W_2 is observed, let $W = \min\{W_1, W_2\}$ denote the overall failure time of a test unit. Then, its cdf and pdf are obtained to be

$$
\begin{aligned}
F(w) &= F(w; \theta) \\
&= 1 - (1 - G_1(w))(1 - G_2(w)) \\
&= \begin{cases} 1 - \exp\left\{ -\left(\dfrac{1}{\theta_{11}} + \dfrac{1}{\theta_{12}} \right) w^\delta \right\}, & \text{if } 0 < w < \tau, \\[2ex] 1 - \exp\left\{ -\left(\dfrac{1}{\theta_{11}} + \dfrac{1}{\theta_{12}} \right) \tau^\delta - \left(\dfrac{1}{\theta_{21}} + \dfrac{1}{\theta_{22}} \right) (w^\delta - \tau^\delta) \right\}, \\ \qquad \text{if } \tau \le w < \infty \end{cases}
\end{aligned}
\tag{6.5}
$$

$$
f(w) = f(w;\theta)
= \begin{cases}
\delta w^{\delta-1}\left(\frac{1}{\theta_{11}}+\frac{1}{\theta_{12}}\right)\exp\left\{-\left(\frac{1}{\theta_{11}}+\frac{1}{\theta_{12}}\right)w^{\delta}\right\}, \\
\qquad\qquad\qquad\qquad\qquad\qquad \text{if } 0 < w < \tau, \\
\delta w^{\delta-1}\left(\frac{1}{\theta_{21}}+\frac{1}{\theta_{22}}\right)\exp\left\{-\left(\frac{1}{\theta_{11}}+\frac{1}{\theta_{12}}\right)\tau^{\delta}\right. \\
\qquad \left.-\left(\frac{1}{\theta_{21}}+\frac{1}{\theta_{22}}\right)(w^{\delta}-\tau^{\delta})\right\}, \quad \text{if } \tau \le w < \infty
\end{cases} \tag{6.6}
$$

respectively, where $\theta = (\theta_1, \theta_2)$ with $\theta_i = (\theta_{i1}, \theta_{i2})$ for $i = 1, 2$. Furthermore let i denote the indicator for the cause of failure. Then, under our assumptions, the joint pdf of (W, C) is given by

$$
f_{w,c}(w,j) = g_j(t)(1 - G_{j'}(t))
= \begin{cases}
\frac{\delta w^{\delta-1}}{\theta_{1j}}\exp\left\{-\left(\frac{1}{\theta_{11}}+\frac{1}{\theta_{12}}\right)w^{\delta}\right\}, \quad \text{if } 0 < w < \tau, \\
\frac{\delta w^{\delta-1}}{\theta_{2j}}\exp\left\{-\left(\frac{1}{\theta_{11}}+\frac{1}{\theta_{12}}\right)\tau^{\delta}\right. \\
\qquad \left.-\left(\frac{1}{\theta_{21}}+\frac{1}{\theta_{22}}\right)(w^{\delta}-\tau^{\delta})\right\}, \quad \text{if } \tau \le w < \infty
\end{cases} \tag{6.7}
$$

for j, $j' = 1, 2$ and $j' \neq j$. The relative risk imposed on a test unit before τ due to failure cause j is denoted by

$$
\pi_{1j} = \Pr[C = j | 0 < W < \tau] = \frac{\theta_{1j}^{-1}}{\theta_{11}^{-1} + \theta_{12}^{-1}}, \quad j = 1, 2. \tag{6.8}
$$

Similarly, the relative risk after τ due to the cause j is denoted by

$$
\pi_{2j} = \Pr[C = j | W \ge \tau] = \frac{\theta_{2j}^{-1}}{\theta_{21}^{-1} + \theta_{22}^{-1}}, \quad j = 1, 2. \tag{6.9}
$$

They are simply the proportion of failure rates in the given time frame. It follows that W and C are independent given the time frame in which a failure has occurred.

6.5 LOG-LIKELIHOOD

Let us now define

n_{1j} = the number of units that fail before τ due to the failure cause j, $j = 1, 2$

n_{2j} = the number of units that fail after τ due to the failure cause j, $j = 1, 2$.

Under the assumption of the cumulative exposure model, the likelihood function of θ based on the Type-I censored sample is

$$L = \prod_{i=1}^{n_{11}} \left[\frac{\delta w_i^{\delta-1}}{\theta_{11}} e^{-w_i^\delta/\theta_{1\bullet}} \right] \prod_{i=1}^{n_{12}} \left[\frac{\delta w_i^{\delta-1}}{\theta_{12}} e^{-w_i^\delta/\theta_{i\bullet}} \right]$$
$$\times \prod_{i=1}^{n_{21}} \left[\frac{\delta w_i^{\delta-1}}{\theta_{21}} e^{-\frac{\tau^\delta}{\theta_{1\bullet}} - \frac{w_i^\delta - \tau^\delta}{\theta_{2\bullet}}} \right] \prod_{i=1}^{n_{22}} \left[\frac{\delta w_i^{\delta-1}}{\theta_{22}} e^{-\frac{\tau^\delta}{\theta_{1\bullet}} - \frac{w_i^\delta - \tau^\delta}{\theta_{2\bullet}}} \right]$$
$$\times e^{-\left\{ \frac{n_c(T^\delta - \tau^\delta)}{\theta_{2\bullet}} + \frac{n_c \tau^\delta}{\theta_{1\bullet}} \right\}}, \tag{6.10}$$

where

$$\frac{1}{\theta_{1\cdot}} = \frac{1}{\theta_{11}} + \frac{1}{\theta_{12}},$$
$$\frac{1}{\theta_{2\cdot}} = \frac{1}{\theta_{21}} + \frac{1}{\theta_{22}},$$
$$n_{1\cdot} = n_{11} + n_{12},$$
$$n_{2\cdot} = n_{21} + n_{22},$$

and

$$n = n_{1\cdot} + n_{2\cdot} + n_c, \quad n \text{ is fixed and known.}$$

Thus, the log-likelihood L of a Type-I censored observation is

$$\log L = -n_{11} \log \theta_{11} + n_{11} \log \delta + (\delta - 1) \sum_{i=1}^{n_{11}} \log w_i - \sum_{i=1}^{n_{11}} \frac{w_i^\delta}{\theta_{1\cdot}}$$
$$- n_{12} \log \theta_{12} + n_{12} \log \delta + (\delta - 1) \sum_{i=1}^{n_{12}} \log w_i - \sum_{i=1}^{n_{12}} \frac{w_i^\delta}{\theta_{1\cdot}}$$

$$
\begin{aligned}
&- n_{21}\log\theta_{21} + n_{21}\log\delta + (\delta-1)\sum_{i=1}^{n_{21}}\log w_i - \sum_{i=1}^{n_{21}}\left(\frac{w_i^\delta - \tau^\delta}{\theta_{2\cdot}}\right)\\
&-\sum_{i=1}^{n_{21}}\frac{\tau^\delta}{\theta_{1\cdot}} - n_{22}\log\theta_{22} + n_{22}\log\delta + (\delta-1)\sum_{i=1}^{n_{22}} w_i\\
&-\sum_{i=1}^{n_{12}}\left(\frac{w_i^\delta - \tau^\delta}{\theta_{2\cdot}}\right) - \sum_{i=1}^{n_{22}}\frac{\tau^\delta}{\theta_{1\cdot}} - n_c\left\{\frac{(T^\delta-\tau^\delta)}{\theta_{2\cdot}} + \frac{\tau^\delta}{\theta_{1\cdot}}\right\} \qquad (6.11)
\end{aligned}
$$

Remark 6.1. The model described above can be easily extended and generalized to accommodate multiple stress levels and multiple competing failure causes.

6.6 FISHER INFORMATION MATRIX

The second partial derivatives of the sample log-likelihood with respect to the model parameters are needed for obtaining Fisher Information matrix. For a single observation, the first partial derivatives of log L with respect to α_j and β_j, $j = 1, 2$, is,

$$
\begin{aligned}
\frac{\partial \log L}{\partial \alpha_j} &= \delta\Bigg(-n_{11} + \frac{1}{\theta_{1j}}\sum_{i=1}^{n_{11}} w_i^\delta + \frac{1}{\theta_{1j}}\sum_{i=1}^{n_{12}} w_i^\delta - n_{21} + \frac{1}{\theta_{2j}}\sum_{i=1}^{n_{21}} w_i^\delta\\
&\quad - \frac{n_{21}}{\theta_{2j}}\tau^\delta + \frac{n_{21}}{\theta_{1j}}\tau^\delta + \frac{1}{\theta_{2j}}\sum_{i=1}^{n_{22}} w_i^\delta - \frac{n_{22}}{\theta_{2j}}\tau^\delta + \frac{n_{22}}{\theta_{1j}}\tau^\delta\\
&\quad + \frac{n_c(T^\delta - \tau^\delta)}{\theta_{2j}} + \frac{n_c\tau^\delta}{\theta_{1j}}\Bigg), \qquad (6.12)\\
\frac{\partial \log L}{\partial \beta_j} &= \delta\Bigg(-n_{11}x_1 + \frac{x_1}{\theta_{1j}}\sum_{i=1}^{n_{11}} w_i^\delta + \frac{x_1}{\theta_{1j}}\sum_{i=1}^{n_{12}} w_i^\delta - n_{21}x_2 + \frac{x_2}{\theta_{2j}}\sum_{i=1}^{n_{21}} w_i^\delta\\
&\quad - \frac{n_{21}\tau^\delta}{\theta_{2j}}x_2 + \frac{n_{21}}{\theta_{1j}}x_1\tau^\delta + \frac{x_2}{\theta_{2j}}\sum_{i=1}^{n_{22}} w_i^\delta - \frac{n_{22}}{\theta_{2j}}\tau^\delta x_2\\
&\quad + \frac{n_{22}\tau^\delta}{\theta_{1j}}x_1 + \frac{n_c x_2(T^\delta - \tau^\delta)}{\theta_{2j}} + \frac{n_c\tau^\delta}{\theta_{1j}}x_1\Bigg). \qquad (6.13)
\end{aligned}
$$

The second partial derivatives are given below:

$$\frac{\partial^2 \log L}{\partial \alpha_j^2} = \delta^2 \left(-\frac{1}{\theta_{1j}} \sum_{i=1}^{n_{11}} w_i^\delta - \frac{1}{\theta_{1j}} \sum_{i=1}^{n_{12}} w_i^\delta - \frac{1}{\theta_{2j}} \sum_{i=1}^{n_{21}} w_i^\delta \right.$$
$$+ \frac{n_{21}\tau^\delta}{\theta_{2j}} - \frac{n_{21}}{\theta_{1j}}\tau^\delta - \frac{1}{\theta_{2j}} \sum_{i=1}^{n_{22}} w_i^\delta + \frac{n_{22}}{\theta_{2j}}\tau^\delta - \frac{n_{22}}{\theta_{1j}}\tau^\delta$$
$$\left. - \frac{n_c(T^\delta - \tau^\delta)}{\theta_{2j}} - \frac{n_c\tau^\delta}{\theta_{1j}} \right), \tag{6.14}$$

$$\frac{\partial^2 \log L}{\partial \beta_j^2} = \delta^2 \left(-\frac{x_1^2}{\theta_{1j}} \sum_{i=1}^{n_{11}} w_i^\delta - \frac{x_1^2}{\theta_{1j}} \sum_{i=1}^{n_{12}} w_i^\delta - \frac{x_2^2}{\theta_{2j}} \sum_{i=1}^{n_{21}} w_i^\delta + \frac{x_2^2 n_{21}\tau^\delta}{\theta_{2j}} \right.$$
$$- \frac{x_1^2 n_{21}}{\theta_{1j}}\tau^\delta - \frac{x_2^2}{\theta_{2j}} \sum_{i=1}^{n_{22}} w_i^\delta + \frac{x_2^2 n_{22}}{\theta_{2j}}\tau^\delta - \frac{x_1^2 n_{22}}{\theta_{1j}}\tau^\delta$$
$$\left. - \frac{n_c x_2^2(T^\delta - \tau^\delta)}{\theta_{2j}} - \frac{n_c x_1^2\tau^\delta}{\theta_{1j}} \right), \tag{6.15}$$

$$\frac{\partial^2 \log L}{\partial \alpha_j \partial \beta_j} = \delta^2 \left(-\frac{x_1}{\theta_{1j}} \sum_{i=1}^{n_{11}} w_i^\delta - \frac{x_1}{\theta_{1j}} \sum_{i=1}^{n_{12}} w_i^\delta - \frac{x_2}{\theta_{2j}} \sum_{i=1}^{n_{21}} w_i^\delta + \frac{x_2 n_{21}\tau^\delta}{\theta_{2j}} \right.$$
$$- \frac{x_1 n_{21}}{\theta_{1j}}\tau^\delta - \frac{x_2}{\theta_{2j}} \sum_{i=1}^{n_{22}} w_i^\delta + \frac{x_2 n_{22}}{\theta_{2j}}\tau^\delta - \frac{x_1 n_{11}}{\theta_{1j}}\tau^\delta$$
$$\left. - \frac{n_c x_2(T^\delta - \tau^\delta)}{\theta_{2j}} - \frac{n_c x_1\tau^\delta}{\theta_{1j}} \right), \tag{6.16}$$

$$\frac{\partial^2 \log L}{\partial \alpha_j \partial \beta_j} = \frac{\partial^2 \log L}{\partial \beta_j \partial \alpha_j}. \tag{6.17}$$

The elements of the Fisher information matrix F, obtained by taking the negative expectations of the second partial and mixed partial derivatives of $\log L$ with respect to α_j and β_j, $j = 1, 2$. These elements have been obtained using the following results with $n_{1\cdot} \equiv n_1$

and $n_{2\cdot} \equiv n_2$:

(a) $n = n_1 + n_2 + n_c$, n is fixed, where

(i) $n_1 \sim \text{Bin}(n, p_1)$,
(ii) $n_2 \sim \text{Bin}(n, p_2)$,
(iii) $n_c \sim \text{Bin}(n, p_3)$,
(iv) $p_1 = P[0 < W < \tau]$

$$= \int_0^\tau \frac{\delta w^{\delta-1}}{\theta_{1\cdot}} e^{-\frac{w^\delta}{\theta_{1\cdot}}} dw \text{ (using (6.6))}$$
$$= 1 - e^{-\tau^\delta/\theta_{1\cdot}},$$

(v) $p_2 = P[\tau < W < T]$

$$= \int_\tau^T \frac{\delta w^{\delta-1}}{\theta_{2\cdot}} e^{-\frac{w^\delta - \tau^\delta}{\theta_{2\cdot}} - \frac{\tau^\delta}{\theta_{1\cdot}}} dw \text{ (using (6.6))}$$
$$= e^{-\frac{\tau^\delta}{\theta_{1\cdot}}}[1 - e^{-(T^\delta - \tau^\delta)/\theta_{2\cdot}}],$$

(vi) $p_3 = 1 - p_1 - p_2$.

(b) $n_{11} \sim \text{Bin}(n_1, \pi_{11})$, where

$$\begin{aligned} E[n_{11}] &= E[E[n_{11} \mid n_1]] \\ &= E\left[n_1 \frac{\theta_{11}^{-1}}{\theta_{11}^{-1} + \theta_{21}^{-1}}\right] \\ &= n\frac{\theta_{11}^{-1}}{\theta_{1\cdot}^{-1}} p_1 \\ &= n\frac{\theta_{11}^{-1}}{\theta_{1\cdot}^{-1}}(1 - e^{-\tau^\delta/\theta_1}). \text{ (using (i) and (iv))} \end{aligned}$$

(c) $n_{12} \sim \text{Bin}(n_2, \pi_{12})$, where

$$\begin{aligned} E[n_{12}] &= E[n_1 - n_{11}] \\ &= E[n_1 - E[n_{11}|n_1]] \end{aligned}$$

$$= n\left[1 - \frac{\theta_{11}^{-1}}{\theta_{1\cdot}^{-1}}\right] p_1$$

$$= n\left[1 - \frac{\theta_{11}^{-1}}{\theta_{1\cdot}^{-1}}\right] (1 - e^{-\tau^\delta/\theta_{1\cdot}}). \text{ (using (b), (i) and (iv))}$$

(d) $n_{21} \sim \text{Bin}(n_1, \pi_{21})$, where

$$\begin{aligned}
E[n_{21}] &= E[E[n_{21} \mid n_2]] \\
&= E\left[n_2 \frac{\theta_{21}^{-1}}{\theta_{21}^{-1} + \theta_{22}^{-1}}\right] \\
&= \frac{\theta_{21}^{-1}}{\theta_{21}^{-1} + \theta_{22}^{-1}} E[n_2] \\
&= n \frac{\theta_{21}^{-1}}{\theta_{21}^{-1} + \theta_{22}^{-1}} p_2 \\
&= n \frac{\theta_{21}^{-1}}{\theta_{2\cdot}^{-1}} e^{-\tau^\delta/\theta_{1\cdot}} [1 - e^{-(T^\delta - \tau^\delta)/\theta_{2\cdot}}]. \text{ (using (ii) and (v))}
\end{aligned}$$

(e) $n_{22} \sim \text{Bin}(n_2, \pi_{22})$

$$\begin{aligned}
E[n_{22}] &= E[n_2 - n_{21}] \\
&= E[n_2 - (n_{21}|n_2)] \\
&= E[n_2 - E[n_{21}|n_2]] \\
&= E\left[n_2 - n_2 \frac{\theta_{21}^{-1}}{\theta_{2\cdot}^{-1}}\right] \\
&= n\left[1 - \frac{\theta_{21}^{-1}}{\theta_{2\cdot}^{-1}}\right] p_2 \\
&= n\left[1 - \frac{\theta_{21}^{-1}}{\theta_{2\cdot}^{-1}}\right] e^{-\tau^\delta/\theta_{1\cdot}} [1 - e^{-(T^\delta - \tau^\delta)/\theta_{2\cdot}}].
\end{aligned}$$

(using (d), (ii) and (v)).

(f)

$$E\left[\sum_{i=1}^{n_{11}} w_i^\delta\right] = E\left[E\left[\sum_{i=1}^{n_{11}} w_i^\delta | n_{11}\right]\right]$$

$$= E\left[\sum_{i=1}^{n_{11}} \int_0^\tau \frac{w_i^\delta \frac{1}{\theta_{11}} \delta w_1^{\delta-1} e^{-w_i^\delta/\theta_{1\cdot}}}{\int_0^\tau \frac{1}{\theta_{11}} \delta w_1^{\delta-1} e^{-w_1^\delta/\theta_{1\cdot}} dw_i} dw_i\right]$$

(using (6.7))

$$= E\left[\sum_{i=1}^{n_{11}} \int_0^\tau \frac{w_i^\delta \frac{1}{\theta_{11}} \delta w_1^{\delta-1} e^{-w_i^\delta/\theta_{1\cdot}}}{\frac{\theta_{1\cdot}}{\theta_{11}}(1 - e^{-\tau^\delta/\theta_{1\cdot}})} dw_i\right]$$

$$= E\left[\sum_{i=1}^{n_{11}} \int_0^\tau \frac{\delta w_i^{2\delta-1} e^{-w_i^\delta/\theta_{1\cdot}}}{\theta_{1\cdot}} dw_i\right]$$

$$= n\frac{\theta_{11}^{-1}}{\theta_{1\cdot}^{-1}}\left[e^{-\tau^\delta/\theta_{1\cdot}}(\tau^\delta + \theta_{1\cdot}) - \theta_{1\cdot}\right].$$

(g)

$$E\left[\sum_{i=1}^{n_{12}} w_i^\delta\right] = E\left[E\left[\sum_{i=1}^{n_{12}} w_i^\delta | n_{12}\right]\right]$$

$$= n\left(1 - \frac{\theta_{11}^{-1}}{\theta_{1\cdot}^{-1}}\right)\left(e^{-\tau^\delta/\theta_{1\cdot}}(\tau^\delta + \theta_{1\cdot}) - \theta_{1\cdot}\right).$$

(using (6.7)).

(h)

$$E\left[\sum_{i=1}^{n_{21}} w_i^\delta\right] = E\left[E\left[\sum_{i=1}^{n_{21}} w_i^\delta | n_{12}\right]\right]$$

$$= E\left[\sum_{i=1}^{n_{21}} \int_\tau^T \frac{w_i^\delta f(w_i, 1) dw_i}{\frac{\delta}{\theta_{21}} \int_\tau^T w_i^{\delta-1} e^{-\frac{w^\delta - \tau^\delta}{\theta_{2\cdot}} - \frac{\tau^\delta}{\theta_{1\cdot}}} dw_i}\right]$$

$$= E\left[\sum_{i=1}^{n_{21}} \int_\tau^T \frac{w_i^\delta f(w_i, 1) dw_i}{\frac{\delta}{\theta_{21}}\left(\frac{\theta_{2\cdot}}{\theta_{21}} e^{-\tau^\delta/\theta_i}[1 - e^{-(T^\delta - \tau^\delta)/\theta_{2\cdot}}]\right)}\right]$$

$$= \frac{n\theta_{21}^{-1}}{\theta_{21}^{-1} + \theta_{22}^{-1}}$$
$$\times \left((\tau^\delta + \theta_{2\cdot})e^{-\tau^\delta/\theta_{1\cdot}} - (T^\delta + \theta_{2\cdot})e^{-\frac{T^\delta - \tau^\delta}{\theta_{2\cdot}} - \frac{\tau^\delta}{\theta_{2\cdot}}} \right).$$

(i)
$$E\left[\sum_{i=1}^{n_{22}} w_i^\delta\right] = E\left[E\sum_{i=1}^{n_{22}} w_i^\delta | n_{22}\right]$$
$$= n\left[1 - \frac{\theta_{21}^{-1}}{\theta_{21}^{-1} + \theta_{22}^{-1}}\right.$$
$$\left.\times \left((\tau^\delta + \theta_{2\cdot})e^{-\tau^\delta/\theta_{1\cdot}} - (T^\delta + \theta_{2\cdot})e^{-\frac{T^\delta - \tau^\delta}{\theta_{2\cdot}} - \frac{\tau^\delta}{\theta_{1\cdot}}} \right)\right].$$
(using (6.7))

Using the results (6.14)–(6.17),

$$E\left[\frac{-\partial \log L}{\partial \alpha_j}\right] = 0, \quad E\left[\frac{-\partial \log L}{\partial \beta_j}\right] = 0, \quad (j = 1, 2),$$

and (a)–(i′) the elements of Fisher information matrix are:

$$E\left[\frac{-\partial^2 \log L}{\partial \alpha_j^2}\right] = n\delta^2 \frac{\theta_{1\cdot}}{\theta_{1j}} A_1(\tau) + n\delta^2 \frac{\theta_{2\cdot}}{\theta_{2j}} A_2(\tau) = P_j, \tag{6.18}$$

$$E\left[\frac{-\partial^2 \log L}{\partial \beta_j^2}\right] = nx_1^2\delta^2 \frac{\theta_{1\cdot}}{\theta_{1j}} A_1(\tau) + nx_2^2\delta^2 \frac{\theta_{2\cdot}}{\theta_{2j}} A_2(\tau) = Q_j, \tag{6.19}$$

$$E\left[\frac{-\partial^2 \log L}{\partial \alpha_j \partial \beta_j}\right] = nx_1\delta^2 \frac{\theta_{1\cdot}}{\theta_{1j}} A_1(\tau) + nx_2\delta^2 \frac{\theta_{2\cdot}}{\theta_{2j}} A_2(\tau) = R_j, \tag{6.20}$$

where

$$A_1(\tau) = 1 - \exp\left\{-\frac{\tau^\delta}{\theta_{1\cdot}}\right\},$$

$$A_2(\tau) = \exp\left\{-\frac{\tau^\delta}{\theta_{1\cdot}}\right\}\left(1 - \exp\left\{-\frac{(T^\delta - \tau^\delta)}{\theta_{2\cdot}}\right\}\right)$$

and $E\left[\frac{\partial^2 \log L}{\partial \alpha_j \partial \beta_j}\right] = E\left[\frac{\partial^2 \log L}{\partial \beta_j \partial \alpha_j}\right]$.

Thus, the Fisher information matrix F is,

$$F = \text{Diag}\{F_1, F_2\},$$

that is, the matrix whose diagonal 2×2 sub matrices F_j, $j = 1, 2$ are:

$$F_j = \begin{bmatrix} P_j & R_j \\ R_j & Q_j \end{bmatrix}.$$

6.7 OPTIMAL PLAN

For any plan, the asymptotic covariance matrix of the maximum likelihood estimates α_j, β_j, $j = 1, 2$ is the inverse of the corresponding Fisher information matrix, i.e.,

$$F^{-1} = \begin{bmatrix} \text{Var}[\hat{\alpha}_1] & \text{Cov}[\hat{\alpha}_1, \hat{\beta}_1] & 0 & 0 \\ \text{Cov}[\hat{\beta}_1, \hat{\alpha}_1] & \text{Var}[\hat{\beta}_1] & 0 & 0 \\ 0 & 0 & \text{Var}[\hat{\alpha}_2] & \text{Cov}[\hat{\alpha}_2, \hat{\beta}_2] \\ 0 & 0 & \text{Cov}[\hat{\beta}_2, \hat{\alpha}_2] & \text{Var}[\hat{\beta}_2] \end{bmatrix}.$$

The sum of asymptotic variances of the MLEs of the log characteristic life at design stress due to all the failure causes is used as an optimality criterion, and the optimal stress change time is determined by minimizing

$$n \sum_{j=1}^{2} \text{AsVar}(\alpha_j + \beta_j x_0). \tag{6.21}$$

6.8 TESTING OF HYPOTHESES

An important inference problem concerning the regression coefficient β_j, $j = 1, 2$ is the test of hypotheses

(i) H_{01}: $\beta_1 = \beta_2 = 0$, against
H_{11}: atleast one of β_j, $j = 1, 2$ is distinct from zero.
(ii) H_{02}: $\beta_1 = 0$, against
H_{12}: $\beta_1 < 0$.
(iii) H_{03}: $\beta_2 = 0$, against
H_{13}: $\beta_2 < 0$.

To test (i), (ii), and (iii), the likelihood ratio statistics used are:

$$\Lambda_1 = -2\log\left[\frac{L(\tilde{\alpha}_1, \tilde{\alpha}_2, 0, 0)}{L(\hat{\alpha}_1, \hat{\alpha}_2, \hat{\beta}_1, \hat{\beta}_2)}\right],$$

$$\Lambda_2 = -2\log\left[\frac{L(\tilde{\alpha}_1, \tilde{\alpha}_2, 0, \tilde{\beta}_2)}{L(\hat{\alpha}_1, \hat{\alpha}_2, \hat{\beta}_1, \hat{\beta}_2)}\right],$$

and

$$\Lambda_3 = -2\log\left[\frac{L(\tilde{\alpha}_1, \tilde{\alpha}_2, \tilde{\beta}_1, 0)}{L(\hat{\alpha}_1, \hat{\alpha}_2, \hat{\beta}_1, \hat{\beta}_2)}\right],$$

where $\tilde{\beta}_j$ is the MLE of β_j under null hypothesis H_{0j}, and $\hat{\beta}_j$ is the unrestricted MLE, $j = 1, 2, 3$. Large values of Λ_1, Λ_2, Λ_3 provide evidence against their respective null hypothesis. The significant levels can be calculated by using the fact that in large samples Λ_1 is approximately distributed as χ_2^2, while Λ_2 and Λ_3 both are approximately distributed as χ_1^2.

6.9 A NUMERICAL EXAMPLE

Example 6.1. The data in Table 6.1 include 30 simulated observations based on

$$n = 30, \quad \delta = 2, \quad \theta_{11} = 6.08, \quad \theta_{12} = 12.03, \quad \theta_{21} = 3.05,$$
$$\theta_{22} = 6.02, \quad T = 2.8, \quad x_0 = 2, \quad x_1 = 3, \quad x_2 = 5.$$

The optimal value of τ obtained is $\tau^* = 2.16$. The following model is fitted to the data

$$\log\theta_{ij}^{\frac{1}{\delta}} = \alpha_j + \beta_j x_i; \quad i = 0, 1, 2; \;\; j = 1, 2.$$

The MLEs $\hat{\alpha}_1$, $\hat{\alpha}_2$, $\hat{\beta}_1$, $\hat{\beta}_2$ are readily obtained using the NMaximize option of *Mathematica 6.0.* The MLEs are

$$\hat{\alpha}_1 = 2.67608, \quad \hat{\alpha}_2 = 1.67246, \quad \hat{\beta}_1 = -0.35588, \quad \hat{\beta}_2 = -0.11461.$$

Table 6.1: Simple step-stress with competing risks of failures simulated data ($n = 30$, $\delta = 2$, $\theta_{11} = 6.08$, $\theta_{12} = 12.03$, $\theta_{21} = 3.05$, $\theta_{22} = 6.02$, $T = 2.8$, $x_0 = 2$, $x_1 = 3$, $x_2 = 5$, $\tau^* = 2.16$)

Before τ^*		After τ^*	
Failure Time	**Failure Cause**	**Failure Time**	**Failure Cause**
0.4359	2	2.1695	1
0.4515	2	2.1874	2
0.4833	2	2.2159	2
0.5154	1	2.2592	2
0.8941	2	2.2602	1
0.9409	1	2.7544	1
0.9954	1	2.3267	2
1.1158	2	2.3465	1
1.1470	1	2.4216	1
1.4844	2	2.7510	1
1.6796	2		

The inverse of Fisher observed information matrix $\hat{F}^{-1}$ are given as

$$\hat{F}^{-1} = \begin{bmatrix} 0.4455 & -0.1010 & 0 & 0 \\ -0.1010 & 0.0242 & 0 & 0 \\ 0 & 0 & 0.3375 & -0.08533 \\ 0 & 0 & -0.0853 & 0.02301 \end{bmatrix}.$$

The observed values of F^{-1}, that is $\hat{F}^{-1}$, were determined by substituting the estimated parameters $\hat{\alpha}_1$, $\hat{\alpha}_2$, $\hat{\beta}_1$, $\hat{\beta}_2$ for the true parameters in the asymptotic covariance matrix. The standard errors of $\hat{\alpha}_1$, $\hat{\alpha}_2$, $\hat{\beta}_1$, and $\hat{\beta}_2$ are obtained by taking square root of the diagonal elements of $\hat{F}^{-1}$. 95% confidence intervals for the parameters are

$$1.36783 \le \alpha_1 \le 3.98433,$$
$$0.53387 \le \alpha_2 \le 2.81105,$$
$$-0.66059 \le \beta_1 \le -0.05116$$

Table 6.2: Tests for parameters in the model $\log \theta_{ij}^{\frac{1}{\delta}} = \alpha_j + \beta_j x_i;\ i = 0, 1, 2;\ j = 1, 2$

Full model $(\alpha_1, \alpha_2, \beta_1, \beta_2)$	Λ	d.f.	$\chi^2_{(0.05)}$
$(\alpha_1, \alpha_2, 0, 0)$	5.4346	2	5.99
$(\tilde{\alpha}_1, \tilde{\alpha}_2, 0, \tilde{\beta}_2)$	4.925	1	3.84
$(\tilde{\alpha}_1, \tilde{\alpha}_2, \tilde{\beta}_1, 0)$	0.5094	1	3.84

and

$$-0.411905 \leq \beta_2 \leq 0.$$

Since β_2 can take negative values only, therefore to turn the confidence interval to sensible one, positive upper bound is replaced by zero.

Table 6.2 shows the likelihood ratio statistics Λ for tests of various submodels against the full model $(\hat{\alpha}_1, \hat{\alpha}_2, \hat{\beta}_1, \hat{\beta}_2)$. The column d.f. indicates the degree of freedom of the Chi-square approximation. The tests based on Λ_2 and Λ_3 lead to the rejection of H_{02} and H_{03} respectively, while that based on Λ_1 leads to acceptance of H_{01}.

6.10 SENSITIVITY ANALYSIS

To use an optimum test plan, one needs information about the values of θ_{11}, θ_{12}, θ_{21}, and θ_{22} which are usually unknown. Therefore, they have to be approximated from experience, similar data, or preliminary test. Incorrect choice of pre-estimates gives a nonoptimal test plan and results in poor estimates of the parameters of life distribution at design stress. The effect of pre-estimates for some selected values of θ_{11}, θ_{12}, θ_{21}, and θ_{22} in terms of relative stress change point, τ^* are presented in Table 6.3.

$$\%\Delta\tau^* = (|(\tau^* - \tau^0)/\tau^*|)100,$$

where τ^* is the optimal stress change point for the plans obtained with the correctly specified values, and τ^0 is the stress change point

Table 6.3: Effects of incorrect pre-estimates of θ_{11}, θ_{12}, θ_{21}, and θ_{22} $\%\Delta\tau^*$ due to $\pm$ 1% to $\pm$5% change in θ_{11}, θ_{12}, θ_{21}, and θ_{22}

Percentage Deviation (%)	θ_{11}	θ_{12}	θ_{21}	θ_{22}	$\%\Delta\tau^*$
+1	6.1408	12.1503	3.0805	6.0802	0.22665
−1	6.0192	11.9097	3.0195	5.9598	0.23221
+2	6.2016	12.2706	3.111	6.1404	0.447874
−2	5.9584	11.7894	2.989	5.8996	0.46999
+3	6.2624	12.3909	3.1415	6.2006	0.66374
−3	5.8976	11.669	2.9589	5.8394	0.713802
+4	6.3232	12.5112	3.172	6.2608	0.87464
−4	5.8368	11.5488	2.928	5.7792	0.963633
+5	6.384	12.6315	3.2025	6.321	1.08044
−5	5.776	11.4285	2.8975	5.719	1.21995

for the plans obtained with mis-specified values. The sensitivity analysis indicates that because these values have a very small effect on the optimal value τ^*, they are not sensitive. Therefore, the optimum plans are robust, and initial estimates have a small effect on optimal values.

6.11 SUMMARY

This chapter deals with the optimum time-censored step-stress ALT plan with competing causes of failure using Khamis–Higgins model which is an alternative to the Weibull cumulative exposure model for the step-stress accelerated life testing. The optimum stress change time is obtained by minimizing the sum of asymptotic variances of the MLEs of the log characteristic lives at design stress due to all the failure causes. Inferential procedures involving design parameters have been discussed and sensitivity analysis carried out.

CHAPTER 7

PRODUCT CONTROL AND ACCELERATED LIFE TESTING

7.1 INTRODUCTION

For both the manufacturer and the customer, reliability is one of the most important characteristics defining the quality of a product or system. The expanding global marketplace resulting in increased competition and higher customer expectations for better quality has put tremendous pressure on today's manufacturers to develop new highly technological and highly reliable products in record time. This has in turn put pressure on the reliability engineers to evaluate the reliabilities of such highly reliable products in short time, while safeguarding producer's as well as consumer's interests. Accelerated life testing coupled with quality control technique of acceptance sampling plans helps in meeting this challenge.

The American National Standards Institute (ANSI) and American Society for Quality Control (ASQC) define quality as "the totality of features and characteristics that bears on its ability to satisfy given needs". This is ensured in quality control through Process Control and Product Control. Process Control aims at controlling and maintaining the quality of the manufactured product so that it conforms to specified quality standards. 'Product Control' means controlling the quality of the product by critical examination at strategic points and this is achieved through 'Acceptance Sampling Plans'. Acceptance sampling is defined in International Organization for Standardization (2002), ISO 3534-2 and its American version

ANSI/ISO/ASQ A3534-2: 2002 (American Society for Quality Control (2002)) as "**Sampling in which decisions are made to accept or not to accept a lot or other grouping of product, materials or service based on sample results**". Acceptance sampling helps manufacture products according to the desired specification limits and hence yields quality assurance.

An acceptance sampling plan (ASP) used to determine the acceptability of a product with respect to its lifetime is called **Life Test Sampling Plan (LSP)** or **Reliability Acceptance Sampling Plan (RASP)**. Such a plan generally specifies the optimal sample size, test statistic, and decision criterion along with costs related to sampling, inspection, warranty, and rejection. Since life testing for high-reliability products under normal operating conditions is impractical, therefore accelerated life tests (ALTs) are incorporated in LSPs and are referred to as **Accelerated Life Test Sampling Plans (ALTSPs)**. Wallace (1985) stressed the need for introducing ALTs to the future plans of MIL-STD-781.

7.2 PRODUCT CONTROL METRICS

In this section, product control indices needed for the study of LSPs have been explained. See Duncan (1986), Mitra (2005) and Schilling and Neubauer (2009) for reference.

Definition 7.1 (Quality Characteristic). One or more elements defining the intended quality level of a product or service constitute ***quality characteristic***.

For instance, a quality restaurant may be one that provides extraordinary cuisine served on the finest china with an ambience of soft music.

Quality characteristic can be categorized in to **structural characteristic, sensory characteristic, time-oriented characteristic, and ethical characteristic**.

Definition 7.2a Structural characteristic includes such element as the length of a part, the weight of a can, the strength of a beam, the viscosity of a fluid etc.

Definition 7.2b Sensory characteristic comprises the taste of good food, the smell of a sweet fragrance, beauty of a model among others.

Definition 7.2c Time-oriented characteristic includes such measures as warranty, reliability, and maintainability.

Definition 7.2d Ethical characteristic includes honesty, courtesy, friendliness etc.

Definition 7.3 (Variables and Attributes). Quality characteristics that are measurable and are expressed on a numerical scale are called **variables**. A quality characteristic that cannot be measured on a numerical scale is expressed as an **attribute.**

The diameter of a bearing expressed in millimeters is a variable, as are the density of a liquid in grams per cubic centimeter and the resistance of a coil in ohms.

Some examples of an attribute are: the smell of cologne that is characterized as either acceptable or not; the color of a fabric that is either acceptable or not.

However, there are some variables that are treated as attributes because it is simpler to measure them this way or because it is difficult to obtain data on them. For instance, the diameter of a bearing is, in theory a variable. However, if the diameter is measured using a go/no-go gage and classified as either conforming or nonconforming (with respect to some established specifications) then the characteristic is expressed as an attribute. The reasons for using a go/no-go gage, as opposed to a micrometer, could be economic; i.e., the time needed to obtain a measurement using a go/no-go gage may be much shorter and consequently less expensive. Alternatively, an inspector may not have enough time to obtain measurements on a numerical scale using a micrometer, so such a variable classification would not be feasible.

Definition 7.4 (Defect and Defective). A **defect** is associated with a quality characteristic that does not meet certain standards. Furthermore, the severity of one or more defects in a product or service may cause it to be unacceptable or **defective**.

The **modern term** for a defect is **nonconformity**, and the term for a defective is a **nonconforming** item.

Definition 7.5 (Process Average). The average defective fraction maintained by the producer in the statistically controlled production process is called process average.

Definition 7.6 (Producer's Risk). This is the risk associated with rejecting (or not accepting) a lot of "good" quality. It is generally desirable to frequently accept lots of this quality level (acceptable quality level (AQL)), for which a numerical value is prescribed.

Definition 7.7. Acceptable Quality Level (AQL) is defined as the maximum percentage or proportion of nonconforming items or number of nonconformities in a lot or batch that can be considered satisfactory as a process average (ANSI/ASQC Standard A2 American Society for Quality Control (1987)).

This is the numerical definition of a good lot, associated with the producer's risk. The AQL value refers to the process average, which, if kept stable at this value, would result in accepting a majority of the lots. Thus, if the producer's risk is 5% for an AQL of 0.02, it means that one considers batches that are 2% nonconforming as good and prefers to reject them no more than 5% of time.

Analogously, the **Acceptable Reliability Level (ARL)** is defined as the reliability level at which the probability of acceptance of lot is a specified high value, say 95%.

Definition 7.8 (Consumer's Risk). This is the risk of accepting a "poor" lot. It is seldom desirable to accept lots with a quality level this poor. A numerical value of this poor quality level (limiting quality level (LQL)) is prescribed.

Definition 7.9. Limiting Quality Level (LQL) is defined as the percentage or proportion of nonconforming items or number of nonconformities in a lot or batch for which the consumer wishes the probability of acceptance to be a specified low value. This is a numerical definition of a poor lot, associated with a consumer's risk.

LQL is sometimes referred to as the **Rejectable Quality Level (RQL)**, **Unacceptable Quality Level (UQL)**, or **Limiting Quality (LQ)**. When LQL is expressed as a percentage of nonconforming items, it may be referred to as the **lot tolerance percent defective (LTPD)**.

Thus, if the consumer's risk is 10% for an LQL of 0.08, this means that batches that are 8% nonconforming are poor and one prefers to accept these batches no more than 10% of the time.

Analogously the **Lot Tolerance Reliability Level (LTRL)** is defined as that level at which probability of lot rejection is a specified high value, say 90%.

Definition 7.10 (Probability of acceptance). Probability that, when using a given acceptance sampling plan, a lot will be accepted when the lot or process is of a specified quality level, p (say) is called probability of acceptance of lot. It is denoted by $P_a(p)$ or $L(p)$.

Definition 7.11. Operating Characteristic (OC curve) is a measure of the performance of a sampling plan, and is a plot of probability of accepting the lot versus proportion nonconforming of the lot. It shows the discriminatory power of the sample plan.

For all sample plans, lots with a low proportion nonconforming have to be accepted most of the time and those with high proportion nonconforming should not be accepted very often. The OC curve indicates the degree to which this objective has been achieved by a sampling plan. The OC curve shows the discriminatory power of a sampling inspection plan. An ideal OC curve is one which rejects all lots of quality worse than some particular value of p and accepts all lots of quality equal to or better than that.

Such curves are of two types:

Type A OC curve: Here, it is assumed that the sample is chosen from an isolated lot of finite size. It plots the probability that a lot will be accepted against the proportion defective in the lot inspected.

Type B OC curve: Here, it is assumed that a stream of lots is produced by the process and that the lot size is large (at least 10 times) compared to the sample size. It plots the proportion of lots that will be accepted against the proportion defective in the producer's process, which gives rise to the lot inspected.

The sample plans are set up with reference to an (AQL, p_0) or ARL. The AQL is the base line requirement for the quality of the producer's product. The producer would like to design a sampling plan such that the OC curve yields a high probability of acceptance at the AQL. On the other side of the OC curve, the consumer wishes to be protected from accepting poor quality from the producer. So the consumer establishes a criterion based on the lot tolerance percent defective (LTPD, p_1) or LTRL. The idea is to only accept poor quality product with a very low probability. The usual way of determining the plan parameters is to consider these two risks simultaneously, which is called the two-point approach.

Definition 7.12. Average Sample Number (ASN) is the average sample size inspected per lot in reaching decisions to accept or not to accept a lot when using a given acceptance sampling plan.

In other words, "The average number of sample units per lot used for making decisions (acceptance or nonacceptance) is ASN".

A plot of ASN against p is called the ASN curve for the plan.

Definition 7.13. Acceptance Number (c) is the maximum allowable number of defective items in a sample.

Definition 7.14 (Specification Limits). Acceptable bounds on individual quality characteristics (say, 2 ± 0.1 cm for the inside diameter, 4 ± 0.2 cm for the outside diameter, and 10 ± 0.5 cm for the length of an axle) are known as **specification limits**.

Specification limits can be of two types:

A **single specification limit** implies only one boundary value for acceptability, either upper U or lower L. Thus, a measurement does not conform to the specification limit if

$$X > U$$

for an **upper specification limit**, or if

$$X < L$$

for a **lower specification limit**.

Double specification limits place both upper and lower boundary values on the acceptability of a measurement. That is, the measurement X is acceptable if and only if

$$L \leq X \leq U.$$

7.3 TYPES OF LSPs (RASPs)

LSPs are classified as attributes sampling plans and variables sampling plans, where the decision can be made on the basis of the number of failures in the former and on the basis of failure times in the latter.

See for e.g., Mitra (2005), Schilling and Neubauer (2009), Stephens (2001) and Wadsworth, Stephens, and Godfrey (2001).

MIL-STD-105E (United States Department of Defense (1989)) is a lot-by-lot standardized sampling plan for attributes. Standardized sampling plans for variables are MIL-STD-414 (United States Department of Defense (1957)). The civilian version of this plan is ANSI/ASQC Z1.9 (American Society for Quality Control (1980)). The international version of the standardized plan is International Organization for Standardization (1981), ISO 3951. ANSI/ASQC Z1.9 (American Society for Quality Control (1980)) is an AQL-based plan in which it is assumed that the distribution of the quality characteristic is normal.

7.4 LSPs (RASPs) BY ATTRIBUTES

If the lifetime distribution is known, the acceptance criteria of an attributes-sampling plan can be converted into other reliability criteria such as mean life, hazard rate, and reliable life (Schilling, 1982, p. 524). For instance, United States Department of Defense (1965) has provided procedures and factors for adapting Military Standard, MIL-STD-105D (United States Department of Defense (1963)) plans, to life and reliability testing when a Weibull distribution of failure times can be assumed. Although characteristics such as mean life can be used to determine the acceptability of the product, the OC curve is given by the corresponding attributes sampling plan of the MIL-STD-105D. The sampling plans described in this report are basically attributes-sampling plans.

The two most frequently used attribute LSPs, viz., single sampling plan and double sampling plan are briefly described in this section.

7.4.1 Single Sampling RASP

A single sampling RASP consists of the number of units on test (n) and the acceptance number (c) such that if there are at most c failures out of n, the lot is accepted. n and c are determined such that the following producer's and consumer's requirements are satisfied:

$$L(p_0) = \sum_{i=0}^{c} \binom{n}{i} p_0^i (1 - p_0)^{n-i} = 1 - \tilde{\alpha} \tag{7.1}$$

and

$$L(p_1) = \sum_{i=0}^{c} \binom{n}{i} p_1^i (1 - p_1)^{n-i} = \tilde{\beta}, \tag{7.2}$$

where $\tilde{\alpha}$ and $\tilde{\beta}$ are respectively producer's risk and consumer's risk.

Generally it may not be possible to obtain a plan which exactly satisfies the above two equations due to discreteness of n and c. Therefore, the smallest n and the corresponding c are determined such that

$$\sum_{i=0}^{c} \binom{n}{i} p_0^i (1 - p_0)^{n-i} \geq 1 - \tilde{\alpha} \tag{7.3}$$

and

$$\sum_{i=0}^{c} \binom{n}{i} p_1^i (1 - p_1)^{n-i} \leq \tilde{\beta}. \tag{7.4}$$

The above inequalities can be solved using the method given by Guenther (1974).

There exists rich literature on formulation of attribute LSPs under normal operating conditions based on single sample. Sobel and Tischendrof (1959) have studied acceptance sampling plans for exponential distribution. These results are extended for the Weibull distribution by Goode and Kao (1961). Gupta and Groll (1961) and Gupta (1962) have studied the sampling plans for the lifetime under the gamma and lognormal distributions, respectively. SSPs for attributes have also been studied by Kantam and Rosaiah (1998) for half logistic distribution, Kantam, Rosaiah, and Rao (2001) for the log-logistic distribution, Rosaiah and Kantam (2005) for inverse Rayleigh distribution, Baklizi (2003) for the Pareto distribution of the second kind when the shape parameter is assumed to be known, Baklizi and EI Masri (2004) for the Birnbaum–Saunders (BS) distribution widely used for the fatigue process, Tsai and Wu (2006) for the generalized Rayleigh distribution, and Balakrishnan, Leiva, and López (2007) for the generalized BS distribution. Rosaiah, Kantam, and Santosh Kumar (2006), Rosaiah, Kantam, and Reddy (2007), Aslam and Shahbaz (2007) have studied economic reliability test plan for generalized log-logistic distribution, Inverse Rayleigh variate, and generalized exponential distribution, respectively. Lio, Tsai, and Wu (2010a, 2010b) have proposed SSP by attributes based on the percentiles of BS distribution and Burr Type-XII distribution. More recently Al-Nasser and Al-Omari (2013), Gui and Zhang (2014) have studied SSPs for attributes, respectively for exponentiated Frétchet distribution and Gompertz distribution.

Example 7.1. Assume that the lifetime of items follows a Weibull distribution (see Section 1.4.2.4), and that the life test is terminated at time t_0. Normally, the producer records the estimated values of the shape parameter of his (her) product from engineering knowledge. If the shape parameter of the Weibull distribution is unknown, it can be estimated from the historical failure time data.

It is assumed that α is an unknown scale parameter and β is a known shape parameter.

The mean lifetime of a Weibull distributed item is given as:

$$\mu = \frac{\alpha}{\beta}\Gamma\left(\frac{1}{\beta}\right). \tag{7.5}$$

The unreliability or fraction nonconforming at time t_0 is obtained using the cdf of the Weibull distribution given in (1.18) and is given by:

$$p = 1 - \exp\left(-\left(\frac{\Gamma(1/\beta)}{\beta}\right)^{\beta}\left(\frac{t_0}{\mu}\right)^{\beta}\right). \tag{7.6}$$

It is convenient to specify the termination time t_0 as a multiple of the specified lifetime, μ_0. Therefore, let $t_0 = a\mu_0$, where a is a positive constant. Then, (7.6) reduces to

$$p = 1 - \exp\left(- \left(\frac{a\Gamma(1/\beta)}{\beta} \right)^{\beta} \left(\frac{\mu_0}{\mu} \right)^{\beta} \right). \tag{7.7}$$

If r_1 is the ARL taken as mean ratio at the producer's risk and r_2 is the LTRL taken as mean ratio at the consumer's risk, then design parameters are obtained as the solutions of the following two inequalities:

$$L\left(p_1 \middle| \frac{\mu}{\mu_0} = r_1 \right) \geq 1 - \tilde{\alpha} \quad \text{and} \quad L\left(p_2 \middle| \frac{\mu}{\mu_0} = r_2 \right) \leq \tilde{\beta}. \tag{7.8}$$

7.4.2 Double Sampling Plan

For a double sampling plan, a random sample is drawn and inspected. A decision amongst acceptance, rejection, or continuing inspection is made. If it is decided to continue, a second sample of items is drawn, and an acceptance or a rejection decision is then made after the inspection of the second sample. If the probability to inspect the second sample is zero, double sampling plan reduces to single sampling plan.

The operation of a double-sampling plan is as follows:

(i) Draw the first random sample of size n_1 from the lot and put them on test for the duration of t_0 units of time. Accept the lot if there are c_1 or smaller number of failures. Reject the lot and terminate the test as soon as (c_2+1) failures are observed, where $c_1 < c_2$.

(ii) If the observed number of failures by t_0 is between $c_1 + 1$ and c_2 (c_2 included), then draw a second sample of size n_2 from the lot and put them on testing for another t_0 units of time. The lot is accepted if the total number of failures from the first and the second samples is no greater than c_2. Otherwise, terminate the test and reject the lot.

The single sampling plan is a special case of double sampling plan when $c_1 = c_2$.

The formulation of double acceptance sampling plans has been studied by different authors. The DASPs have been developed for Rayleigh, log-logistic, Weibull and general life distribution by Aslam (2007), Aslam and Jun (2010), Aslam, Jun, and Ahmad (2009, 2010), respectively. Srinivasa Rao, Ghitany, and Kantam (2009) have developed DASPs based on truncated life tests following a Marshall–Olkin extended Lomax distribution. Srinivasa Rao (2011) has proposed a DASP based on truncated life tests when the lifetime of a product follows the Marshall–Olkin extended exponential distribution Aslam *et al.* (2011) have developed a double-sampling procedure for Burr Type-XII distribution percentiles.

Example 7.2. Assume that the lifetime of items follows a Weibull distribution (see Section 1.4.2.4), and that the life test is terminated at time t_0.

Since the percentiles provide more useful information about the quality deterioration for products as compared to the mean lifetime, therefore the 100qth-percentile (see (1.20)) denoted by t_q is used.

Take $\delta_q^0 = \frac{t}{t_q^0}$, the cdf of the Weibull distribution (see (1.18)) gives the fraction nonconforming and can be represented as:

$$p = 1 - \exp\{(\delta_q^0)^\beta(\ln(1-q))\}. \tag{7.9}$$

Denote the Double Acceptance Sampling plan as $(n_1, n_2, c_1, c_2, \delta_q^0)$, then the probability of acceptance of lot is given by:

$$L(p) = \sum_{i=0}^{c_1} \binom{n_1}{i} p^i(1-p)^{n_1-i} + \sum_{x=c_1+1}^{c_2} \binom{n_1}{x} p^x(1-p)^{n_1-x} \left[\sum_{i=0}^{c_2-x} \binom{n_2}{i} p^i(1-p)^{n_2-i}\right], \tag{7.10}$$

where p is the probability that an item fails before time t, given a specified 100qth percentile lifetime t_q^0.

Using the two-point approach, the parameters of the Double Acceptance Sampling plan are obtained as solutions of the following two inequalities simultaneously:

$$L\big(p_1 \mid \delta_q^0 = r_1\big) \geq 1 - \tilde{\alpha} \quad \text{and} \quad L(p_2 \mid \delta_q^0 = r_2) \leq \tilde{\beta}, \tag{7.11}$$

where p_1 is the probability of failure corresponding to the producer's risk, ARL, p_2 is the failure probability corresponding to consumer's risk, LTRL, r_1 is the percentile ratio at the producer's risk α (or the percentile ratio corresponding to ARL) and r_2 is the percentile ratio at the consumer's risk β (or the percentile ratio corresponding to LTRL).

7.4.3 Other LSPs By Attributes

There are other attribute life test sampling procedures. These include sequential sampling plans based on the sequential probability ratio test developed by Wald (1947), repetitive group sampling (RGS) plan (Sherman, 1965), skip lot sampling plans (SkSP) (Dodge, 1955a), and chain sampling plans (ChSP) (Dodge, 1955b). These plans are described below briefly:

In **sequential sampling plan** at each phase, based on the cumulative inspection results, a decision is made to either accept the lot, reject the lot, or continue sampling. Sequential sampling is usually an item-by-item inspection process, even though it is possible to have groups of items at any given phase.

In a **RGS plan**, a random sample is drawn, then, according to a fixed criterion, the lot is either accepted, rejected, or the sample is completely disregarded, and one begins all over again with a new sample. This is continued until the fixed criterion tells us to either accept or reject the lot. It is not needed to keep track of how many times the sample has been repeated.

SkSP is followed under the assumption that there is a continuous flow of lots from the production process and lots are offered for inspection one by one in the order of production and the production process is capable of producing units whose process quality level is stable.

The operating procedure of Skip Lot sampling plan for the truncated life test has the following steps:

(a) Start with normal inspection, using reference plan (single sampling plan is used as reference plan).
(b) When i consecutive lots are accepted on normal inspection, switch to skipping inspection and inspect only a fraction, f of the lots.
(c) When a lot is rejected on skipping inspection, switch to normal inspection.
(d) Screen each rejected lot and correct or replace all defective units found.

This results in reduction of the average number of items inspected while maintaining a probability of acceptance at least as high as a lot-by-lot inspection would yield. The skip-lot sampling plan is one of the sampling schemes widely used in the industry for the inspection purpose. The main advantage of the SkSP scheme is to provide the inspection of the product at a low cost (Hsu, 1980). More details about the procedure and applications of SkSP schemes can be found in Aslam *et al.* (2015) and the references therein.

In **chain sampling plan**, the acceptance or rejection of a lot is based not only on the sample from that lot, but also on sample results from the past lots. It is, therefore, assumed that the lots should have the same quality and that they come from a steady flow of items which are produced by a single source. The operating procedure for the plan is as follows:

1. For each lot, select a sample of n units and test each unit for conformance to the specified requirements.
2. Accept the lot if the observed number of defectives d is zero, reject the lot if $d \geq 2$.
3. Accept the lot if d is one and no defective units are found in the immediately preceding i samples of size n.

The chain sampling plan is used for tests that are costly thereby necessitating the need for small samples and results in better OC curves as compared to other sampling plans. If the sample size is small, the acceptance number is usually small and sometimes it is zero resulting in the entire OC curve being convex. Consequently, even for extremely good lots with a low proportion nonconforming, the probability of lot acceptance decreases rapidly from one. So from a producer's point of view, there is a chance that good lots will be rejected more frequently than they should be. The chain sampling plan achieves a desirable OC curve for zero acceptance number when the proportion nonconforming is small.

Group acceptance sampling plans in which a multiple number of items as group can be tested simultaneously in a tester have also been studied in the literature (see for e.g., Aslam and Jun, 2009a, 2009b).

Some other attribute sampling procedures have also been standardized. Skip lot plans related to the International Organization for Standardization (1989), ISO 2859-1 are contained in International Organization for Standardization (1991), ISO 2859-3.

7.5 LSPs BY VARIABLES

Variable sampling plans for reliability are similar to acceptance sampling plans (Schilling and Neubauer, 2009). Inspection procedures by variables are based on the measurement of a variable quality characteristic, and the decision to accept or reject the lot is a function of these measurements (as opposed to the number of defectives). Variables inspection is applicable whenever the testing of individual items involves measurement and the form of the distribution is known. Since inspection by variables, when it is applicable, makes greater use of the information concerning the lot than does inspection by attributes, variables plans require smaller sample sizes than attributes plans furnishing the same protection. Variables sampling plans pertain to a single quality characteristic, and it is usually assumed that measurements of this quality characteristic are independent, and identically distributed.

There are **two** basic types of variable sampling plans. The **first** deals with controlling a process parameter such as the mean or standard deviation. The desirable settings of the parameters are such that certain conditions regarding the mean lot quality and the corresponding probability of lot acceptance are satisfied. Plans can be designed for single or double specification limits and a process standard deviation that is known or unknown.

The **second** type of plan deals with controlling the proportion of the product that is nonconforming. Here as well, single or double sampling specification limits and a process variability that is known or unknown can be used to determine the plans.

In this section variable, sampling plans for a process parameter, and for proportion nonconforming are briefly described in Sections 7.5.1 and 7.5.2, respectively.

7.5.1 Variable Sampling Plans for a Process Parameter

Variable sampling plans for a process parameter are used when either the average quality of the product or process or the variability of the quality is of concern. For example mean life of a lamp, and the standard deviation of an electrical test on semiconductors for use in a ballistic missile. Some plans in this category involve:

- **Hypothesis testing on process parameters** which are one- or two-sided tests depending on whether the parameter is to be controlled against specifications on one or both sides.
- **Acceptance control chart** which is a natural extension of a standard test of hypothesis on individual lots is to plot the results for successive lots in control chart form. This serves to allow trend and runs analysis on a continuing series of lots and affords the acceptance control engineer all the advantages of the control chart technique. The critical value for the test serves as the acceptance control limit (ACL). Lots which plot inside the ACL are accepted. Those which plot outside are rejected. This idea was first proposed by Freund (1957) in a celebrated paper which later won the Brumbaugh Award from the American Society for Quality Control as the best technical contribution of the year.
- **Sequential sampling plans for a process parameter** which are used on variable data in a manner analogous to sequential

plans for attributes. Sequential sampling requires one of three decisions to be made after each unit is inspected: Accept the lot, do not accept the lot, or inspect another piece. This process is continued until one of the first two decisions is made. As in attribute sequential testing, a cumulative sum of the characteristic is plotted as a function of the number of units inspected.

The variable SSP for a process parameter is briefly described in Section 7.5.1.1.

7.5.1.1 *Single Sampling Plan for a Process Parameter*

Standard statistical tests of hypothesis form the basis for the methodology of single sampling by variables for process parameters such as mean life, quantile life, hazard rate or reliable life. In fact, such plans are simply tests of hypotheses. They are used as one- or two-sided tests depending on whether the parameter is to be controlled against specifications on one or both sides. Sample sizes are critical in acceptance sampling applications and may be determined from the appropriate power or OC curves of the test. The power curve shows probability of rejection plotted against hypothetical values of process parameter. Its complement is the OC curve. The lot is accepted or rejected in the same way as the null hypothesis is accepted or rejected. See Schilling and Neubauer (2009) for reference.

7.5.2 Single Sampling by Variables for Proportion Nonconforming

The basic idea of variable sampling for proportion nonconforming is to show that the sample results are sufficiently far within the specification limit(s) to assure the acceptability of lot with reasonable probability.

Let X be a quality characteristic measurable on a continuous scale. If an upper limit, U, is specified, then the product is qualified as defective when $X > U$. Similarly if a lower limit, L, is specified, then the product is qualified as defective when $X < L$.

The operating procedure for variable single sampling plans for proportion nonconforming is based on:

7.5.2.1 $\bar{X}$-method, k-method and M-method

$\bar{X}$-method

$\bar{X}$-method requires that the sample size and an acceptance constant k be specified and that σ be known. An acceptance limit $\bar{X}_a$ for $\bar{X}$ is set a distance $k\sigma$ within the specification limit. The procedure is shown in Table 7.1.

k-method

The $\bar{X}$ method is a special case of what is called the k method. The procedure involved in k method is shown in Table 7.2. The more

Table 7.1: $\bar{X}$-method

Lower Specification Limit	Upper Specification Limit
1. Set $\bar{X}_a = L + k\sigma$. 2. Select a random sample of size n. 3. Compute $\bar{X}$. 4. If $\bar{X} \geq \bar{X}_a$, accept the lot; if $\bar{X} < \bar{X}_a$, reject the lot.	1. Set $\bar{X}_a = U - k\sigma$. 2. Select a random sample of size n. 3. Compute $\bar{X}$. 4. If $\bar{X} \leq \bar{X}_a$, accept the lot; if $\bar{X} > \bar{X}_a$, reject the lot.

Table 7.2: k-Method (Given n, k)

Lower Specification Limit	Upper Specification Limit
1. Select a random sample of size n. 2. Compute $Z_L = (\bar{X} - L)/\sigma$, for σ known or $Z_L = (\bar{X} - L)/s$, for σ unknown. 3. If $Z_L \geq k$, accept the lot; if $Z_L < k$, reject the lot. 4. Equivalently, if $\bar{X} - k\sigma \geq L$, accept the lot; if $\bar{X} - k\sigma < L$, reject the lot. If σ is unknown, use appropriate values of n and k with $\bar{X} - ks$ as above.	1. Select a random sample of size n. 2. Compute $Z_U = (\bar{X} - L)/\sigma$, for σ known or $Z_U = (U - \bar{X})/s$, for σ unknown. 3. If $Z_U \geq k$, accept the lot; if $Z_U < k$, reject the lot. 4. Equivalently, if $\bar{X} + k\sigma \leq U$, accept the lot; if $\bar{X} + k\sigma > U$, reject the lot. If σ is unknown, use appropriate values of n and k with $\bar{X} + ks$ as above.

general k method may be used when the standard deviation is not known simply by substituting the sample standard deviation

$$s = \sqrt{\frac{\sum_{i=1}^{n}(X_i - \bar{X})^2}{(n-1)}}, \tag{7.12}$$

for σ in the known standard deviation procedure and choosing an appropriate value of k and sample size, n, for the unknown standard deviation case.

It can be seen that the $\bar{X}$ method is a special case of the k method, since for a lower specification limit acceptance would occur if

$$\bar{X} - k\sigma \geq L \Rightarrow \bar{X} \geq L + k\sigma \Rightarrow \bar{X} \geq \bar{X}_a.$$

Also, $\bar{X}_a$ is a fixed constant only if σ is known and so the k method is the only alternative for the case of unknown standard deviation.

Schneider's (1989) approach

Schneider (1989) has used acceptance/rejection decision that employs the k-method with the sample average $\bar{X}$ replaced by MLE, $\hat{\mu}$, of location parameter, and σ replaced by MLE, $\hat{\sigma}$, of scale parameter of the log-lifetime distribution. Failure-censored sampling plans for lognormal and Weibull life distributions are considered. The lower specification limit L is specified. Define the test statistic

$$T = \hat{\mu} - k\hat{\sigma}. \tag{7.13}$$

The lot is rejected if $T < \ln L$, and is accepted otherwise. Large sample theory is used to obtain the sample size and the acceptability constant for a given degree of censoring and given two points on the OC curve, $(p_\alpha, 1-\alpha)$, and (p_β, β).

M-method

It is often desirable to base lot acceptance on the estimates of the proportion nonconforming in the lot. An estimate of this sort can be made under attribute inspection by dividing the number of nonconformances or defectives d found in the sample size n, to obtain an unbiased estimate of the proportion nonconforming in the lot. This estimate is then compared with the constant c/n to determine lot acceptance. For variable inspection a procedure has been developed by Lieberman and Resnikoff (1955), which involves the use of a uniform minimum variance unbiased estimate of the lot proportion nonconforming p, where $p = P[X < L]$ in case of lower specification limit and $p = P[X > U]$ in case of upper specification limit, in the acceptance sampling criteria. Let

$$M = \int_{k\sqrt{n/(n-1)}}^{\infty} \frac{1}{\sqrt{2\pi}} \mathrm{e}^{-t^2/2} \mathrm{d}t, \sigma \text{ known} \tag{7.14}$$

be the threshold number, related to k from the k-method, with which p is compared to decide whether to accept or reject the lot in case of variable inspection. Table 7.3 gives the comparison of the k and M-methods for variables and attributes when the lot is accepted based on upper specification limit. Table 7.4 presents procedure for application of the M-method when the standard deviation σ is known.

When the standard deviation is unknown, the noncentral t-distribution is involved in the estimation procedure, which leads to inclusion of values from the incomplete beta function in the estimate of p (Schilling and Neubauer, 2009).

Example 7.3. Let the quality characteristic be normal with process standard deviation known and upper specification limit, U.

Table 7.3: Decision criteria for acceptance (sample of size n)

	Attributes	**Variables**
k-Method	$d \leq c$	$z = \dfrac{U - \bar{X}}{\sigma} \geq k$
M-Method	$p = \dfrac{d}{n} \leq \dfrac{c}{n} = M$	$p \leq M$

Table 7.4: M-method (σ known, given n, M)

Lower Specification Limit	Upper Specification Limit
1. Select a random sample of size n. 2. Compute $\bar{X}$. 3. Compute $$Q_L = \frac{\bar{X} - L}{\sigma}\sqrt{\frac{n}{n-1}}$$ and obtain $$\hat{p}_L = \int_{Q_L}^{\infty} \frac{1}{\sqrt{2\pi}} e^{-\frac{t^2}{2}} dt$$ from cumulative standard normal tables. This gives a minimum variance unbiased estimate of p. 4. If $\hat{p}_L \leq M$ accept; $\hat{p}_L > M$ reject.	1. Select a random sample of size n. 2. Compute $\bar{X}$. 3. Compute $$Q_U = \frac{U - \bar{X}}{\sigma}\sqrt{\frac{n}{n-1}}$$ and obtain $$\hat{p}_U = \int_{Q_U}^{\infty} \frac{1}{\sqrt{2\pi}} e^{-\frac{t^2}{2}} dt$$ from cumulative standard normal tables. This gives a minimum variance unbiased estimate of p. 4. If $\hat{p}_U \leq M$ accept; $\hat{p}_U > M$ reject.

The fraction defective in a lot is given by:

$$p = 1 - \Phi(v) = \Phi(-v), \tag{7.15}$$

with

$$v = \frac{U - \mu}{\sigma}, \tag{7.16}$$

and $\Phi(\cdot)$ is the cdf of standard normal distribution.

When the standard deviation σ is known, a single variables sampling plan is specified by its sample size, n, and its acceptability constant, k. A sample of size n, drawn at random from the lot, is inspected, and the lot is sentenced by the decision rule: accept the lot when $\bar{X} + k\sigma \leq U$, and reject the lot when $\bar{X} + k\sigma > U$ (see Table 7.2(4)).

Under the conditions stated, the probability of acceptance, P_a, of a lot is given by

$$P_a = \Phi(w), \tag{7.17}$$

with

$$w = \sqrt{n}\frac{U - k\sigma - \mu}{\sigma},$$

Eliminating U between (7.15) and (7.17), the auxiliary variables v and w are found to be related to n and k by:

$$\frac{w}{\sqrt{n}} + k = v$$

or

$$w = (v - k)\sqrt{n}. \tag{7.18}$$

To construct the OC curve, one can start with a set of values of p, obtain v by (7.15), then w by (7.18), and finally P_a by (7.17). Or alternatively one can start with a set of values of P_a, and find the corresponding p's by (7.17), (7.18), and (7.15), in that order.

If a single point (p, P_a) is prescribed, w and v are fixed and any set of values (n, k) satisfying (7.18) with those values of w and v will specify a sampling plan with an OC curve passing through that point.

When two points, (p_1, P_1) and (p_2, P_2), on the OC curve are prescribed, (7.18) must be simultaneously satisfied by (v_1, w_1) and by (v_2, w_2); solving for (n, k) one obtains

$$n = \left[\frac{w_1 - w_2}{v_1 - v_2}\right]^2, \tag{7.19}$$

and

$$k = \frac{v_1 w_2 - v_2 w_1}{w_2 - w_1}. \tag{7.20}$$

These formulae were derived by Wallis (1948) in the paper in which variable sampling was originally proposed (Hamaker, 1979).

For ***k*-method** if the lower specification limit L is specified, a similar procedure is adopted, i.e., and the lot is sentenced by the decision rule:

accept the lot when $\bar{X} - k\sigma \geq L$, else reject the lot when $\bar{X} - k\sigma < L$ (see Table 7.2(4)).

Using ***M*-method** and two-point approach, if an upper specification limit, U, is given, a sample of size n is selected, the sample mean $\bar{X}$ is found. The value of Q_U is computed and the percent of the distribution ($\hat{p}_U$) that does not exceed this value is found (see Table 7.4(3)). The maximum allowable percent nonconforming, M using (7.14) is found using k in (7.20). If $\hat{p}_U \leq M$, the lot is accepted. Otherwise, the lot is rejected.

The procedure for using ***M*-method** with a lower specification limit is similar to that for an upper specification limit.

Remark 7.1. Besides the situation highlighted in Example 7.3, the variable life test single sampling plans under normal operating conditions have been studied by many authors, for example, Epstein and Sobel (1953, 1955), Epstein (1954), Aroian (1964), Bulgren and Hewette (1973), Kockerlakota and Balakrishnan (1986), Fairbanks (1988), Balasooriya (1995), Balasooriya and Saw (1998) for exponential distribution, Hatter and Moore (1980), Fertig and Mann (1980) and Schneider (1989) for Weibull distribution. Suresh and Ramanathan (1997) have derived acceptance sampling plans by variables for a class of symmetric distributions. Balasooriya, Saw, and Gadag (2000) have presented progressively failure-censored sampling plans and Balasooriya and Low (2004) have extended their results to the case with competing causes of failure.

7.5.3 Variable Repetitive Group Sampling Plan (VRGSP) for Proportion Nonconforming

Balamurali and Jun (2006) have proposed RGS procedure for the application of variables inspection. It has already been pointed out in (Section 7.4.3) that Sherman (1965) has proposed RGS plan for attributes.

Example 7.4. Suppose that the quality characteristic of interest has the upper specification limit U and follows a normal distribution with known standard deviation σ. Then the two operating procedures of the variables RGS plan are explained below:

***k*-method**

Step 1. Take a random sample of size n, say $(X_1, X_2, \ldots, X_n)$ and compute

$$\nu = \frac{(U - \bar{X})}{\sigma}$$

Step 2. Accept the lot if $\nu \geq k_a$ and reject the lot if $\nu < k_r$. If $k_r \leq \nu < k_a$, then repeat the Steps 1 and 2. $(k_a > k_r)$.

Thus, the proposed variable RGS plan is characterized by three parameters, namely n, k_a and k_r.

If $k_a = k_r$, then the proposed plan is reduced to the **variables single sampling plan**.

***M*-method**

Step 1. Take a random sample of size n, say $(X_1, X_2, \ldots, X_n)$ and compute

$$Q_u = \frac{(U - \bar{X})}{\sigma}\sqrt{\frac{n}{n-1}}.$$

Step 2. Also determine the minimum variance unbiased estimate of the fraction nonconforming, p, as

$$\hat{p}_u = \int_{Q_u}^{\infty} \frac{1}{\sqrt{2\pi}} \mathrm{e}^{-t^2/2} \mathrm{d}t.$$

Step 3. Accept the lot if $\hat{p}_u \leq M_a$ and reject the lot if $\hat{p}_u > M_r$.

If $M_a \leq \hat{p}_u < M_r$, then repeat the Steps 1 and 2 $(M_a < M_r)$.

Thus, the parameters of the variable RGS plan under the M-method are the sample size n, acceptable criterion M_a and M_r.

Jun *et al.* (2010) have proposed a variable group sampling plan under failure-censored reliability tests for Weibull distribution.

7.5.4 Other VRASPs for Proportion Nonconforming

There are other types of RASPs by variables for proportion nonconforming existing in the literature such as Double Sampling Plans (Schilling and Neubauer, 2009), sequential sampling plans for proportion nonconforming (Schilling and Neubauer, 2009), chain sampling plans (Govindaraju and Balamurali, 1998; Balamurali and Usha, 2013). Aslam *et al.* (2015) have designed skip-lot variable sampling plan that allows re-inspection whenever decision cannot be made on the basis of original inspection.

7.6 LSPs BASED ON A COST MODEL

Costs related to LSPs are test cost, accept cost, reject cost, and also warranty cost since almost all products sold to the consumer nowadays carry a warranty of some kind. These are explained below briefly:

Definition 7.15. **Test cost** refers to costs associated with the length of test duration and the number of test items.

Definition 7.16. **Accept cost** is the external failure cost for the products in the accepted lot.

Definition 7.17. **Reject cost** means scrap or reprocessing costs for products in the rejected lot.

Definition 7.18. **Warranty** is defined as a contractual obligation incurred by a manufacturer in connection with the sale of a product under which the manufacturer is required to ensure the product functions properly during the warranty period.

This definition is due to Blischke and Murthy (1994).

Definition 7.19. **Warranty costs** are additional costs incurred on the manufacturers as all failures under warranty need to be either rectified or compensated by the manufacturer.

Therefore, in designing acceptance sampling plans for a product sold under warranty, it is worthwhile to consider the costs associated with the warranty policies.

7.6.1 Types of Warranty Policies

There are different types of warranty policies. They can be divided in to different categories (Blischke and Murthy, 1994, 1996) as indicated below:

- Involving reliability improvement or not;
- Single item or group of items (cumulative warranty);
- One-dimensional (limit on age) or two-dimensional (limits on age and usage);
- Renewing or nonrenewing;
- Base or extended.

In this work, one-dimensional warranties are considered.

Failure-free and pro-rata rebate are the two common types of warranty policies.

Definition 7.20. A failure-free warranty (FFW) policy obligates the manufacturer to maintain the product free of charge during the warranty period.

The FFW is also referred to as **free replacement warranty (FRW)**.

Failure-free policies can be further divided into two categories, namely, renewing and nonrenewing

- Renewing policy — if an item fails within the warranty time, it is replaced by a new item with a new warranty. In effect, warranty begins anew with each replacement.
- Nonrenewing policy — replacements of a failed item do not alter the original warranty.

Definition 7.21. A pro-rata warranty (PRW) policy obligates the manufacturer to refund a fraction of the purchase price if the product fails within the warranty period.

Occasionally, FRW and PRW can be combined as one policy called a **general rebate policy (GRP)** to provide consumers with more choices (Thomas and Rao, 1999; Murthy and Blischke, 2000).

7.6.2 Warranty Cost Under a Free Replacement Policy

For a warranty period of length $\tau > 0$ and assumed constant cost c' to replace or repair the item, the unit cost of warranty claims to the manufacturer (C_w) for a product that fails at time Y is

$$C_w(Y|c', \tau) = \begin{cases} c' & \text{if } 0 < Y \leq \tau \\ 0, & \text{if } Y > \tau. \end{cases} \tag{7.21}$$

7.6.3 Warranty Cost Under a Pro-Rata Replacement Policy

The **PRW** policy, which involves a sharing of the repair or replacement cost by the manufacturer and customer is based on some product age dependent formula, the most common being linear. Here for a warranty interval $(0, \tau]$ and constant $c' > 0$, the unit cost of repair or replacement to the manufacturer is given by a cost, which is proportional to the age of the product. Such PRW policies are applied to items that are significantly influenced by aging, such as tires.

$$C_w(Y|c', \tau) = \begin{cases} \dfrac{c'(\tau - y)}{\tau}, & \text{if } 0 < y \leq \tau \\ 0, & \text{if } y > \tau. \end{cases} \tag{7.22}$$

These policies gained interest as the efforts in developing warranty models for predicting manufacturing capability and costs evolved. Other policies can be constructed by combining the FRW and PRW policies into a single plan.

7.6.4 Warranty Cost Under a Combination — Free and Pro-Rata Replacement Policy

Here failures in $(0, \tau_1)$ cost c', but between τ_1 and τ_2 the cost is prorated based on the age of the unit (Mamer, 1982); thus

$$C_w(Y|c', \tau_1, \tau_2) = \begin{cases} c', & \text{if } Y < \tau_1 \\ \dfrac{c'(\tau_2 - y)}{\tau_2 - \tau_1}, & \text{if } \tau_1 \leq Y < \tau_2 \\ 0, & \text{if } y > \tau_2. \end{cases} \tag{7.23}$$

7.6.5 Expected Total Cost for the Sampling Plan

The optimal sampling plan for the life testing decision problem can be derived using the aforementioned relevant costs. (7.23) is used for illustrative purpose.

Example 7.5. Let X denotes the lifetime of a product. Suppose there is a lot of N products each with lifetimes distributed with failure density function $f(x|\theta)$ (life distribution function $F(x|\theta)$), where θ is a parameter (in general, vector). According to the past experiences of selling similar products, it is plausible to estimate θ, and then conduct a sampling test so as to determine the acceptance or rejection of this lot of products. Let c_1 is the cost of putting an item on test from a set of n products randomly sampled out of the lot; c_2 is the testing cost per unit time; and c_3 is the cost of rejecting an item when the whole lot has been rejected. Let the periods of FRW and PRW are pre-determined as τ_1 and τ_2, respectively. However, the PRW of the producer between τ_1 and τ_2 is assumed to vary in time instead of being constant. As a result, the expected warranty cost per product item using (7.23) is given by:

$$E[C_w(X|c', \tau_1, \tau_2, \theta)] = c'\left(\int_0^{\tau_1} f(x|\theta)\mathrm{d}x + \int_{\tau_1}^{\tau_2} \frac{(\tau_2 - x)}{\tau_2 - \tau_1} f(x|\theta)\mathrm{d}x\right). \tag{7.24}$$

Suppose that the test stops at censoring time t_0, and the lot is accepted if the number of failures is less than r (say), else it is rejected. Then, the expected cost resulting from selling the acceptance lot is given by:

$$W(n, r, t_0) = (N - n)E[C_w(X|c', \tau_1, \tau_2, \theta)]P[X_{(r)} > t_0], \tag{7.25}$$

and the expected rejection cost resulting from rejecting the lot is given by:

$$R(n, r, t_0) = (N - n)c_3(1 - P[X_{(r)} > t_0]), \tag{7.26}$$

where $X_{(r)}$ is the rth order statistic of lifetimes in the set of randomly sampled n products.

Because the sampling cost and the testing cost can both be obtained by simply multiplying the unit cost of sampling with the sample size, and multiplying the unit cost of testing with the censoring time, i.e., nc_1 and c_2t_0, respectively, the expected total cost for the sampling plan with the decision making parameters is given by:

$$E\text{Cost}(n, r, t_0) = nc_1 + c_2t_0 + W(n, r, t_0) + R(n, r, t_0). \quad (7.27)$$

7.6.6 Optimal LSP

The optimal LSP (n^*, r^*, t_0^*) can be determined as:

$$(n^*, r^*, t_0^*) = \arg \inf_{(n,r,t_0)} E\text{Cost}(n, r, t_0). \quad (7.28)$$

See for reference Kwon (1996), and Nechwal *et al.* (2009).

7.7 OPTIMUM ACCELERATED LIFE TESTS SAMPLING PLANS (ALTSPs)

When one needs to test the acceptability of high-reliability products, it is impractical to use LSPs that are designed under normal operating conditions since such products last for a very long time. ALTSPs help in deciding acceptance/rejection of such products.

An **optimum ALTSP** safeguards producer's and consumer's interests by determining optimum total number of test units to be tested under accelerated condition, the number of test units at each stress level, maximum allowable number of nonconforming items, and other experimental variables, and thus helps in obtaining accurate reliability estimates.

7.8 SENSITIVITY ANALYSIS

The use of an optimum test plan requires information about the design parameters. Incorrect choice of these gives a poor estimate of the mean or quantile life at design stress. The sensitivity analysis identifies the sensitive parameters which need to be estimated with special care for the purpose of minimizing the risk of obtaining erroneous solutions. If a small change in any parameter value does not result in relatively large change in the optimum plan, then the proposed test plan is robust, and the initial values have a small effect on optimal values.

7.9 DESIGN OF ALTSP AND PALTSP MODELS: A REVIEW

The existing literature on design of optimum ALTSP models are based on single sampling plans under continuous inspection. The sampling plans are based on acceptance of a product on a lot-by-lot basis. Such plans are used when grouping the product in lots follow naturally. This is in contrast to items produced on a continuous basis (such as on a conveyor belt, where successive operations are conducted on the item) where the formation of lots is not feasible. In this section, ALTSP and PALTSP models studied by different authors have been briefly described.

7.9.1 Constant-Stress ALTSP Models

Yum and Kim (1990) are the first to integrate accelerated life testing with LSP under constant-stress scheme for the exponential distribution based on two stress levels. Hsieh (1994) has extended their work and obtained sampling plans that minimize the total censoring number.

Bai, Kim, and Chun (1993) have designed failure-censored ALTSPs for lognormal and Weibull distributions in which optimum total sample size, sample proportions allocated to each stress and lot acceptability constants are determined when the test uses two prescribed levels of stress higher than use condition with a given degree of censoring at each stress. Bai, Chun, and Kim (1995) have devised failure-censored constant-stress ALTSPs for Weibull distribution under expected test time constraint. A cost model for an ALTSP has been constructed by Lin and Chiu (1995). Seo, Jung, and Kim (2009) have designed optimum time-censored and failure-censored constant-stress ALTSPs for Weibull distribution with non constant shape parameter. Optimum constant-stress ALTSPs for the Weibull distribution under Type-I progressive interval censoring with random removals have been formulated by Ding, Yang, and Tse (2010).

7.9.2 Time-varying Stress ALTSP Models

Optimum failure-censored step-stress ALTSPs for Weibull distribution have been formulated by Chung, Seo, and Yun (2006). Srivastava and Sharma (2014a) have studied time-censored ramp-stress ALTSP using variable single sampling scheme for controlling the process parameter. Srivastava and Sharma (2015a) have also studied time censored ramp-stress ALTSP using single sampling plan by variables for proportion nonconforming.

7.9.3 RASP With Acceleration Factor

Kim and Yum (2009, 2011) have formulated RASPs with known acceleration factor for the Weibull life distribution under Type-I and hybrid censored data.

7.9.4 Constant-Stress PALTSP Models

Srivastava and Sharma (2014c) have proposed PALTSP model for Burr Type-XII distribution using tampered failure rate model with constant-stress setting.

7.9.5 Time-Varying Stress PALTSP Models

Srivastava and Sharma (2014b) have designed time-censored step-stress PALTSP using tampered failure rate model and warranty. Optimum time-censored step-stress PALTSP with competing causes of failures using tampered failure rate model has also been studied by Srivastava and Sharma (2015b).

CHAPTER 8

OPTIMUM TIME-CENSORED RAMP-STRESS ALTSPs

8.1 INTRODUCTION

In the present chapter, quality control technique of acceptance sampling plan is aligned with accelerated life testing under ramp-stress loading for inducing early failures in high reliability items. The log-logistic distribution found appropriate for insulating materials has been used for the purpose.

Sections 8.2–8.5 of the chapter highlight basic assumptions, life distribution under ramp-test, test procedure, and lot acceptance test procedure, respectively. The test is conducted under fully accelerated environmental condition using continuous inspection resulting in exact failure times. In fully accelerated life testing, all the test units are run at an accelerated condition. The formulation of optimum ramp-stress accelerated life test sampling plans (ALTSPs) for log-logistic distribution under Type-I censoring has been explained in Section 8.6. Variable single sampling plan (SSP) for proportion nonconforming has been used based on Schneider's (1989) approach to k-method (see Section 7.5.2). In Section 8.7, optimum design of time-censored step-stress ALTSP under variable SSP for controlling process parameter based on median lifetime of log-logistic distribution has been discussed. The optimal plans consist in finding optimum sample size, sample proportion allocated to each stress, and stress rate factor such that producer's and consumer's interests are safeguarded. Variance optimality criterion is used when expected cost per lot is not taken into consideration, and bilevel programming approach (see Appendix C.1) has been used in cost optimization problems. The methods developed have been illustrated using some numerical examples, and sensitivity analyses carried out in the context of ramp-stress ALTSP based on variable SSP for proportion nonconforming.

8.1.1 Notation

X	product lifetime
Y	$\ln X$
ψ_i	stress rate ψ at stress level i, $\psi_i > 0$, $i = 1, 2$
ξ_i	stress rate factor ξ at ith stress level; $\xi_1 < \xi_2 = 1$; $\xi_i = \psi_i/\psi_2$, $i = 1, 2$
ϕ_1, ϕ_2	proportion of sample allocated to stress rates ψ_1 and ψ_2, respectively, $\phi_1 + \phi_2 = 1$
$s(x)$	stress at failure time x
s_0	stress level under normal operating condition or design constant stress
ϕ_1^*	optimum proportion of sample allocated to stress rate ψ_1
N	total number of test items in an ALT
n_i	total number of test items allocated to stress level i; $i = 1, 2$, $n_1 = n\phi_1, n_2 = n\phi_2$, $n = n\phi_1 + n\phi_2 = n_1 + n_2$
z	$(\ln(x) - \mu(\xi))/\sigma$; standardized log failure time under stress rate factor ξ
τ'	$(\ln(\tau) - \mu(\xi))/\sigma$; standardized log censoring time under stress rate factor ξ
α, λ	shape and scale parameters of log-logistic life distribution
γ_0, γ_1	parameters of the inverse power law, $\gamma_1 > 0$, $-\infty < \gamma_0 < \infty$
α', η	shape and scale parameters of log-logistic life distribution under ramp-test

μ, σ	location, scale parameter of the transformed distribution
$\mu(\xi)$	location parameter under stress rate factor

$$\xi = \ln(\eta(\xi)) = \left(\frac{1}{1+\gamma_1}\right)\left(\gamma_0 + \gamma_1 \ln\left(\frac{s_0}{\xi\psi_2}\right) + \ln(1+\gamma_1)\right)$$

L, L'	lower specification limit on X, Y
T	test statistic used to decide the lot acceptability
$\tilde{\alpha}, \tilde{\beta}$	producer's risk and consumer's risk, respectively; $0 < \tilde{\alpha} < 1$, $0 < \tilde{\beta} < 1$
$p_{\tilde{\alpha}}, p_{\tilde{\beta}}$	fraction nonconforming to be accepted with probability atleast $1 - \tilde{\alpha}$, and rejected with probability $\tilde{\beta}$
P_1	probability that a unit will fail by time τ at ξ_1 and is given by $1 - \{1 + [\tau/((e^{\gamma_0}(s_0/(\psi_2\xi_1))^{\gamma_1}(1+\gamma_1))^{1/(1+\gamma_1)})]^{\alpha(1+\gamma_1)}\}^{-1}$
P_2	probability that a unit will fail by time τ at ξ_2 and is given by $1 - \{1 + [\tau/((e^{\gamma_0}(s_0/(\psi_2))^{\gamma_1}(1+\gamma_1))^{1/(1+\gamma_1)})]^{\alpha(1+\gamma_1)}\}^{-1}$
$\Phi(\cdot)$	cdf of standard normal distribution
$H(\cdot)$	cdf of logistic distribution
w_p	quantile p of standard normal distribution; $\Phi(w_p) = p$
z_p	quantile p of standard logistic distribution; $z_p = -\ln((1-p)/p)$
K	lot acceptability constant
C_s	cost of sampling and putting an item on test
C_τ	cost per unit of test time
C_{re}	cost of rejecting an item
C_w	cost associated with an external failure

8.2 ASSUMPTIONS

1. The stress rates ψ_1 and ψ_2 ($\psi_1 < \psi_2$) are used in a ramp-test.
2. At any constant stress, s, the product life, X, has a log-logistic distribution. The log-logistic life distribution has been found to

be appropriate for high reliability components. The hazard rate function of log-logistic distribution is given by

$$h^*(x) = \frac{\frac{\alpha}{\lambda}\left(\frac{x}{\lambda}\right)^{\alpha-1}}{1+\left(\frac{x}{\lambda}\right)^{\alpha}}, \quad t \geq 0, \ \alpha > 0, \ \lambda > 0 \tag{8.1}$$

which is always decreasing with time if $\alpha \leq 1$; first increasing, then decreasing with time if $\alpha > 1$.

3. For the effect of changing stress, the linear cumulative exposure model holds.
4. The inverse power law holds for $\lambda(s)$ where λ is a linear function of a (possibly transformed) stress s, i.e.,

$$\lambda(s) = e^{\gamma_0}\left(\frac{s_0}{s}\right)^{\gamma_1}, \tag{8.2}$$

where parameters γ_0 and γ_1 are the characteristics of the product.

5. At any constant-stress s, the lifetime of a unit follows a log-logistic distribution with cdf

$$G(x;\alpha,\lambda) = \frac{\left(\frac{x}{\lambda}\right)^{\alpha}}{1+\left(\frac{x}{\lambda}\right)^{\alpha}}, \quad x \geq 0, \ \alpha > 0, \ \lambda > 0, \tag{8.3}$$

where α and λ are shape and scale parameters, respectively, of the distribution, and the inverse power law (see (8.2)) holds for $\lambda(s)$.

6. The lifetimes of test units are statistically independent.
7. The stress applied to test units is continuously increased with constant rate ψ from zero.

8.3 LOG-LOGISTIC LIFE DISTRIBUTION UNDER RAMP-STRESS TEST

The cdf and pdf of log-logistic distribution is given by:

$$F(x;\alpha,\lambda) = 1 - \left(1+\left(\frac{x}{\lambda}\right)^{\alpha}\right)^{-1}, \quad x \geq 0, \ \alpha > 0, \ \lambda > 0, \tag{8.4}$$

$$f(x;\alpha,\lambda) = \left(\frac{\alpha}{\lambda}\right)\left(\frac{x}{\lambda}\right)^{\alpha-1}\left(1+\left(\frac{x}{\lambda}\right)^{\alpha}\right)^{-2}, \quad x \geq 0,\ \alpha > 0,\ \lambda > 0, \tag{8.5}$$

where α and λ are shape and scale parameters, respectively.

The stress at failure time x is (from assumption #7) $s(x) = \psi x$.

From the linear cumulative exposure model and the inverse power law, the cdf of the lifetime X of a unit tested under stress rate ψ is:

$$G(x) = F(\varepsilon(x)), \tag{8.6}$$

where $F(\cdot)$ is the assumed cdf (see (8.4)) with the scale parameter λ set equal to one,

$$\varepsilon(x) = \int_0^x \frac{1}{\lambda(s(u))} du \tag{8.7}$$

is the cumulative exposure (damage) model. Hence, the cdf and pdf, respectively of log-logistic distribution reduces to

$$G(x;\alpha,\eta) = \frac{\left(\int_0^x \frac{du}{\lambda(s(u))}\right)^{\alpha}}{1+\left(\int_0^x \frac{du}{\lambda(s(u))}\right)^{\alpha}}, \text{ (using (8.4), (8.6), and (8.7))}$$

$$= \frac{\left(\frac{x}{\eta}\right)^{\alpha'}}{1+\left(\frac{x}{\eta}\right)^{\alpha'}}, \tag{8.8}$$

$$g(x;\alpha,\eta) = \left(\frac{\alpha'}{\eta}\right)\left(\frac{x}{\eta}\right)^{\alpha'}\left(1+\left(\frac{x}{\eta}\right)^{\alpha'}\right)^{-2}, \tag{8.9}$$

where

$$\alpha' = \alpha(1+\gamma_1), \tag{8.10}$$

$$\eta = \left(e^{\gamma_0}\left(\frac{s_0}{\xi}\right)^{\gamma_1}(1+\gamma_1)\right)^{\frac{1}{1+\gamma_1}}, \tag{8.11}$$

$G(\cdot)$ is the log-logistic distribution with scale parameter η and shape parameter α', η is a function of ξ.

8.3.1 Log-Lifetime Distribution

The logistic $Y = \ln(X)$ is used instead of using the actual lifetime X. Thus, Y follows logistic distribution with location parameter

$$\mu = \ln(\eta), \tag{8.12}$$

and scale parameter

$$\sigma = (\alpha')^{-1}. \tag{8.13}$$

The pdf and cdf for Y, respectively are

$$h(y;\mu,\sigma) = \frac{\exp\left\{\frac{y-\mu}{\sigma}\right\}}{\sigma\left(1+\exp\left\{\frac{y-\mu}{\sigma}\right\}\right)^2}, \qquad -\infty < y < \infty, \tag{8.14}$$

$$H(y;\mu,\sigma) = 1 - \left(1+\exp\left\{\frac{y-\mu}{\sigma}\right\}\right)^{-1}, \qquad -\infty < y < \infty. \tag{8.15}$$

8.4 LIFE TEST PROCEDURE

1. Out of the total of n ($n = n\phi_1 + n\phi_2 = n_1 + n_2$) items, n_1 items randomly chosen are allocated to stress rate ψ_1 and the remaining n_2 items are allocated to stress rate ψ_2.
2. Under **Type-I Censoring**, the test is continued until:
   (a) all the test units fail, or
   (b) a prescribed censoring time,

   whichever occurs earlier.

8.5 LOT ACCEPTANCE SAMPLING PROCEDURE

The following lot acceptance sampling procedure is considered.

1. Samples of n items are randomly selected from the lot and are tested according to the test procedure explained in Section 8.4.
2. Maximum likelihood estimators (MLEs) $\hat{\mu}$ and $\hat{\sigma}$ of location parameter μ_0 and scale parameter σ_0 at use condition, respectively are obtained from the test data.

3. Assume that one-sided lower specification limit L^* is assigned to the lifetime of a product. The lower specification limit on Y is $L' = \ln(L^*)$. The value of test statistic $T = \hat{\mu}_0 - k\hat{\sigma}_0$ is compared with L', and the lot is accepted if $T \geq L'$ and rejected otherwise. In this chapter, T is denoted by T'. The sample size n and acceptability constant k are to be determined according to the optimality criterion and constraints, i.e., lots with fraction nonconforming $p \leq p_{\tilde{\alpha}}$ are accepted with probability at least $1-\tilde{\alpha}$ and lots with $p > p_{\tilde{\beta}}$, are rejected with probability at least $1 - \tilde{\beta}$.

8.6 OPTIMUM TIME-CENSORED RAMP-STRESS ALTSP UNDER VARIABLE SSP FOR PROPORTION NONCONFORMING

In this section, an optimum time-censored ramp-stress ALTSP for the log-logistic distribution designed by Srivastava and Sharma (2015a) has been explained.

8.6.1 Log-Likelihood

The log-likelihood of an observation y at stress rate ψ is obtained. The stress rate factor ξ at the ith stress level is

$$\xi_i = \frac{\psi_i}{\psi_2}, \quad i = 1, 2. \tag{8.16}$$

For the low stress rate factor, ψ_1, $\xi_1 = \frac{\psi_1}{\psi_2}$. and for the high stress rate, ψ_2, $\xi_2 = 1$.

Define the indicator function $I \equiv I(z)$ in terms of the censoring time τ' by

$$I = \begin{cases} 1, & \text{if } z \leq \tau', \text{ failure observed by time } \tau' \\ 0, & \text{if } z > \tau', \text{ censored at time } \tau'. \end{cases} \tag{8.17}$$

The log-likelihood of a single observation y is,

$$L(\gamma_0, \gamma_1, \alpha) = L = I(z)(-\ln(\sigma) - z - 2\ln(1 + e^{-z})) - \bar{I}(z)\ln(1 + e^{\tau'}). \tag{8.18}$$

Let the log-likelihood of unit j be L_j then, log-likelihood, L_0, for n independent observations is $L_0 = L_1 + \cdots + L_n$.

The maximum likelihood estimates $\hat{\gamma}_0, \hat{\gamma}_1$, and $\hat{\alpha}$ are the parameter values that maximize L_0.

8.6.2 Fisher Information Matrix

The Fisher information matrix is obtained by taking expectations of the negative of the second partial derivatives of the log likelihood function with respect to γ_0, γ_1, and α. The first derivatives are

$$\frac{\partial L}{\partial \gamma_0} = \alpha \cdot \left(I(z)\left(1 - \frac{2e^{-z}}{1+e^{-z}}\right) + \bar{I}(z)\left(\frac{1}{1+e^{-\tau'}}\right)\right), \tag{8.19}$$

$$\frac{\partial L}{\partial \gamma_1} = \frac{1}{(1+\gamma_1)} \cdot \left(I(z)\left(1 - \left(1 - \frac{2e^{-z}}{(1+e^{-z})}\right)(z - \alpha T_1)\right) - \bar{I}(z)\frac{1}{(1+e^{-\tau'})}(\tau' - \alpha T_1)\right), \tag{8.20}$$

$$\frac{\partial L}{\partial \alpha} = \frac{1}{\alpha} \cdot \left(I(z)\left(1 - z + \frac{2ze^{-z}}{(1+e^{-z})}\right) - \bar{I}(z)\frac{\tau'}{(1+e^{-\tau'})}\right). \tag{8.21}$$

The elements of the Fisher information matrix for an observation at stress rate, ψ_i, $i = 1, 2$, are the negative expectations of these second partial derivatives. The calculation of elements of Fisher information matrix is shown in Appendix B.1.

The Fisher information matrix for an item with ramp rate, ξ, is:

$$F_\xi(\gamma_0, \gamma_1, \alpha) = \begin{bmatrix} E\left[-\frac{\partial^2 L}{\partial \gamma_0^2}\right] & E\left[-\frac{\partial^2 L}{\partial \gamma_0 \partial \gamma_1}\right] & E\left[-\frac{\partial^2 L}{\partial \gamma_0 \partial \alpha}\right] \\ E\left[-\frac{\partial^2 L}{\partial \gamma_0 \partial \gamma_1}\right] & E\left[-\frac{\partial^2 L}{\partial \gamma_1^2}\right] & E\left[-\frac{\partial^2 L}{\partial \gamma_1 \partial \alpha}\right] \\ E\left[-\frac{\partial^2 L}{\partial \gamma_0 \partial \alpha}\right] & E\left[-\frac{\partial^2 L}{\partial \gamma_1 \partial \alpha}\right] & E\left[-\frac{\partial^2 L}{\partial \alpha^2}\right] \end{bmatrix}. \tag{8.22}$$

Since $n\phi_1$ units are tested under stress rate factor $\xi_1 = \psi_1/\psi_2$, and $n\phi_2$ remaining units are tested under stress rate factor $\xi_2 = 1$, the Fisher information matrix for the plan with a sample of n independent items at two stress levels is

$$F = n\phi_1 F_{\xi_1}(\gamma_0, \gamma_1, \alpha) + n\phi_2 F_{\xi_2}(\gamma_0, \gamma_1, \alpha), \tag{8.23}$$

where $n = n\phi_1 + n\phi_2$, $\phi_1 + \phi_2 = 1$.

8.6.3 Asymptotic Variance of the Test Statistic

The asymptotic variance–covariance matrix of the maximum likelihood estimates $\hat{\gamma}_0$, $\hat{\gamma}_1$, and $\hat{\alpha}$ is the inverse of the corresponding Fisher information matrix, i.e.,

$$\Sigma = \begin{bmatrix} \mathrm{Var}[\hat{\gamma}_0] & \mathrm{Cov}[\hat{\gamma}_0, \hat{\gamma}_1] & \mathrm{Cov}[\hat{\gamma}_0, \hat{\alpha}] \\ \mathrm{Cov}[\hat{\gamma}_1, \hat{\gamma}_0] & \mathrm{Var}[\hat{\gamma}_1] & \mathrm{Cov}[\hat{\gamma}_1, \hat{\alpha}] \\ \mathrm{Cov}[\hat{\alpha}, \hat{\gamma}_0] & \mathrm{Cov}[\hat{\alpha}, \hat{\gamma}_1] & \mathrm{Var}[\hat{\alpha}] \end{bmatrix} = F^{-1}, \tag{8.24}$$

using time-censored data.

The asymptotic variance of the test statistic T' is given by

$$\mathrm{AsVar}(T') = \mathrm{AsVar}(\hat{\mu}_0) - 2k\ \mathrm{Ascov}(\hat{\mu}_0, \hat{\sigma}_0) + k^2\mathrm{AsVar}(\hat{\sigma}_0) = \frac{V}{n}, \tag{8.25}$$

where V is obtained in Appendix B.2.

8.6.4 Operating Characteristic (OC) Function

Based on the asymptotic distribution theory, the test statistic T' is asymptotically normally distributed with mean $\mu_0 - k\sigma_0$ and variance $= V/n$. The standardized variate,

$$U = \sqrt{n}\left[T' - (\mu_0 - \mathrm{k}\sigma_0)\right]/\sqrt{V} \sim N(0,1) \text{ as } n \to \infty,$$

where $\mathrm{AsVar}(T') = V/n$.

The OC curve is obtained by plotting L_p against the fraction nonconforming, p, where

$$L_p = P[T' \geq L'] = 1 - \Phi\left[U \geq \frac{(\sigma_0(z_p + k))\sqrt{n}}{\sqrt{V}}\right], \tag{8.26}$$

and $z_p = (L' - \mu_0)/\sigma_0$ is the quantile of the standard logistic distribution corresponding to the fraction nonconforming p, and $\Phi(\cdot)$ is the cdf of standard normal distribution.

8.6.5 Optimal Plans

The optimal plan is obtained using variance optimality criterion. It consists in finding out optimal sample size, lot acceptability constant, low stress rate, and sample proportion allocated to each stress such that producer's risk and consumer's risk are safeguarded by minimizing $V = n\,\text{AsVar}(T')$ (see (8.25)). The high stress rate ψ_2 is known from technical considerations.

8.7 OPTIMUM TIME-CENSORED RAMP-STRESS ALTSP UNDER VARIABLE SSP FOR THE PROCESS PARAMETER BASED ON MEDIAN LIFETIME

In this section, an optimum time-censored ramp-stress ALTSP based on variable single sampling plan for controlling process parameter due to Srivastava and Sharma (2014a) has been discussed. The plan involves hypothesis testing approach for testing whether the median lifetime of the product under study has achieved a required level or not.

The producer may be interested in testing whether the null hypothesis

$$H_0 : \mu \geq \mu_1 \quad (\text{or } \eta \geq \eta_1),$$

is admissible, where $\mu = \log(\eta)$, and μ is the minimum acceptable log of median lifetime, whereas the consumer might also specify some particular value μ_2 less than μ_1, which can be regarded as a clearly rejectable log of median lifetime, and consider the alternative hypothesis

$$H_1 : \mu \leq \mu_2 \quad (\text{or } \eta \leq \eta_2).$$

Since normal distribution belongs to exponential family of distributions, and $\mu = \log(\eta)$ is a monotone increasing function of η, therefore the test based on $\hat{\mu}$ is a **uniformly most powerful (UMP) size-**$\tilde{\alpha}$ test of $H_0 : \mu \geq \mu_1$ versus $H_1 : \mu \leq \mu_2$ (Lehmann, 1986; Mood, Graybill, and Boes, 1974).

The UMP size-$\tilde{\alpha}$ test is:

Reject H_0 if $\hat{\mu} < c$, where c is given as a solution to $P_{H_0}[\hat{\mu} < c] = \tilde{\alpha}$, and $\hat{\mu}$ is asymptotically normally distributed. Now,

$$\tilde{\alpha} = P_{H_0}[\hat{\mu} < c] = \Phi\left(\frac{\sqrt{n}(c-\mu)}{\sqrt{\text{AsVar}(\hat{\mu})}}\right); \tag{8.27}$$

where $\frac{\sqrt{n}(c-\mu)}{\sqrt{\text{AsVar}(\hat{\mu})}} = z_{\tilde{\alpha}}$ is the critical value of standard normal distribution such that the size of critical region is $\tilde{\alpha}$.

The estimates of the design parameters are obtained using the method of maximum likelihood (see Section 8.7.1). The asymptotic variance of the test statistic for deciding about lot disposition is obtained in Section 8.7.2. The behavior of UMP life test is described by OC curve (see Section 8.7.3). The optimal plan consists in finding optimum sample size, lot acceptability constant, sample proportions allocated to each stress, and stress rate factor by minimizing the expected total cost per lot. Bilevel programming approach is used for this purpose.

8.7.1 Log-Likelihood

Just like in 8.6.1 the log-likelihood of a single observation y is,

$$\begin{aligned} L(\gamma_0, \gamma_1, \alpha) = L = I(z)(-\log\sigma - z - 2\log(1+e^{-z})) \\ - \bar{I}(z)\log(1+e^{\tau'}), \end{aligned} \tag{8.28}$$

Let the log-likelihood of unit j be L_j, then log-likelihood, L_0, for n independent observations is $L_0 = L_1 + \cdots + L_n$.

The maximum likelihood estimates $\hat{\gamma}_0, \hat{\gamma}_1$, and $\hat{\alpha}$ are the parameter values that maximize L_0.

8.7.2 Asymptotic Variance of the Test Statistic

Based on the asymptotic distribution theory of MLE, $\hat{\mu}$ which is the function of $(\hat{\gamma}_0, \hat{\gamma}_1, \hat{\alpha})$, say $w(\hat{\gamma}_0, \hat{\gamma}_1, \hat{\alpha})$ is asymptotically normally distributed with mean μ, and asymptotic variance $\text{AsVar}(\hat{\mu}) = \sigma_0/n$, i.e.,

$$\sqrt{n}(\hat{\mu} - \mu) \sim N(0, \sigma_0), \quad \text{where } \mu = \log(\eta), \tag{8.29}$$

and

$$\begin{aligned} \text{AsVar}(\hat{\mu}) &\equiv \text{AsVar}(w(\hat{\gamma}_0, \hat{\gamma}_1, \hat{\alpha})) \\ &= \underset{\sim}{W}' F^{-1} \underset{\sim}{W}, \quad \text{where } \underset{\sim}{W}' = \left[\frac{\partial w}{\partial \hat{\gamma}_0} \; \frac{\partial w}{\partial \hat{\gamma}_1} \; \frac{\partial w}{\partial \hat{\alpha}} \right] \\ &\text{with } w \equiv w(\hat{\gamma}_0, \hat{\gamma}_1, \hat{\alpha}) \end{aligned} \tag{8.30}$$

and Fisher Information Matrix, F is given in (8.22).

8.7.3 OC Function

The behavior of UMP life test can be described by its OC curve which plots the probabilities of lot acceptance versus log median lifetime, μ. The OC function $G_n(\mu) \equiv G(\mu; n, \phi, \xi_1)$ is given by

$$\begin{aligned} G_n(\mu) = P[\hat{\mu} \geq c|\mu] &= \text{P}\left[Z \geq \frac{\sqrt{n}(c-\mu)}{\sqrt{\sigma_0}} \middle| \mu \right], \\ &\text{where } Z \sim N(0,1) \text{ as } n \to \infty, \\ &= 1 - \Phi\left[\frac{\sqrt{n}(c-\mu)}{\sqrt{\sigma_0}} \right], \quad \mu > 0. \end{aligned} \tag{8.31}$$

The lot acceptability constant c is calculated by taking average of the lot acceptability constant w.r.t. producer's risk c_1 and consumer's risk c_2, where $c_1 = \frac{z_{1-\tilde{\alpha}}\sqrt{\sigma_0}}{\sqrt{n}} + \mu_1$ and $c_2 = \frac{z_{\tilde{\beta}}\sqrt{\sigma_0}}{\sqrt{n}} + \mu_2$. Thus, $c = \left(\frac{c_1+c_2}{2}\right)$.

8.7.4 Optimal Plan

The optimal plan is determined by finding out optimum sample size, lot acceptability constant, sample proportion allocated to each stress, and low stress rate by minimizing the expected total cost per lot comprising warranty costs with respect to acceptance or rejection of the lot, sampling cost, and testing cost such that producer's risk and consumer's risk are safeguarded. The high stress rate ψ_2 is known from technical considerations. Bilevel programming approach is used for this purpose.

8.8 COST STRUCTURE

In this section, cost structure that takes into consideration testing cost, sampling cost, and warranty costs based on rejection and acceptance of a lot for adoption of an optimal test planning has been obtained.

The company takes responsibility for the product warranty once the batch has been accepted by the company; and the periods of FRW and PRW (see Section 7.6) are predetermined as c_f, and c_{pw}, respectively. However, the pro-rata warranty of the producer between c_f, and c_{pw} is assumed to vary in time instead of being constant. As a result, the warranty costs of the company could be expressed by a piecewise function of life time X which is given as:

$$\text{Cost}_w(X) = \begin{cases} C_w, & \text{if } x < c_f \\ C_w \dfrac{(c_{pw} - x)}{(c_{pw} - c_f)}, & \text{if } c_f \le x < c_{pw} \\ 0, & \text{if } x \ge c_{pw}, \end{cases} \tag{8.32}$$

where X is the lifetime of the product.

8.8.1 Cost Structure under Variable SSP for Proportion Nonconforming

Since X follows log-logistic distribution, therefore the expected warranty cost per product is given by

$$\begin{aligned} &E(\gamma_0, \gamma_1, \alpha, \phi_1, \phi_2, \xi_1) \\ &\quad = C_w \int_0^{c_f} f(x)dx + C_w \int_{c_f}^{c_p} \frac{(c_{pw} - x)}{(c_{pw} - c_f)} f(x)dx. \end{aligned} \tag{8.33}$$

Thus, the expected cost resulting from selling the acceptance lot is given by

$$\begin{aligned} &W_1(\gamma_0, \gamma_1, \alpha, \phi_1, \phi_2, \xi_1) \\ &\quad = (N - n)\, E(\gamma_0, \gamma_1, \alpha, \phi_1, \phi_2, \xi_1) P[T \geq L' | p_{\tilde{\alpha}}], \end{aligned} \tag{8.34}$$

and the expected rejection cost resulting from the rejecting the lot is given by

$$R_1(\gamma_0, \gamma_1, \alpha, \phi_1, \phi_2, \xi_1) = (N - n) C_{re} P[T < L' | p_{\tilde{\alpha}}], \tag{8.35}$$

where T is taken as T' in time-censored case, and T'' in failure-censored case.

Since the sampling cost C_s, and the testing cost C_τ can both be obtained by simply multiplying the unit cost of sampling with the sample size, and multiplying the unit cost of testing with the censoring time, i.e., 'nC_s', and 'τC_τ', respectively, therefore, the expected total cost per lot for the given sampling plan using time-censored data is given by

$$\begin{aligned} &T\text{Cost}'(n \mid \gamma_0, \gamma_1, \alpha, \phi_1, \phi_2, \xi_1) \\ &\quad = nC_s + \tau C_\tau + W_1(\gamma_0, \gamma_1, \alpha, \phi_1, \phi_2, \xi_1) \\ &\qquad + R_1(\gamma_0, \gamma_1, \alpha, \phi_1, \phi_2, \xi_1). \end{aligned} \tag{8.36}$$

8.8.2 Cost Structure under Variable SSP for the Process Parameter based on Median Lifetime

The expected warranty cost resulting from selling the acceptance lot is given by

$$W_2(\gamma_0, \gamma_1, \alpha, \phi_1, \phi_2, \xi_1) = (N-n)E(\gamma_0, \gamma_1, \alpha, \phi_1, \phi_2, \xi_1)P[\hat{\mu} \geq c|\mu_1], \quad (8.37)$$

and the expected rejection cost resulting from the rejecting the lot is given by

$$R_2(\gamma_0, \gamma_1, \alpha, \phi_1, \phi_2, \xi_1) = (N-n)C_{re}P[\hat{\mu} < c|\mu_1]. \quad (8.38)$$

Therefore, the expected total cost per lot for the given sampling plan using time-censored data is given by

$$\begin{aligned} &T\text{Cost}''(n \mid \gamma_0, \gamma_1, \alpha, \phi_1, \phi_2, \xi_1) \\ &\quad = nC_s + \tau C_\tau + W_2(\gamma_0, \gamma_1, \alpha, \phi_1, \phi_2, \xi_1) \\ &\qquad + R_2(\gamma_0, \gamma_1, \alpha, \phi_1, \phi_2, \xi_1). \end{aligned} \quad (8.39)$$

8.9 DESIGN OF OPTIMAL SAMPLING PLANS

In this section, the optimal sampling plans under ramp-stress based on variable SSP for proportion nonconforming and variable SSP for the process parameter based on median lifetime have been devised.

8.9.1 Design of Optimization Problem for Time-Censored Ramp-Stress ALTSP under Variable SSP for Proportion Nonconforming

In determining an optimal sampling plan (see 8.6.5) for two given points $(p_{\tilde{\alpha}}, 1-\tilde{\alpha})$ and $(p_{\tilde{\beta}}, \tilde{\beta})$ on the OC curve, the two equations:

$$w_{\tilde{\alpha}} - \frac{(\sigma_0(z_{p_{\tilde{\alpha}}} + k))\sqrt{n}}{\sqrt{V}} = 0, \quad (8.40)$$

$$w_{1-\tilde{\beta}} - \frac{(\sigma_0(z_{p_{\tilde{\beta}}} + k))\sqrt{n}}{\sqrt{V}} = 0 \tag{8.41}$$

have to be solved for n^* and k^*. Thus,

$$n^* = \left(\frac{w_{\tilde{\alpha}} - w_{1-\tilde{\beta}}}{z_{p_{\tilde{\alpha}}} - z_{p_{\tilde{\beta}}}} \right) \frac{V}{\sigma_0^2} \tag{8.42}$$

and

$$k^* = \frac{w_{1-\tilde{\beta}} z_{P_{\tilde{\alpha}}} - w_{\tilde{\alpha}} z_{P_{\tilde{\beta}}}}{w_{\tilde{\alpha}} - w_{1-\tilde{\beta}}}. \tag{8.43}$$

The value of k^* is determined by the two points $(p_{\tilde{\alpha}}, 1 - \tilde{\alpha})$ and $(p_{\tilde{\beta}}, \tilde{\beta})$ on the OC curve and n^* depends on these two points and V. See Schneider (1989) for reference.

To minimize total sample size n, it is reasonable to design the test plan so that V is minimized.

Thus, the optimal design problem for obtaining optimal sampling plan under ramp-test with time-censored data is formulated as:

$$\begin{aligned} &\text{Minimize } V \\ \text{s.t. } &0 \le \phi_i \le 1, \ i = 1, 2 \\ &\sum_{i=1}^{2} \phi_i = 1, \\ &\phi_i P_i \ge 0.1, \ i = 1, 2, \\ &0 < \xi_1 < 1. \end{aligned} \tag{8.44}$$

Since the sample size is unknown while minimizing V, therefore instead of giving minimum mean number of failures (MMNF) a pre-specified value in advance, 0.1, say, is assigned to the ratio of MMNF and the sample size. Another suitable number for MMNF can be specified according to cost, time, and precision implications (Meeker and Hahn, 1985; Yang, 1994).

8.9.2 Design of Optimization Problem for Time-Censored Ramp-Stress ALTSP under Variable SSP for the Process Parameter based on Median Lifetime

The cost optimization problem is formulated using bilevel programming approach (see Appendix C.1) for determining optimum sampling plan under ramp-test with time-censored data. In the first stage, optimum low stress rate and sample proportion allocated to each stress level are obtained by minimizing $n/|F|$, i.e., D-optimality based criterion; followed by the second stage in which optimal sample size has been obtained by minimizing the expected total cost per lot of the sampling plan. To obtain an optimal total sample size, it is necessary to make sufficiently low size of risk that the producer and consumer are willing to accept, say at most $\tilde{\alpha}$ and $\tilde{\beta}$, ($\tilde{\alpha}, \tilde{\beta} < 0.5$ where $0 < \tilde{\alpha} < 1, 0 < \tilde{\beta} < 1$), evaluated at the corresponding acceptable and rejectable μ_1 and μ_2. Thus, the minimum sample size is determined by optimally designing ramp-stress ALT so that the constraints $G_n(\mu_1) \geq 1 - \tilde{\alpha}$ and $G_n(\mu_2) \leq \tilde{\beta}$ are satisfied. *Mathematica 8.0* has been used to formulate the optimal plan.

The optimization problem can be formulated as:

$$\begin{cases} & \underset{\phi_1,\xi_1}{\text{Min}} \dfrac{n}{|F|} \\ \text{s.t.} & 0 \leq \phi_i \leq 1, \ i = 1, 2, \\ & \displaystyle\sum_{i=1}^{2} \phi_i = 1, \\ & 0 < \xi_1 < 1, \\ & \text{where } (\phi_1^*, \xi_1^*) \text{ solves} \\ & \begin{cases} & \underset{n}{\text{Min}} \ TCost''(n \mid \gamma_0, \gamma_1, \alpha, \phi_1, \phi_2, \xi_1) \\ \text{s.t.} & G_n(\mu_2) \leq \tilde{\beta}, \\ & N \geq n, \ n \in \mathrm{Z}^+. \end{cases} \end{cases} \tag{8.45}$$

Let,

$$\Psi_1 = \{(\phi_1, \xi_1) : 0 \le \phi_i \le 1, i = 1, 2, \sum_{i=1}^{2} \phi_i = 1, 0 < \xi_1 < 1\},$$

$$\Psi_2 = \{(n) : G_{n,c}(\mu_2) \le \tilde{\beta}, N \ge n, n \in Z^+\},$$

and

$$(\phi_1^*, \xi_1^*) = \underset{\phi_1, \xi_1}{\text{Argmin}} \left\{ \frac{n}{|F|} : (\phi_1, \xi_1) \in \Psi_1 \right\}.$$

Then, the optimization problem reduces to:

$$\underset{n}{\text{Min}} \left\{ T\text{Cost}''(n \mid \gamma_0, \gamma_1, \alpha, \phi_1, \phi_2, \xi_1) : n \in \Omega_2, (\phi_1, \xi_1) = (\phi_1^*, \xi_1^*) \right\}. \tag{8.46}$$

Similarly optimization problem for time-censored ramp-stress ALTSP under variable SSP for proportion non-conforming with cost consideration can be formulated subject to the constraints that the producer's and consumer's risks are safeguarded.

8.10 NUMERICAL EXAMPLES

In this section, the methods described in preceding sections have been explained using numerical examples. The lifetime distribution is assumed to be log-logistic, the inverse power law between lifetime and the voltage stress hold.

8.10.1 Numerical Example on Time-Censored Ramp-Stress ALTSP under Variable SSP for Proportion Nonconforming

The determination of time-censored optimal sampling plan under ramp-test based on SSP for proportion nonconforming depends on α, γ_0, γ_1, s_0, ψ_2, and τ. One must obtain their values from experience, similar data, or a preliminary test. $(p_{\tilde{\alpha}}, 1-\tilde{\alpha})$ and $(p_{\tilde{\beta}}, \tilde{\beta})$ are chosen by taking into consideration producer's and consumer's interests.

The following hypothetical data set has been used:

$$\alpha = 1.26, \quad \gamma_0 = 1.4, \quad \gamma_1 = 0.7, \quad s_0 = 20, \quad \psi_2 = 10, \quad \tau = 20.$$

The following steps yield an optimum ramp-stress ALTSP.

Step 1 (OC Curve):

Choose two points $(p_{\tilde{\alpha}}, 1-\tilde{\alpha})$ and $(p_{\tilde{\beta}}, \tilde{\beta})$ on the OC curve as (0.01, 0.95) and (0.10, 0.10), respectively.

Step 2 (Acceptability Constant):

Compute the acceptability constant k. Since $w_{\tilde{\alpha}} = -1.645$, $w_{1-\tilde{\beta}} = 1.282$, $z_{P\tilde{\alpha}} = 4.595$, and $z_{P_{\tilde{\beta}}} = -2.197$, therefore $k^* = 3.247$.

Step 3 (Optimal Plan):

Determine (ϕ_1^*, ξ_1^*) which minimize V evaluated at $k^* = 3.247$. Let the high stress rate be 10 kV/second.

Consider the following hypothetical data set:

$$\gamma_0 = 1.4, \quad \gamma_1 = 0.7, \quad \alpha = 1.26, \quad \psi_2 = 10, \quad s_0 = 20, \quad \tau = 20$$

for obtaining the optimal values.

Optimal allocation and optimal stress rate factor are obtained by minimizing nAsVar(T), using NMinimize option of *Mathematica 7.0*. These are obtained as $\xi_1^* = 0.567$, $\phi_1^* = 0.459$ and $\phi_2^* = 0.541$.

Thus, the optimum low stress rate is $\psi_1^* = \xi_1^*\psi_2 = 5.67 \approx$ 6 KV/sec. Optimal sample size is 35 and optimal number of units to be tested at low stress rate ψ_1 is 16, and at high stress rate ψ_2 is 19.

Step 4 (Simulated data):

Simulate Type-I censored sample from a simple ramp-test using log-logistic model with high stress rate $\psi_2 = 1$, and low stress rate $\psi_1 = \psi_1^*$. The simulated data set given in Table 8.1 includes 35 simulated observations for low and high stress rates.

Step 5 (Graphical Goodness of Fit):

Figure 8.1 shows the associated log-logistic probability plot at low stress rate. The failure times in Table 8.1 at low stress are arranged in increasing order and are ranked from $1, 2, \ldots, i, \ldots, n$. The plot is obtained by taking these ranked failure times on X-axis, and their corresponding ordinate values, viz., $\eta((1-(i/n+1))^{-1}-1)^{1/\alpha'}$ for failure times, where $\xi_1^* = 0.567$ on Y-axis (Grosh, 1989). These

Table 8.1: Simulated Data: Type-I Censored sample on a simple ramp-test using log-logistic model ($\alpha = 1.26$, $\gamma_0 = 1.4$, $\gamma_1 = 0.7$, $s_0 = 20$, $\psi_2 = 10$, $\tau = 20$, $k^* = 3.247$)

Step-Stress	Failure Times	Number of Censored Units
$\xi_1^* = 0.567$ **(Low Stress)**	1.974, 0.809, 2.284, 1.725, 2.432, 1.895, 0.851, 1.465, 2.112, 1.261, 2.117, 1.567, 2.377, 2.376, 2.048	1
$\xi_2 = 1$ **(High Stress)**	1.891, 2.034, 1.327, 2.287, 0.535, 1.808, 1.381, 0.954, 1.629, 2.001, 1.598, 0.722, 0.177, 2.018, 1.299, 1.509, 1.333, 2.144, 2.142	0

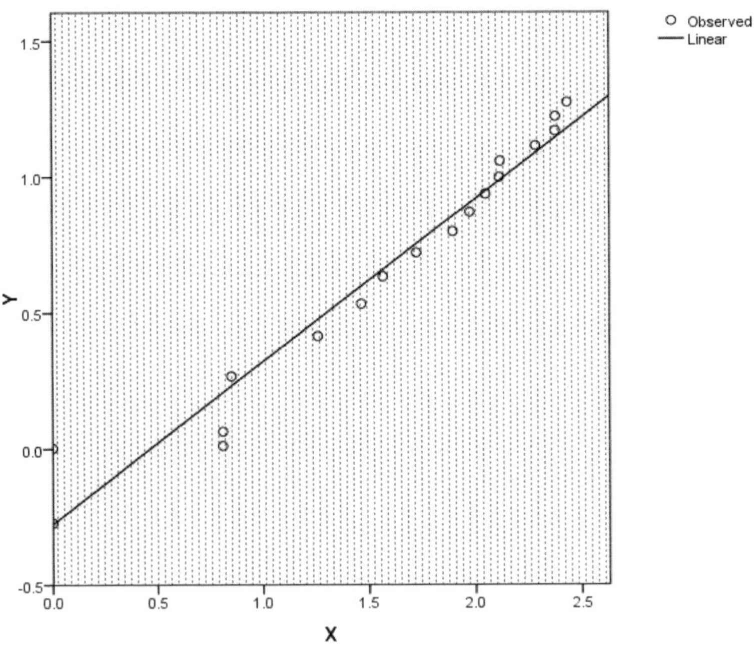

Figure 8.1: The log-logistic probability plot at low stress for the simulated data ($\alpha = 1.26$, $\gamma_0 = 1.4$, $\gamma_1 = 0.7$, $s_0 = 20$, $\psi_2 = 10$, $\tau = 20$, $k^* = 3.247$)

ordinate values are obtained using (8.4). The plotted points tend to follow straight line, which is substantiated by fitting straight line to these points resulting in high value of coefficient of determination $r^2 = 0.988$. The log-logistic distribution, therefore, appears to describe the data adequately.

Similarly, Figure 8.2 shows the associated log-logistic probability plot at high stress. The failure times in Table 8.1 at high stress are arranged in increasing order and are ranked from $1, 2, \ldots, i, \ldots, n$. The plot is obtained by taking these ranked failure times on X-axis, and their corresponding ordinate values, viz., $\eta((1-(i/n+1))^{-1}-1)^{1/\alpha'}$ for failure times, where $\xi_2 = 1$ on Y-axis, (Grosh, 1989). These ordinate values are obtained using (8.4). The plotted points tend to follow straight line, which is substantiated by fitting straight line to these points resulting in high value of coefficient of

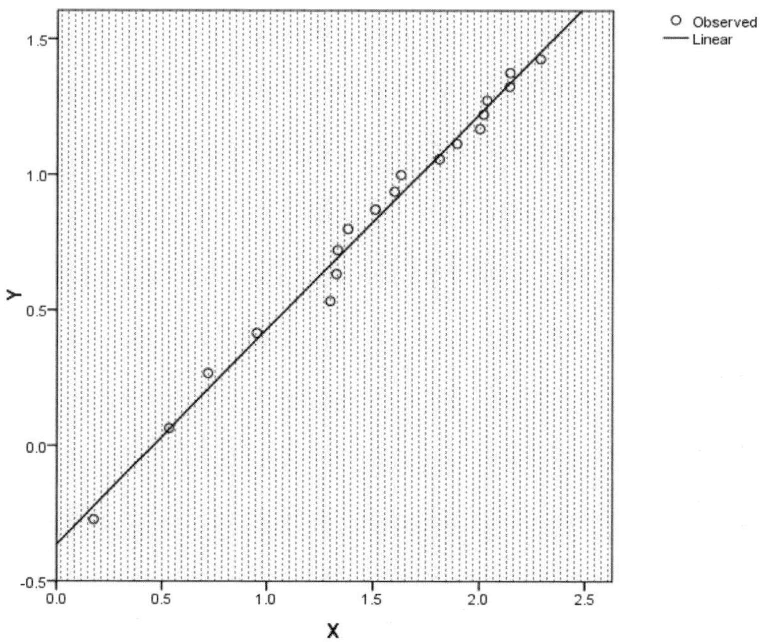

Figure 8.2: The log-logistic probability plot at high stress for the simulated data ($\alpha = 1.26$, $\gamma_0 = 1.4$, $\gamma_1 = 0.7$, $s_0 = 20$, $\psi_2 = 10$, $\tau = 20$, $k^* = 3.247$)

determination $r^2 = 0.955$. The log-logistic distribution, therefore, appears to describe the data adequately.

Step 6 (MLEs of the design parameters):

The MLEs of the design parameters obtained using simulated data in Table 8.1 are

$$\hat{\gamma}_0 = 1.770, \quad \hat{\gamma}_1 = 1.089, \quad \hat{\alpha} = 1.278.$$

These are obtained using the NMaximize option of *Mathematica 7.0.*

Step 7 (Decision Criterion):

Let, the lower specification limit be $L^* = 1.7$ hours, say, indicating that items with lifetime smaller than $\ln L^* = L'$ are nonconforming.

Using the estimated values of the design parameters obtained in step 6, MLEs $\hat{\mu}$ and $\hat{\sigma}$ are computed. For given lower specification limit L', $\hat{\mu}_0 - k^*\hat{\sigma}_0 = \hat{\mu}_0 - 3.247\ \hat{\sigma}_0 = 0.554 > 0.531$, therefore, the lot is rejected.

Figure 8.3 depicts the OC curve obtained using $V^* = 0.523$, $n^* = 35$ and the two points $(p_{\tilde{\alpha}}, 1-\tilde{\alpha})$ and $(p_{\tilde{\beta}}, \tilde{\beta})$ as (0.01, 0.95) and (0.10, 0.10), where z_p varies with p.

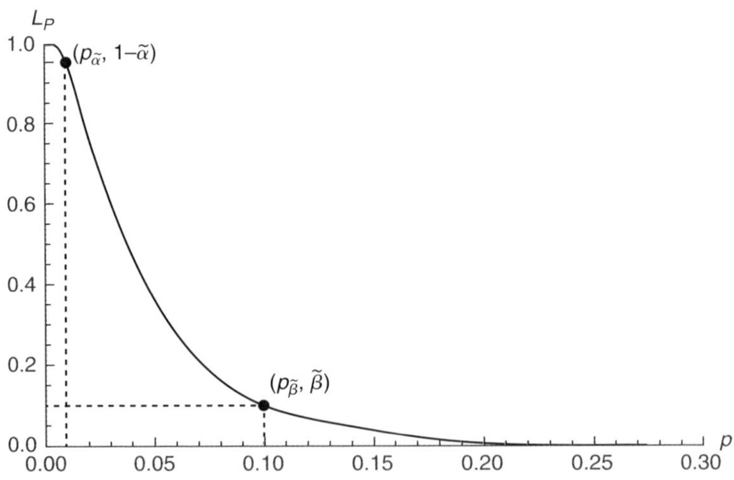

Figure 8.3: OC curve under time-censored variable SSP for proportion nonconforming ($\alpha = 1.26$, $\gamma_0 = 1.4$, $\gamma_1 = 0.7$, $s_0 = 20$, $\psi_2 = 10$, $\tau = 20$, $k^* = 3.247$)

Table 8.2: Optimal ALTSP under time-censored variable SSP for proportion nonconforming when $p_{\tilde{\alpha}}$ and $p_{\tilde{\beta}}$ change ($\alpha = 1.26$, $\gamma_0 = 1.4$, $\gamma_1 = 0.7$, $s_0 = 20$, $\psi_2 = 10$, $\tau = 20$)

$1-\tilde{\alpha}$	$\tilde{\beta}$	$p_{\tilde{\alpha}}$	$p_{\tilde{\beta}}$	k^*	ξ_1^*	ϕ_1^*	ϕ_2^*	V^*	n^*	n_1^*	n_2^*
0.95	0.05	0.001	0.01	5.751	0.413	0.429	0.571	11	101	44	58
			0.05	4.926	0.481	0.429	0.571	9	28	12	16
			0.10	4.552	0.506	0.435	0.565	8	18	8	10
		0.01	0.05	3.770	0.547	0.450	0.550	6	113	51	62
			0.10	3.396	0.562	0.457	0.543	5	47	22	26
			0.15	3.165	0.569	0.461	0.539	5	31	14	17
	0.10	0.001	0.01	5.607	0.425	0.428	0.572	11	77	33	44
			0.05	4.680	0.498	0.433	0.567	8	20	9	12
			0.10	4.260	0.523	0.440	0.560	7	13	6	7
		0.01	0.05	3.667	0.551	0.452	0.548	6	86	39	47
			0.10	3.247	0.567	0.459	0.541	5	36	16	19
			0.15	2.987	0.575	0.464	0.536	5	23	11	12
0.99	0.05	0.001	0.01	5.553	0.430	0.427	0.573	10	140	60	80
			0.05	4.586	0.504	0.434	0.566	8	37	16	21
			0.10	4.148	0.529	0.442	0.558	7	23	10	13
		0.01	0.05	3.628	0.553	0.452	0.548	6	157	71	86
			0.10	3.190	0.569	0.460	0.540	5	65	30	35
			0.15	2.919	0.577	0.465	0.535	5	41	19	22
	0.10	0.001	0.01	5.416	0.442	0.427	0.573	10	112	48	64
			0.05	4.352	0.518	0.438	0.562	7	28	12	16
			0.10	3.870	0.543	0.448	0.552	6	17	8	10
		0.01	0.05	3.531	0.557	0.454	0.546	6	126	57	69
			0.10	3.049	0.573	0.463	0.537	5	51	24	27
			0.15	2.751	0.581	0.467	0.533	4	32	15	17

Table 8.2 presents optimum sampling plan for various values of $p_{\tilde{\alpha}}$ and $p_{\tilde{\beta}}$ when $(1-\tilde{\alpha}, \tilde{\beta}) = (0.95, 0.05)$, (0.95, 0.10), (0.99, 0.05), and (0.99, 0.10). It is observed that for a given $p_{\tilde{\alpha}}$ as $p_{\tilde{\beta}}$ increases n^* and V^* decrease.

Figures 8.4–8.11 obtained using Table 8.2 depict the OC curve of the optimum sampling plan for various values of $p_{\tilde{\alpha}}$ and $p_{\tilde{\beta}}$ when

$$(1-\tilde{\alpha}, \tilde{\beta}) = (0.95, 0.05), (0.95, 0.10), (0.99, 0.05) \text{ and } (0.99, 0.10).$$

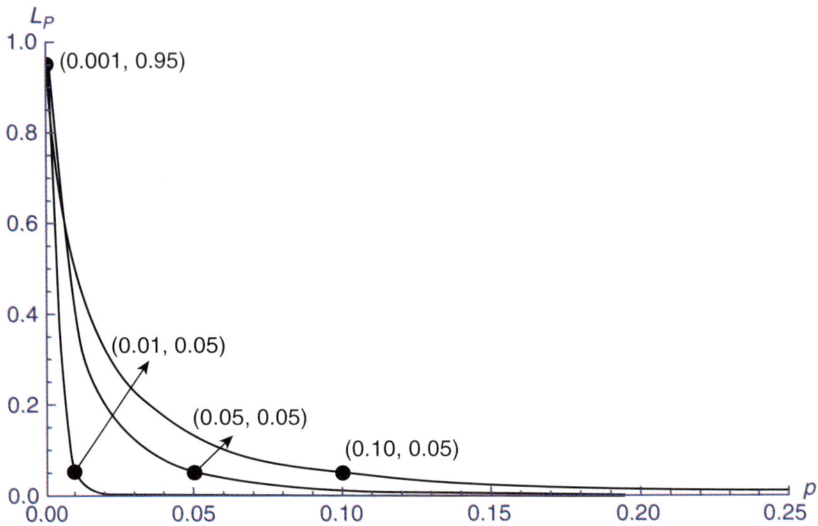

Figure 8.4: OC curves for $(p_{\tilde{\alpha}}, 1 - \tilde{\alpha})$ as (0.001, 0.95) and $(p_{\tilde{\beta}}, \tilde{\beta})$ as (0.01, 0.05), (0.05, 0.05), (0.10, 0.05)

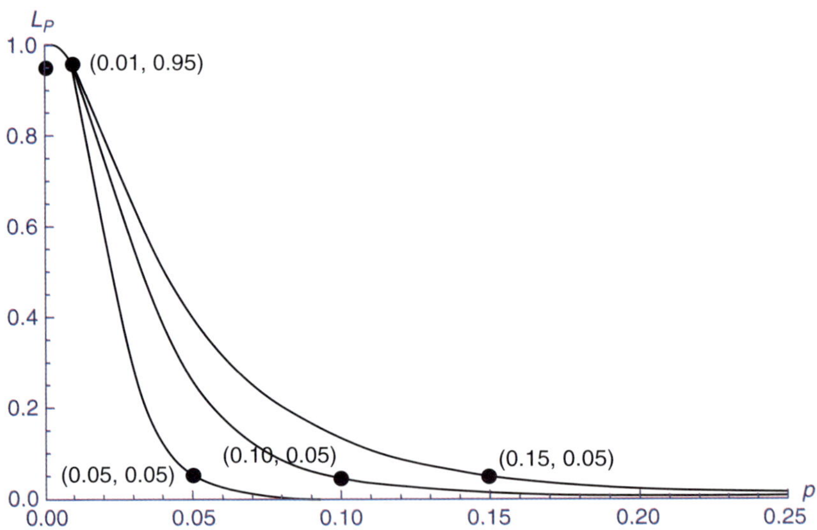

Figure 8.5: OC curves for $(p_{\tilde{\alpha}}, 1 - \tilde{\alpha})$ as (0.01, 0.95) and $(p_{\tilde{\beta}}, \tilde{\beta})$ as (0.05, 0.05), (0.10, 0.05), (0.15, 0.05)

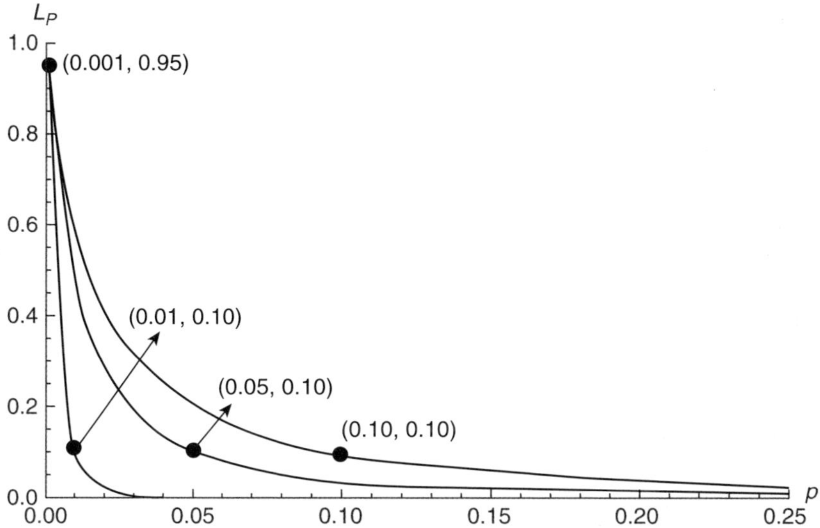

Figure 8.6: OC curves for $(p_{\tilde{\alpha}}, 1 - \tilde{\alpha})$ as (0.001, 0.95) and $(p_{\tilde{\beta}}, \tilde{\beta})$ as (0.01, 0.10), (0.05, 0.10), (0.10, 0.10)

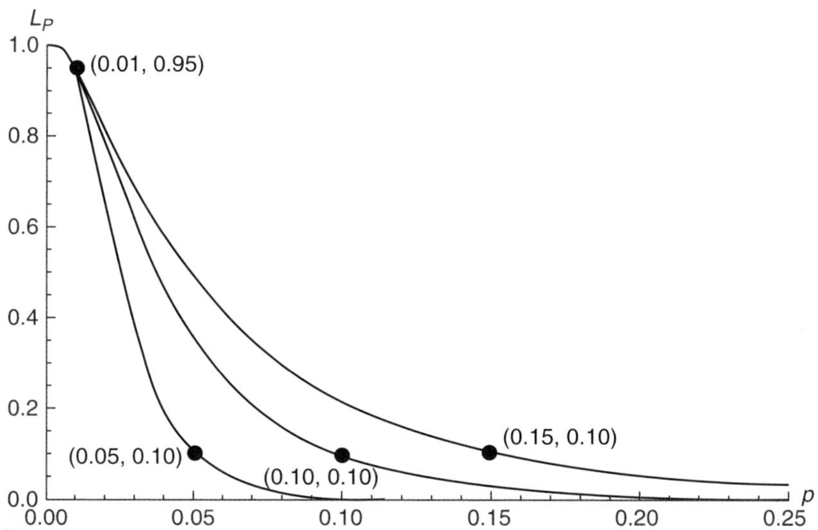

Figure 8.7: OC curves for $(p_{\tilde{\alpha}}, 1 - \tilde{\alpha})$ as (0.01, 0.95) and $(p_{\tilde{\beta}}, \tilde{\beta})$ as (0.05, 0.10), (0.10, 0.10), (0.15, 0.10)

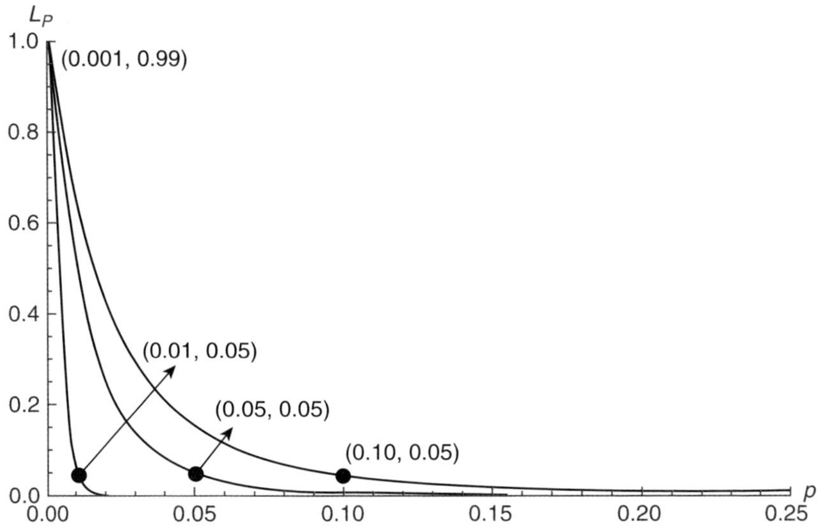

Figure 8.8: OC curves for $(p_{\tilde{\alpha}}, 1 - \tilde{\alpha})$ as (0.001, 0.99) and $(p_{\tilde{\beta}}, \tilde{\beta})$ as (0.01, 0.05), (0.05, 0.05), (0.10, 0.05)

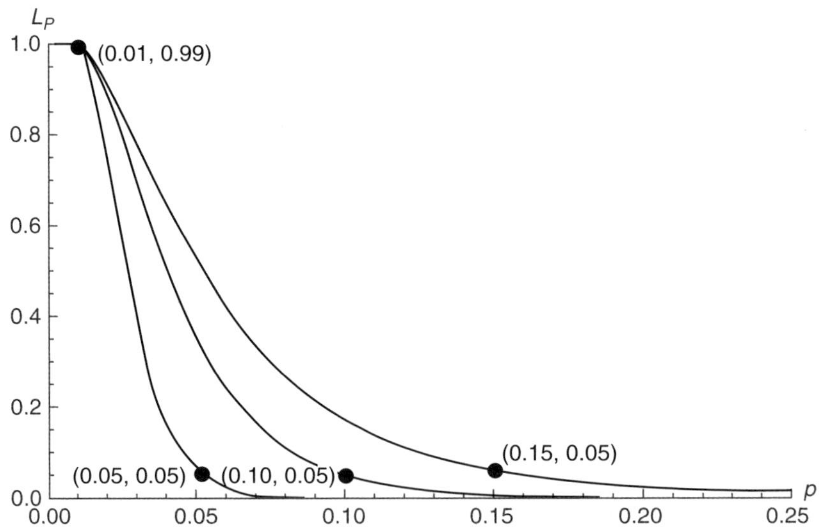

Figure 8.9: OC curves for $(p_{\tilde{\alpha}}, 1 - \tilde{\alpha})$ as (0.01, 0.99) and $(p_{\tilde{\beta}}, \tilde{\beta})$ as (0.05, 0.05), (0.10, 0.05), (0.15, 0.05)

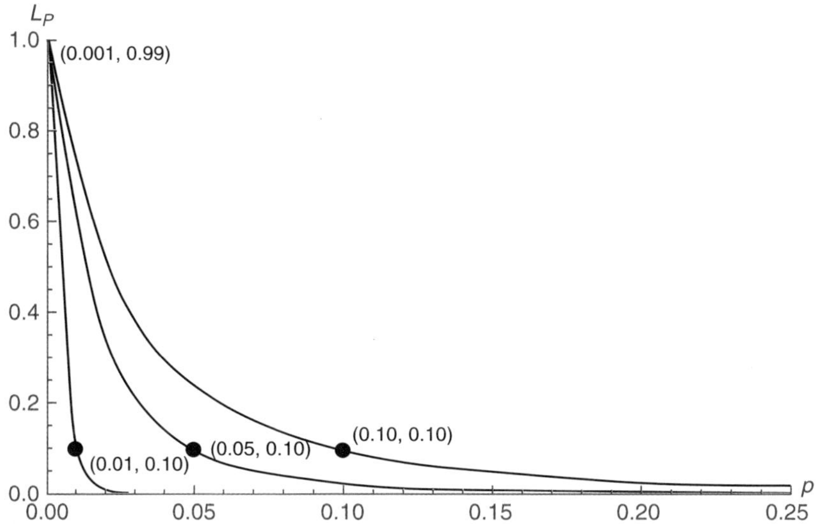

Figure 8.10: OC curves for $(p_{\tilde{\alpha}}, 1 - \tilde{\alpha})$ as (0.001, 0.99) and $(p_{\tilde{\beta}}, \tilde{\beta})$ as (0.01, 0.10), (0.05, 0.10), (0.10, 0.10)

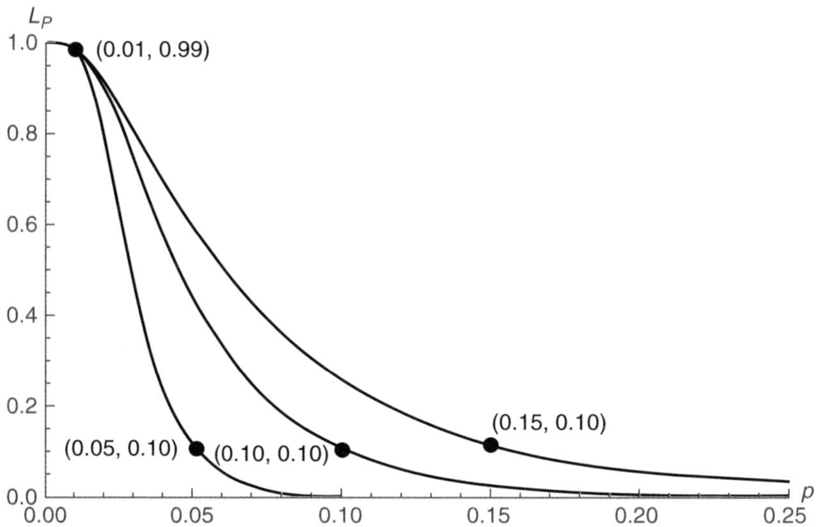

Figure 8.11: OC curves for $(p_{\tilde{\alpha}}, 1 - \tilde{\alpha})$ as (0.01, 0.99) and $(p_{\tilde{\beta}}, \tilde{\beta})$ as (0.05, 0.10), (0.10, 0.10), (0.15, 0.10)

8.10.2 Numerical Example on Time-Censored Ramp-Stress under Variable SSP for the Process Parameter based on Median Lifetime

Here, the method is explained by incorporating various cost factors. Bilevel programming is used to obtain optimal plan.

ABC Corporation is an IC manufacturing company. Since IC products are relatively expensive, it is worthwhile to conduct a more cost saving life testing program in which an appropriate sample size can be found out to determine the acceptance or rejection of a batch of IC products with the objective of cost minimization for the company. However, given that various relevant costs would influence the profits of ABC Corporation in the process of the life test, it would be essential for the managers to select the appropriate decision parameters regarding the accelerated life test sampling plan. See for example, Huang, Hseish, and Ho (2008) and Seo, Jung, and Kim (2009).

The determination of time-censored optimal sampling plan under ramp-stress based on SSP by variable for process parameter depends on α, γ_0, γ_1, s_0, ψ_2, τ, and various cost factors, viz., C_s, C_{re}, C_w, and C_τ. One must obtain their values from experience, similar data, or a preliminary test. $(p_{\tilde{\alpha}}, 1-\tilde{\alpha})$ and $(p_{\tilde{\beta}}, \tilde{\beta})$ are chosen taking in to consideration producer's and consumer's interests.

Steps 1–3

Same as in 8.10.1.

Step 4 (Cost Structure):

Let the product lot size N be 200 and the manager assesses various cost factors, viz., C_s, C_{re}, C_w, and C_τ (in appropriate monetary units) with regard to the lot of IC products. The components are sold under the general rebate warranty policy with $c_f = 10$ and $c_{pw} = 45$ time units.

Thus, the optimal proportion of the sample allocated to low stress, and optimal low stress rate are $\xi_1^* = 0.530$, $\phi_1^* = 0.364$, and $\phi_2^* = 0.635$. The value of optimal sample size n^* is 25 and the optimum lot acceptability constant c^* is 7.934. The optimal cost is 2,149 monetary units, for example dollar.

Step 5 (Decision Criterion):

An agreement between the producer and the consumer is considered to obtain an optimal failure-censored acceptance sampling plan such that the respective probabilities of rejecting a good item and accepting a bad item are at most $\tilde{\alpha} = 0.01$ and $\tilde{\beta} = 0.10$, respectively. In addition, a lot of items are acceptable if the log of median lifetime of items is longer than or equal to $\mu_1 = 10$ days (say, i.e., 240 hours). When the log of median lifetime of items reduces to $\mu_2 = 5$ days (say, i.e., 120 hours) or less, the lot considered is rejected. In this case, the ratio of mean is μ_2/μ_1 is considered as 0.5.

Thus, the manager must accept a lot if $\hat{\mu} \geq 7.934$. The optimal number of samples at low stress rate $n\phi_1^* = 9$ and at high stress rate $n\phi_2^* = 14$.

Table 8.3 presents optimum sample size, lot acceptability constant, and optimum total cost for selected values of $\tilde{\alpha}$, $\tilde{\beta}$, and μ_2/μ_1. From Table 8.3, it is observed that as μ_2/μ_1 increases, the optimal acceptability constant and sample size increase. This is quite evident because, if the probabilities of classifying good lots as bad ones and bad lot as good ones reduce, more empirical information is needed to judge the lots.

Figure 8.12 depicts the plot of $G_n(\mu_1)$ against different values of μ_1 for given pairs $(\tilde{\alpha}, \tilde{\beta}) = (0.01, 0.05)$, (0.01, 0.10), (0.05, 0.05), and (0.05, 0.10).

8.11 SENSITIVITY ANALYSES

Sensitivity analysis identifies the sensitive parameters which need to be estimated with special care for the purpose of minimizing the risk of obtaining an erroneous optimal solution. In this section, sensitivity analyses with respect to ramp-stress ALTSP based on variable SSP for proportion nonconforming have been conducted. The percentage

Table 8.3: Optimal sample size and the corresponding acceptability constant for selected values of $\tilde{\alpha}$, $\tilde{\beta}$, μ_2/μ_1, $\alpha = 3.5$, $\gamma_0 = 5$, $\gamma_1 = 1.2$, $s_0 = 25$, $\psi_2 = 12$, $\tau = 24$, $C_s = 25$, $C_{re} = 120$, $C_w = 250$, $C_\tau = 50$)

$\tilde{\alpha}$	$\tilde{\beta}$	μ_2/μ_1	c_1	c_2	C	n^*	$T\text{Cost}''(n^*)$
0.01	0.05	0.5	12.060	3.543	7.802	22	2081
		0.6	11.735	4.773	8.254	31	2352
		0.7	11.425	5.993	8.709	46	2838
		0.8	11.123	7.206	9.165	74	3813
		0.9	10.811	8.427	9.619	142	5758
	0.10	0.5	11.933	3.935	7.934	25	2149
		0.6	11.657	5.087	8.372	34	2450
		0.7	11.353	6.255	8.804	51	2989
		0.8	11.061	7.416	9.238	83	4061
		0.9	10.762	8.580	9.671	161	5998
0.05	0.05	0.5	11.610	3.390	7.500	18	1963
		0.6	11.366	4.634	8.000	25	2182
		0.7	11.139	5.861	8.500	36	2573
		0.8	10.897	7.103	9.000	58	3368
		0.9	10.500	8.500	9.500	187	5966
	0.10	0.5	11.528	3.810	7.669	20	2025
		0.6	11.291	4.994	8.143	28	2271
		0.7	11.067	6.169	8.618	41	2711
		0.8	11.061	7.416	9.238	83	4061
		0.9	10.762	8.580	9.671	161	5998

deviation (PD), of an optimal setting is measured by

$$\text{PD} = \left(\frac{|Z^{**} - Z^*|}{Z^*}\right) \times 100,$$

where Z^* is the optimal asymptotic variance of T obtained with the given design parameters, and Z^{**} is the one obtained when the parameter is misspecified. Let n^{**} be the optimal sample size resulting from the misspecified parameter.

The effects of % change in the pre-estimated parameters α, γ_0, and γ_1 in terms of the relative change in asymptotic variance of test statistic, T', are presented for time-censored data set in Table 8.4.

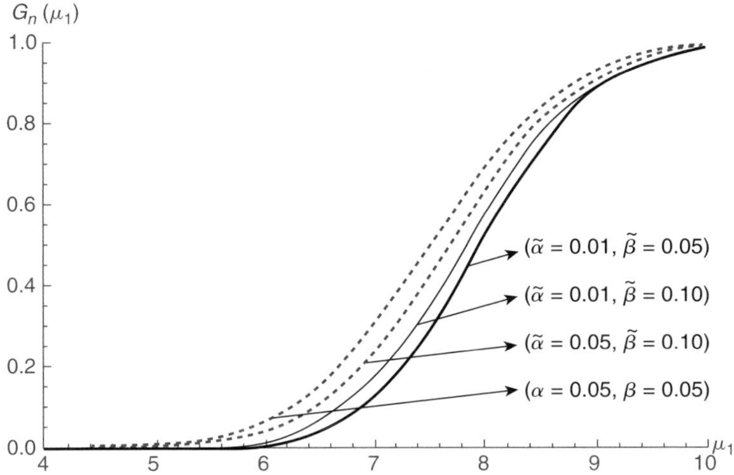

Figure 8.12: UMP-OC curve under time-censored variable SSP for the Process Parameter based on median lifetime ($\alpha = 3.5$, $\gamma_0 = 5$, $\gamma_1 = 1.2$, $s_0 = 25$, $\psi_2 = 12$, $\tau = 24$, $C_s = 25$, $C_{re} = 120$, $C_w = 250$, $C_\tau = 50$)

Table 8.4 shows the optimal test sampling plans for various deviations from the design parameter estimates. It is found that if the absolute relative changes in asymptotic variance are not too far from the true values the increase is small. The result of sensitivity analysis shows that the plan is robust.

8.12 SUMMARY

In this chapter, the formulation of optimum ramp-stress ALTSPs for log-logistic distribution based on variable single sampling plans for controlling proportion nonconforming as well as process parameter have been described. The data type is time-censored. The test is conducted under a fully accelerated environmental condition and mode of inspection used is continuous resulting in exact failure times. Schneider's (1989) approach to k-method is used in variable SSP for proportion nonconforming, and hypothesis testing approach is used for controlling the process parameter based on median life time in variable SSP for process parameter. The optimal plans consist in finding optimum sample size, sample proportion allocated

Table 8.4: Optimal ALTSP when design parameter estimates change ($\alpha = 1.26$, $\gamma_0 = 1.4$, $\gamma_1 = 0.7$, $s_0 = 20$, $\psi_2 = 10$, $\tau = 20$, $Z^* = 0.5230$, $k^* = 3.247$, $n^* = 35$)

% Change	α	γ_0	γ_1	ξ_1^*	ψ_1^*	ϕ_1^*	ϕ_2^*	Z^{**}	n^*	n_1^*	n_2^*	PD
5	1.323	1.470	0.735	0.512	5.122	0.436	0.564	5.244	37	16	21	0.258
−5	1.197	1.330	0.665	0.634	6.337	0.506	0.494	5.208	34	17	17	0.419
6	1.336	1.484	0.742	0.502	5.025	0.433	0.567	5.245	38	16	21	0.285
−6	1.184	1.316	0.658	0.649	6.488	0.520	0.480	5.203	34	18	16	0.516
7	1.348	1.498	0.749	0.493	4.931	0.431	0.569	5.246	38	16	22	0.303
−7	1.172	1.302	0.651	0.665	6.646	0.536	0.464	5.198	33	18	16	0.617
8	1.361	1.512	0.756	0.484	4.842	0.428	0.572	5.247	38	16	22	0.312
−8	1.159	1.288	0.644	0.681	6.811	0.556	0.444	5.193	33	18	15	0.720
9	1.373	1.526	0.763	0.475	4.755	0.426	0.574	5.247	39	16	22	0.310
−9	1.147	1.274	0.637	0.698	6.983	0.579	0.421	5.187	33	19	14	0.826
10	1.134	1.260	0.630	0.467	4.671	0.425	0.575	5.246	39	17	22	0.299
−10	1.134	1.260	0.630	0.71637	7.164	0.606	0.394	5.181	33	20	13	0.934

to each stress, and stress rate factor such that producer's and consumer's interests are safeguarded. Variance optimality criterion is used when cost is not taken into consideration, and bilevel programming approach is used in cost optimization problem. The methods developed have been illustrated using numerical examples, and results of sensitivity analyses carried out with respect to ramp-stress ALTSP based on variable SSP for proportion nonconforming show that the optimal plans are robust to small deviations from the baseline parameter estimates.

CHAPTER 9

OPTIMUM TIME-CENSORED STEP-STRESS PALTSP WITH WARRANTY USING TAMPERED FAILURE RATE MODEL

9.1 INTRODUCTION

The present chapter deals with optimum design of partially accelerated life test sampling plan (PALTSP) using tampered failure rate (TFR) model in step-stress setting.

Under step-stress PALT, a test item is first tested under normal operating conditions and if it does not fail, then it is tested under accelerated conditions until it fails or is censored. In the literature, step-stress PALT has been studied extensively (see Section 1.19.2).

In this chapter, acceptance sampling plan has been coupled with PALT in order to facilitate formulation of decision rules for accepting or rejecting a lot of high reliability items. The design of optimum time-censored step-stress PALTSP has been explained using hazard acceleration and the Burr Type-XII life distribution. The TFR model proposed by Bhattacharyya and Soejoeti (1989) has been used for the purpose (see Section 1.12.2.1). The Burr Type-XII life distribution has been found appropriate for accelerated life testing experiments (see Soliman, 2005). Variable single sampling plan for proportion nonconforming based on Schneider's (1989) approach to k-method has been used (see Section 7.5.2).

The optimum sampling plan consists in finding optimum sample size and optimum stress change point by minimizing expected total cost per lot comprising warranty costs based on general rebate warranty policy (see Section 7.6) with respect to acceptance or rejection of the lot, sampling cost, and testing cost such that the producer's and consumer's interest are safeguarded. Bilevel programming approach is used for the purpose. The method developed has been explained by an example, and sensitivity analysis carried out.

9.1.1 Notation

n^*	total number of test items in a step-stress PALT ($n = n_u + n_a + n_c$)
n_u	number of items failed at use condition
n_a	number of items failed at accelerated condition
n_c	number of censored items
α	scale parameter ($\alpha > 0$)
c, k	shape parameter $c(>0)$, $k(>0)$, (k is known)
A	acceleration factor ($A > 1$)
η_1	censoring time of a step-stress PALT in case of Type-I censoring
T	lifetime of an item at use condition
T_1	test statistic used to decide the acceptability of lot
Y	log-lifetime of an item, $Y = \ln T$
Y_i	observed value of the total log-lifetime Y_i of item i, $i = 1, 2, \ldots, n$
Y^*	total log-life time of an item under TFR model
L^*	lower specification limit on T
L'	$\ln L^*$, Lower specification limit on Y
τ_1	change time in a step-stress PALT, that is $\tau_1 < \eta_1$
P_u	probability of an item fails at use condition in a step PALT, that is, $P_u = (1 - (1 + e^{\tau'})^{-k})$
P_a	probability of an item fails at accelerated condition in a step PALT, that is, $P_a = (1 + e^{\tau'})^{k(A-1)}((1 + e^{\tau'})^{-kA} - (1 + e^{\eta'})^{-kA})$
P_c	$P_c = (1 - P_a - P_u)$ $= ((1+e^{\tau'})^{-k} - (1+e^{\tau'})^{k(A-1)}((1+e^{\tau'})^{-kA} - (1+e^{\eta'})^{-kA}))$
$F_1(\cdot)$	cumulative distribution function (cdf) of the Burr Type-XII distribution

$F(\cdot)$	probability density function (pdf) of the Burr Type-XII distribution
$G(\cdot)$	cdf of $Y = \ln T$
$G(\cdot)$	pdf of $Y = \ln T$
$G_1(\cdot)$	cdf of Y^*
$G_1(\cdot)$	pdf of Y^*
$R(y)$	reliability functions at time y
$H(y)$	hazard (failure) rate at time y
C_s	cost of sampling and sending n items to the testing ground
C_{re}	rejection cost per item, when the whole lot has been rejected
C_η	testing cost per unit time
C_w	warranty cost incurred within w_f
z, τ', η'	$((y-\mu)/\sigma), ((\tau-\mu)/\sigma)$ and $((\eta-\mu)/\sigma)$ are the standardized log-lifetime, standardized log stress change point time, and standardized log censoring time of a unit, respectively.
$p_{\tilde{\alpha}}, p_{\tilde{\beta}}$	fraction nonconforming to be accepted with probability atleast $1-\tilde{\alpha}$, and rejected with probability $\tilde{\beta}$
k_1	lot acceptability constant

9.2 THE MODEL

In this section, formulation of an optimum time-censored step-stress PALT plan for the Burr Type-XII distribution has been discussed (Srivastava and Sharma, 2014b).

9.2.1 Assumptions

1. The total lifetime T of an item is Burr Type-XII life distribution.
2. The lifetime of an item tested at both use and accelerated condition follows Burr Type-XII distribution.
3. The lifetimes of test items are independently and identically distributed random variables.
4. TFR model under step-stress set-up is assumed.

5. Under Type-I censoring, the test terminates when the censoring time 'η' is reached.

9.2.2 Burr Type-XII Life Distribution

The pdf and cdf, respectively of Burr Type-XII distribution are

$$f(t;c,k,\alpha) = \left(\frac{kc}{\alpha}\right)\left(\frac{t}{\alpha}\right)^{c-1}\left(1+\left(\frac{t}{\alpha}\right)^{c}\right)^{-(k+1)}, \quad t \geq 0 \qquad (9.1)$$

and

$$F(t;c,k,\alpha) = 1 - \left(1+\left(\frac{t}{\alpha}\right)^{c}\right)^{-k}, \quad t \geq 0. \qquad (9.2)$$

See Section 1.4.2.9 for details.

9.2.3 Log-Lifetime Distribution

Let for convenience, $Y = \ln T$ then, pdf and cdf of Y, respectively are

$$g_Y(y;\mu,k,\sigma) = \frac{k\exp\{\frac{y-\mu}{\sigma}\}}{\sigma\left(1+\exp\{\frac{y-\mu}{\sigma}\}\right)^{(k+1)}},$$
$$-\infty < y < \infty,\ -\infty < \mu < \infty,\ k > 0,\ \sigma > 0, \qquad (9.3)$$

$$G_Y(y;\mu,k,\sigma) = 1 - \left(1+\exp\left\{\frac{y-\mu}{\sigma}\right\}\right)^{-k},$$
$$-\infty < y < \infty,\ -\infty < \mu < \infty,\ k > 0,\ \sigma > 0, \qquad (9.4)$$

where,

$$\mu = \ln(\alpha) \qquad (9.5)$$

is location parameter, and

$$\sigma = (c)^{-1} \qquad (9.6)$$

is the scale parameter.

The survival function and hazard function are given, respectively by

$$\bar{G}_Y(y) = \left(1 + \exp\left\{\frac{y-\mu}{\sigma}\right\}\right)^{-k},$$
$$-\infty < y < \infty,\ -\infty < \mu < \infty,\ k > 0,\ \sigma > 0, \quad (9.7)$$

$$h(y) = \frac{k \exp\{\frac{y-\mu}{\sigma}\}}{\sigma\left(1 + \exp\{\frac{y-\mu}{\sigma}\}\right)},$$
$$-\infty < y < \infty,\ -\infty < \mu < \infty,\ k > 0,\ \sigma > 0. \quad (9.8)$$

9.2.4 Test Procedure

In the step-stress PALT,

1. All n items are first tested at use condition.
2. If any item out of n items does not fail at use condition by time τ_1, then it is put on accelerated condition and run until it fails or until pre-specified censoring time η_1; whichever occurs earlier.

9.2.5 Lot Acceptance Sampling Procedure

1. Samples of n items are randomly selected from the lot and are tested according to the test procedure in Section 9.2.4.
2. Maximum likelihood estimators (MLEs) $\hat{\mu}$ and $\hat{\sigma}$ of location parameter μ and scale parameter σ, respectively are obtained from the PALT data.
3. The decision about acceptance/rejection of a lot is made using Schneider's approach (Schneider, 1989) to k-method wherein the sample average is replaced by ML estimate, $\hat{\mu}$, of location parameter, μ, and standard deviation is replaced by ML estimate, $\hat{\sigma}$, of scale parameter, σ. In this chapter, k-method has been called k_1-method. The value of test statistic

$$T_1 = \hat{\mu} - k_1\hat{\sigma}$$

is compared with L', and the lot is accepted if $T_1 \geq L'$; and rejected otherwise, where the lower specification limit $L' = \ln L^*$. The sample size n and acceptability constant k_1 are determined according to the D-optimality based criterion. Lots with fraction nonconforming $p \leq p_{\tilde{\alpha}}$ are accepted with probability at least $1 - \tilde{\alpha}$, and lots with $p > p_{\tilde{\beta}}$ are rejected with probability at least $1 - \tilde{\beta}$.

9.2.6 Life Distribution under TFR Model

Consider a step-stress PALT in which an item is tested under normal operating conditions until a fixed time τ_1. If it does not fail, the stress is increased at time τ_1, and held constant till the censoring time η_1. Let $\tau = \ln \tau_1$ and $\eta = \ln \eta_1$. It is assumed that the effect of changing stress is to multiply the initial failure rate function $h(\cdot)$ by an acceleration factor $A(>1)$ subsequent to the change point τ. Let $h^*(y)$ be the failure rate function of the step-stress log-lifetime Y^*, then the TFR model is defined as

$$h^*(y) = \begin{cases} h(y), & \text{if } y \leq \tau \\ A \cdot h(y), & \text{if } y > \tau. \end{cases} \tag{9.9}$$

The acceleration factor, $A(>1)$, depends on the stress levels, and possibly on τ but it is plausible to assume that it does not depend on y (Bhattacharyya and Soejoeti, 1989). The survival function corresponding to (9.9) is given by

$$\bar{G}_1(y) = \begin{cases} \left(1 + \exp\left\{\dfrac{y-\mu}{\sigma}\right\}\right)^{-k}, & \text{if } y \leq \tau \\ \left(1 + \exp\left\{\dfrac{\tau-\mu}{\sigma}\right\}\right)^{k(A-1)} & \\ \qquad \left(1 + \exp\left\{\dfrac{y-\mu}{\sigma}\right\}\right)^{-kA}, & \text{if } y > \tau, \end{cases} \tag{9.10}$$

where $A > 1$ is an acceleration factor (see Bhattacharyya and Soejoeti, 1989).

Therefore, the pdf and cdf of total log-lifetime Y^* of an item is given by

$$g_1(y) = \begin{cases} g_1'(y) = \dfrac{k\exp\{\frac{y-\mu}{\sigma}\}}{\sigma\left(1+\exp\{\frac{y-\mu}{\sigma}\}\right)^{(k+1)}}, & \text{if } y \le \tau \\[2ex] g_1''(y) = \dfrac{kA\exp\{\frac{y-\mu}{\sigma}\}}{\sigma\left(1+\exp\{\frac{y-\mu}{\sigma}\}\right)^{(kA+1)}} \\ \qquad\qquad \left(1+\exp\{\frac{\tau-\mu}{\sigma}\}\right)^{-k(A-1)}, & \text{if } y > \tau \end{cases} \tag{9.11}$$

and

$$G_1(y) = \begin{cases} 1 - \dfrac{1}{\left(1+\exp\{\frac{y-\mu}{\sigma}\}\right)^{k}}, & \text{if } y \le \tau \\[2ex] 1 - \dfrac{1}{\begin{pmatrix} \left(1+\exp\{\frac{\tau-\mu}{\sigma}\}\right)^{-k(A-1)} \\ \left(1+\exp\{\frac{y-\mu}{\sigma}\}\right)^{kA} \end{pmatrix}}, & \text{if } y > \tau. \end{cases} \tag{9.12}$$

9.3 PARAMETER ESTIMATION

Since MLEs are asymptotically normally distributed and asymptotically efficient in many cases, therefore maximum likelihood method is used to estimate α, c, and A, from the test data.

Since the lifetimes $Y_1, \ldots, Y_n$ of n test items are independent and identically distributed random variables, therefore it is sufficient to consider the likelihood function for a single observation.

The likelihood function for the lifetimes of n test items is given by:

$$\begin{aligned}
L &= \prod_{i=1}^{n} L_i(A, \alpha, c, y_i, \delta_{1i}, \delta_{2i}) \\
&= \prod_{i=1}^{n} (g_1'(y_i))^{\delta_{1i}} (g_1''(y_i))^{\delta_{2i}} (\bar{G}_1(\eta))^{\bar{\delta}_{1i}\bar{\delta}_{2i}} \\
&= \prod_{i=1}^{n} \left(\frac{k \exp\{\frac{y_i-\mu}{\sigma}\}}{\sigma\left(1+\exp\{\frac{y_i-\mu}{\sigma}\}\right)^{(k+1)}} \right)^{\delta_{1i}} \\
&\times \left(\frac{kA \exp\{\frac{y_i-\mu}{\sigma}\}}{\sigma\left(1+\exp\{\frac{y_i-\mu}{\sigma}\}\right)^{(kA+1)} \left(1+\exp\{\frac{\tau-\mu}{\sigma}\}\right)^{-k(A-1)}} \right)^{\delta_{2i}} \\
&\times \left(\left(1+\exp\left\{\frac{\eta-\mu}{\sigma}\right\}\right)^{-kA} \left(1+\exp\left\{\frac{\tau-\mu}{\sigma}\right\}\right)^{k(A-1)} \right)^{\bar{\delta}_{1i}\bar{\delta}_{2i}},
\end{aligned} \tag{9.13}$$

where

$$\delta_{1i} = \begin{cases} 1, & y_i \le \tau, i = 1, 2, \ldots, n_u \\ 0, & \text{otherwise} \end{cases}$$

$$\delta_{2i} = \begin{cases} 1, & \tau < y_i \le \eta, \;\; i = n_u + 1, \ldots, n_u + n_a \\ 0, & \text{otherwise} \end{cases}$$

and

$$\bar{\delta}_{1i} = 1 - \delta_{1i}, \quad \bar{\delta}_{2i} = 1 - \delta_{2i}. \tag{9.14}$$

On maximizing the natural logarithm of the above likelihood function, the MLE of A, α, and c can be obtained.

9.4 FISHER INFORMATION MATRIX

The Fisher information F is obtained by taking expectation of the negative of second partial derivatives of $\ln L(A, \alpha, c)$ with respect to A, α, and c.

The Fisher information matrix for an observation is:

$$F = \begin{bmatrix} E\left[-\frac{\partial^2 \ln L}{\partial A^2}\right] & E\left[-\frac{\partial^2 \ln L}{\partial A \partial \alpha}\right] & E\left[-\frac{\partial^2 \ln L}{\partial A \partial c}\right] \\ E\left[-\frac{\partial^2 \ln L}{\partial A \partial \alpha}\right] & E\left[-\frac{\partial^2 \ln L}{\partial \alpha^2}\right] & E\left[-\frac{\partial^2 \ln L}{\partial \alpha \partial c}\right] \\ E\left[-\frac{\partial^2 \ln L}{\partial A \partial c}\right] & E\left[-\frac{\partial^2 \ln L}{\partial \alpha \partial c}\right] & E\left[-\frac{\partial^2 \ln L}{\partial c^2}\right] \end{bmatrix}. \tag{9.15}$$

The calculation of elements of Fisher information matrix is shown in Appendix B.3.

9.5 ASYMPTOTIC VARIANCE OF TEST STATISTIC

For any plan, the asymptotic variance–covariance matrix of the MLEs $\hat{A}, \hat{\alpha}, \hat{c}$ is the inverse of the corresponding Fisher information matrix, that is

$$\sum = F^{-1} = \begin{bmatrix} \mathrm{Var}(\hat{A}) & \mathrm{Cov}(\hat{A}, \hat{\alpha}) & \mathrm{Cov}(\hat{A}, \hat{c}) \\ \mathrm{Cov}(\hat{A}, \hat{\alpha}) & \mathrm{Var}(\hat{\alpha}) & \mathrm{Cov}(\hat{\alpha}, \hat{c}) \\ \mathrm{Cov}(\hat{A}, \hat{c}) & \mathrm{Cov}(\hat{\alpha}, \hat{c}) & \mathrm{Var}(\hat{c}) \end{bmatrix}, \tag{9.16}$$

where F is the Fisher information matrix.

The asymptotic variance of the test statistic T_1 is given by

$$\mathrm{AsVar}(T_1) = \mathrm{AsVar}(\hat{\mu}) - 2k_1\,\mathrm{Ascov}(\hat{\mu}, \hat{\sigma}) + k_1^2\,\mathrm{AsVar}(\hat{\sigma}) = \frac{V}{n}, \tag{9.17}$$

where V is obtained in Appendix B.4.

9.6 OPERATING CHARACTERISTIC (OC) FUNCTION

Based on the asymptotic distribution theory,

$$\frac{T_1 - (\mu - k_1\sigma)}{\sqrt{\mathrm{AsVar}(T_1)}} \sim N(0,1) \quad \text{as } n \to \infty. \tag{9.18}$$

The standardized variate,

$$U = \frac{(T_1 - (\mu - k_1\sigma))\sqrt{n}}{\sqrt{V}} \sim N(0,1) \quad \text{as } n \to \infty. \tag{9.19}$$

The OC curve is obtained by plotting L_p against the fraction nonconforming, p, where

$$L_p = P[T_1 \geq L' \,|\, p] = 1 - \Phi\left[\frac{(\sigma(u_P + k_1))\sqrt{n}}{\sqrt{V}}\right], \tag{9.20}$$

$$u_p = \frac{(L' - \mu)}{\sigma} \tag{9.21}$$

is the quantile obtained using (9.4) corresponding to the fraction nonconforming p, and $\Phi(\cdot)$ is the standard normal distribution function. Therefore, the following equations have to be solved for k_1^* in order to obtain an optimal sampling plan for two given points $(p_{\tilde{\alpha}},\ 1-\tilde{\alpha})$ and $(p_{\tilde{\beta}},\ \tilde{\beta})$ on the OC curve,

$$z_{\tilde{\alpha}} = \frac{(\sigma(u_{P_{\tilde{\alpha}}} + k_1))\sqrt{n}}{\sqrt{V}}, \tag{9.22}$$

$$z_{1-\tilde{\beta}} - \frac{(\sigma(u_{P_{\tilde{\beta}}} + k_1))\sqrt{n}}{\sqrt{V}} = 0, \tag{9.23}$$

where $z_{\tilde{\alpha}}$ and $z_{1-\tilde{\beta}}$ denote the quantiles of the standard normal distribution, $u_{P_{\tilde{\alpha}}}$ and $u_{P_{\tilde{\beta}}}$ are the quantiles obtained using (9.4). Thus,

$$k_1^* = \frac{z_{1-\tilde{\beta}} u_{P_{\tilde{\alpha}}} - z_{\tilde{\alpha}} u_{P_{\tilde{\beta}}}}{z_{\tilde{\alpha}} - z_{1-\tilde{\beta}}}. \tag{9.24}$$

9.7 COST CONSIDERATION

The company takes responsibility for the product warranty once the batch has been accepted by it; and the periods of FRW and PRW are pre-determined as w_f and w_p, respectively. However, the pro-rata warranty of the producer between w_f and w_p is assumed to vary in time instead of being constant. Consequently, the warranty cost of the company could be expressed by a piecewise function of T, which is given by (9.25). It is assumed that after an accept/reject decision, products in the accepted lot are sold under the warranty policy and those in the rejected lot are scrapped or reprocessed.

9.7.1 Expected Warranty Cost Per Item

According to the general rebate warranty policy (see Section 1.14), the cost of accepting an item with lifetime T is

$$\mathrm{Cost}_w(T) = \begin{cases} C_w, & \text{if } T < w_f \\ \dfrac{C_w(w_p - T)}{(w_p - w_f)}, & \text{if } w_f \leq T < w_p \\ 0, & \text{if } T \geq w_p. \end{cases} \tag{9.25}$$

Thus, the expected warranty cost per item is given by

$$\begin{aligned} &E[\mathrm{Cost}_w(T \mid \alpha, c, k, \eta, \tau)] \\ &\quad = C_w \int_0^{w_f} f(t)dt + C_w \int_{w_f}^{w_p} \frac{(w_p - t)}{(w_p - w_f)} f(t)dt. \end{aligned} \tag{9.26}$$

9.7.2 Expected Total Cost Per Lot

The expected warranty cost from selling the accepted lot is

$$\begin{aligned} &W(A, \alpha, c, k, \eta, \tau) \\ &\quad = (N - n)E[\mathrm{Cost}_w(T \mid \alpha, c, k, \eta, \tau)]P[T_1 \geq L' | p \leq p_{\tilde{\alpha}}], \end{aligned} \tag{9.27}$$

and the expected rejection cost that results from rejecting the lot is given by

$$R(A, \alpha, c, k, \eta, \tau) = (N - n)C_{re}P[T_1 < L' | p \leq p_{\tilde{\alpha}}]. \tag{9.28}$$

Therefore, the expected total cost per lot for the given sampling plan is

$$\begin{aligned} &T\text{Cost}(n \mid A, \alpha, c, k, \eta, \tau) \\ &\quad = nC_s + \eta C_\eta + W(A, \alpha, c, k, \eta, \tau) + R(A, \alpha, c, k, \eta, \tau), \end{aligned} \tag{9.29}$$

where the sampling costs and the testing costs are obtained by multiplication of unit cost of sampling with the sample size and multiplying unit cost of testing with the censoring time.

9.8 FORMULATION OF AN OPTIMIZATION PROBLEM

The optimal plan has been obtained using bilevel programming approach (See Appendix C.1). In the first stage, optimum stress change point is obtained by minimizing $n/|F|$, i.e., D-optimality based criterion; followed by the second stage in which optimal sample size has been obtained by minimizing the expected total cost per lot of the sampling plan. To obtain an optimal total sample size, it is necessary to have sufficiently low size of risk that the producer and consumer are willing to accept, say at most $\tilde{\alpha}$ and $\tilde{\beta}$, respectively evaluated at the corresponding acceptable and rejectable nonconforming proportions $p_{\tilde{\alpha}}$ and $p_{\tilde{\beta}}$, where $0 < \tilde{\alpha} < 1$, $0 < \tilde{\beta} < 1$. Thus, the optimum sample size is determined by minimizing the expected total cost per lot subject to the constraints $P[T_1 \geq L'|p \leq p_{\tilde{\alpha}}] \geq 1 - \tilde{\alpha}$ and $P[T_1 \geq L'|p > p_{\tilde{\beta}}] \leq \tilde{\beta}$.

The Fisher information matrix F is a function of A, α, c, k, η, and τ. Thus, the optimal design problem can be formulated as a nonlinear optimization problem:

$$\begin{cases} & \underset{\tau}{\text{Min}}\ \dfrac{n}{|F|} \\ \text{s.t.} & \tau > 0, \tau < \eta, \\ & \text{where } \tau^* \text{ solves} \\ & \begin{cases} & \underset{n}{\text{Min}}\ T\text{Cost}(n \mid A, \alpha, c, k, \eta, \tau^*) \\ \text{s.t.} & P[T_1 \geq L'|p \leq p_{\tilde{\alpha}}] \geq 1 - \tilde{\alpha}, \\ & P[T_1 \geq L'|p > p_{\tilde{\beta}}] \leq \tilde{\beta}, \\ & N \geq n,\ n \in Z^+. \end{cases} \end{cases} \tag{9.30}$$

Let Ψ_1 and Ψ_2 be the solution sets defined as:

$$\Psi_1 = \{\tau : \tau > 0, \tau < \eta\}, \tag{9.31}$$

$$\begin{aligned}\Psi_2 = \{n : P[T_1 \geq L'|p \leq p_{\tilde{\alpha}}] \geq 1 - \tilde{\alpha},\ \ P[T_1 \geq L'|p > p_{\tilde{\beta}}]\\ \leq \tilde{\beta}, N \geq n, n \in Z^+\},\end{aligned} \tag{9.32}$$

and

$$\tau^* = \underset{\tau}{\text{argmin}}\left\{\frac{n}{|F|} : \tau \in \Psi_1\right\}. \tag{9.33}$$

Then, the optimization problem reduces to:

$$\underset{n}{\text{Min}}\,\{T\text{Cost}(n \,|\, A, \alpha, c, k, \eta, \tau^*) : n \in \Psi_2, \tau^* \in \Psi_1\}. \tag{9.34}$$

Since, the optimum criterion depends on k, α, c, A, and η; one must obtain their values from experience, similar data, or a preliminary test to obtain the optimum values of τ and n.

9.8.1 Algorithm

The optimal plan has been obtained by writing codes using *Mathematica* software.

Step 1: Initialization
Initialize the hypothetical data set (A, α, c, k, and η).

Step 2: Variance–Covariance Matrix (F^{-1})

Steps 2.1: Write codes to obtain the Fisher Information Matrix F by taking expectation of the negative of second partial derivatives of $\ln L$ with respect to A, α, and c.

Steps 2.2: Calculate F^{-1} using F in Step 2.1.

Step 3: First Stage Algorithm
The first stage consists in obtaining optimal stress change point using D-optimality based criterion.

Steps 3.1: Minimize $\dfrac{n}{|F|}$ which satisfy $\tau > 0, \tau < \eta$ for the optimal value of τ, i.e., $\underset{\tau}{\text{Min}}\{n/|F| : \tau > 0, \tau < \eta\}$ using NMinimize option of *Mathematica* software.

Step 4: Second Stage Algorithm

At the second stage, decision-maker inclusively considers the optimal value of stress change point τ^* found in the first stage to obtain optimal sample size by minimizing expected total cost per lot subject to producer's and consumer's requirement again using NMinimize option of *Mathematica* software.

Steps 4.1: Minimize $T\text{Cost}(n \,|\, A, \alpha, c, k, \eta, \tau^*)$ which satisfy

$$
\begin{aligned}
&P[T_1 \geq L'|p \leq p_{\tilde{\alpha}}] \geq \ 1 - \tilde{\alpha},\\
&P[T_1 \geq L'|p > p_{\tilde{\beta}}] \leq \tilde{\beta},\\
&N \geq n,\ n \in \mathrm{Z}^+,
\end{aligned}
$$

for the optimal value of the sample size n^*, i.e.,

$$\min_{n} \{T\text{Cost}(n \,|\, A, \alpha, c, k, \eta, \tau^*) : n \in \Psi_2, \tau^* \in \Psi_1\},$$

where Ψ_1 and Ψ_2 are defined in (9.31) and (9.32).

9.9 NUMERICAL EXAMPLE

In this section, a hypothetical step-stress PALTSP experiment is considered to illustrate the method described in this chapter. Let, electronic components used in communications equipment be manufactured in batches of $N = 2{,}000$. The components are sold under the general rebate warranty policy with $w_f = 100$ and $w_p = 450$ time units. From past experience, it is found that the lifetimes of the components possess a Burr Type-XII distribution with parameters (α, c), and A is the acceleration factor. The estimated costs are $C_s = 250$, $C_w = 2500$, $C_\eta = 500$, and $C_{re} = 1200$ (in appropriate monetary units).

Step 1: OC Curve

Choose two points $(p_{\tilde{\alpha}}, 1 - \tilde{\alpha})$ and $(p_{\tilde{\beta}}, \tilde{\beta})$ on the OC curve as (0.01, 0.99) and (0.10, 0.01), respectively.

Step 2: Optimal Lot Acceptability Constant
Compute the acceptability constant k_1, using (9.24). Since $z_{\tilde{\alpha}} = -2.32635$, $z_{1-\tilde{\beta}} = 2.32635$, $u_{p_{\tilde{\alpha}}} = -4.77828$, and $u_{p_{\tilde{\beta}}} = -2.38847$,

$$k_1^* = \frac{(2.32635 \times (-4.77828)) - ((-2.32635) \times (-2.38847))}{(-2.32635 - 2.32635)}$$
$$= 3.58337.$$

Step 3: Optimal Plan
Assume $A = 2.17$, $\alpha = 0.24$, $c = 1.6$, $k = 1.2$, and $\eta = 1.9$, then the optimal stress change point τ obtained by using NMinimize option of *Mathematica* 8.0 is $\tau^* = 0.57658$. The expected total cost and sample size of the optimal sampling plan is $(T\text{Cost}^*, n^*) = (322{,}058, 256)$.

Step 4: Simulated Data
The data in Table 9.1 gives 256 simulated observations based on data $A = 2.17$, $\alpha = 0.24$, $c = 1.6$, $k = 1.2$, $\eta = 1.9$, and $\tau^* = 0.57658$.

Thus, the total number of units tested at use condition is $n_u = 219$, and at accelerated condition it is $n_a = 36$, and $n_c = 1$.

Step 5: Graphical Goodness of Fit
If the data represent a sample from a Burr with $k > 1$, a probability plot of data when $k = 1$ will be concave up, if the data are from a Burr with $k < 1$, a probability plot of data when $k = 1$ will be concave down. Thus, for a probability plot using $k = 1$ that appears to be concave up, the values of k greater than 1 until the plot with best linearity is obtained should be tried. If the plot using $k = 1$ appears to be concave down, the values of k less than one should be attempted. Thus, the Burr distribution is a general purpose distribution that can be applied to all types of failure data, complete or with any censored pattern. See Zimmer (1998) for details.

The failure times at normal use and at accelerated condition given in Table 9.1 are both arranged in increasing

Table 9.1: Simulated data: Type-I censored sample on a simple ramp-test using log-lifetime model ($A = 2.17$, $\alpha = 0.24$, $c = 1.6$, $k = 1.2$, $\eta = 1.9$, $\tau^* = 0.57658$, $n_u = 219$, $n_a = 36$, $n_c = 1$)

Step-Stress	Failure Times							
Use Condition	0.00283	0.00915	0.01161	0.01894	0.02014	0.02516	0.02549	0.02580
	0.02657	0.02690	0.02703	0.02713	0.02757	0.02876	0.03239	0.03282
	0.03329	0.03491	0.03848	0.03994	0.04053	0.04153	0.04200	0.04525
	0.04608	0.04665	0.04753	0.04755	0.04968	0.05209	0.05274	0.05642
	0.05990	0.06145	0.06241	0.06251	0.06344	0.06522	0.06808	0.06981
	0.06998	0.07006	0.07122	0.07312	0.07404	0.07596	0.07826	0.08170
	0.08232	0.08444	0.08459	0.08573	0.08596	0.08650	0.08688	0.08741
	0.09214	0.09229	0.09324	0.09439	0.09536	0.10001	0.10006	0.10446
	0.10776	0.10844	0.10945	0.11059	0.11300	0.11437	0.11571	0.11662
	0.11768	0.11901	0.13076	0.13163	0.13165	0.13221	0.13452	0.14270
	0.14284	0.14436	0.14678	0.14732	0.15263	0.15472	0.15648	0.15660
	0.15735	0.15737	0.15815	0.15989	0.16179	0.16206	0.16394	0.16510
	0.16619	0.16692	0.16811	0.16826	0.16920	0.17096	0.17163	0.17550
	0.17884	0.18170	0.18237	0.18785	0.18864	0.18864	0.18919	0.20130
	0.21967	0.24422	0.25882	0.26131	0.26487	0.26693	0.26769	0.26790
	0.26868	0.27180	0.27658	0.27728	0.28540	0.28944	0.29121	0.29791
	0.30025	0.30213	0.30442	0.30547	0.31524	0.31966	0.32019	0.32334
	0.33913	0.34211	0.34258	0.34279	0.34283	0.34320	0.34518	0.34621

	0.34684	0.34786	0.34892	0.35036	0.35094	0.35115	0.35287	0.35329
	0.35338	0.35368	0.35460	0.36143	0.36232	0.36427	0.36708	0.36979
	0.37240	0.37508	0.37681	0.38239	0.38337	0.38477	0.39317	0.39425
	0.39667	0.40029	0.40294	0.40636	0.41184	0.41611	0.41617	0.42373
	0.43409	0.43586	0.43808	0.44175	0.44296	0.44580	0.44717	0.44747
	0.44758	0.44941	0.45263	0.46282	0.46394	0.46586	0.46919	0.48118
	0.48567	0.48926	0.49600	0.49872	0.49959	0.50226	0.50986	0.51174
	0.51756	0.52413	0.52592	0.52592	0.52840	0.53471	0.53964	0.54161
	0.54240	0.55315	0.55499	0.55760	0.55977	0.56287	0.56594	0.56629
	0.56723	0.56735	0.56966					
Accelerated	0.59341	0.61228	0.61634	0.63482	0.63661	0.63854	0.63952	0.64320
Condition	0.65171	0.65512	0.66952	0.67729	0.69128	0.69536	0.69931	0.70272
	0.73002	0.73043	0.74619	0.76953	0.79165	0.79175	0.80127	0.81722
	0.84613	0.87897	0.92822	0.99774	1.00149	1.03479	1.03647	1.11834
	1.15986	1.17971	1.52557	1.60843				

order, and are ranked. The plot is obtained by taking these ranked failure times on X-axis, and their corresponding ordinate values, viz,

$$\mu + \sigma \ln\left(\left(1 - \frac{i}{(n+1)}\right)^{-1/k} - 1\right)$$

at normal operating condition, and

$$\mu + \sigma \ln\left(\left(\left(1 - \frac{i}{(n+1)}\right) \times \left(1 + \exp\left\{\frac{\tau - \mu}{\sigma}\right\}\right)^{-k(A-1)}\right)^{\frac{-1}{kA}} - 1\right)$$

at accelerated condition, where $\tau^* = 0.57658$ on Y-axis, (Grosh, 1989). These ordinate values are obtained using (9.12). The plotted points tend to follow straight line which is substantiated by fitting straight line to these points resulting in high value of coefficient of determination $\boldsymbol{r^2} = \mathbf{0.977}$. Therefore, for the simulated data under Type-I censoring, the Burr Type-XII distribution appears to describe the data adequately.

Step 6: MLEs of the Design Parameters

The MLEs of the model parameters α and c obtained using simulated data given in Table 9.1 are $\hat{\alpha} = 0.11012$ and $\hat{c} = 2.42496$.

These are obtained using the NMaximize option of *Mathematica* 8.0.

Step 7: Decision Criterion

Consider the lower specification limit as L^* indicating that items with lifetime smaller than $\ln L^*$ are nonconforming.

Using the estimated values of the design parameters obtained in step 6, MLEs $\hat{\mu}$ and $\hat{\sigma}$ are computed. For given lower specification limit L', $\hat{\mu} - k_1\hat{\sigma} = \hat{\mu} - 3.58337\hat{\sigma} > L'$, lot is accepted; otherwise rejected.

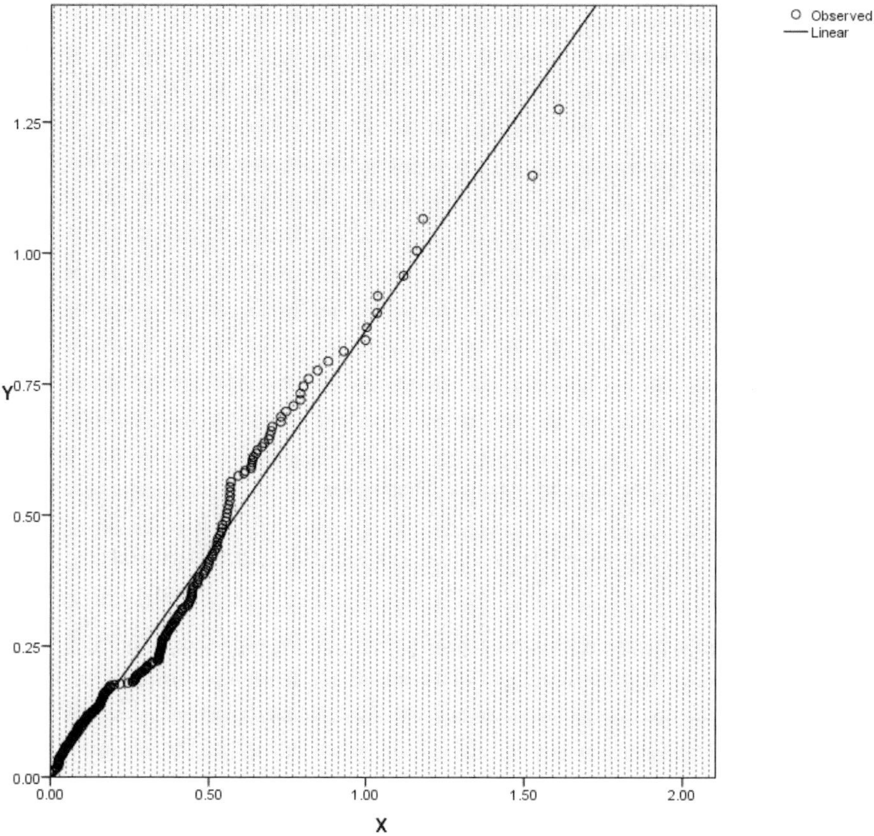

Figure 9.1: Log-lifetime distribution probability plot for the simulated data under step-stress PALT

Figure 9.2 depicts the OC curve obtained from u_P, V, n^*, and the two points $(p_{\tilde{\alpha}}, 1-\tilde{\alpha})$ and $(p_{\tilde{\beta}}, \tilde{\beta})$ taken as (0.01, 0.99) and (0.10, 0.01), respectively.

Tables 9.2 and 9.3 present optimum sampling plans for various values of $p_{\tilde{\alpha}}$ and $p_{\tilde{\beta}}$ when $(1-\tilde{\alpha},\ \tilde{\beta}) = (0.95, 0.10)$ and (0.99, 0.01). It is observed that for given $p_{\tilde{\alpha}}$ as $p_{\tilde{\beta}}$ increases, n^* and V^* decrease. Table 9.4 shows that the increase in censoring time leads to increase in optimal stress change point.

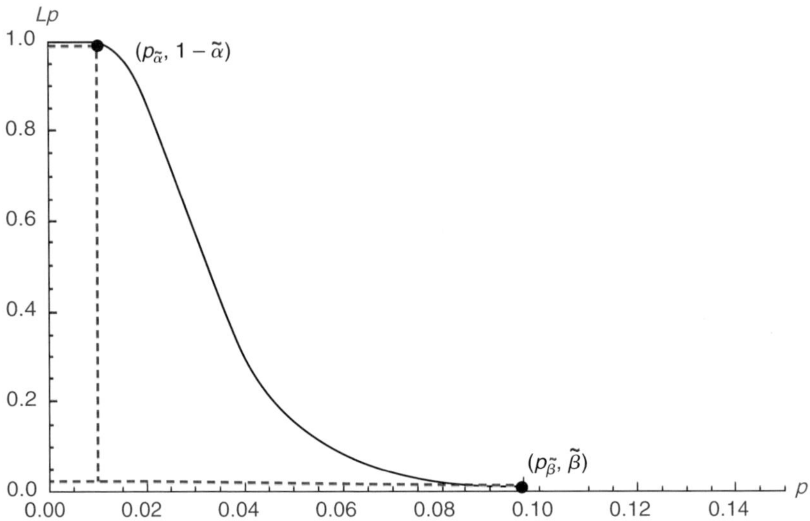

Figure 9.2: OC curve with $(p_{\tilde{\alpha}}, 1 - \tilde{\alpha})$ as (0.01, 0.99) and $(p_{\tilde{\beta}}, \tilde{\beta})$ as (0.10, 0.01)

Table 9.2: Optimal ALTSP ($A = 2.17$, $\alpha = 0.24$, $c = 1.6$, $k = 1.2$, $\eta = 1.9$, $1 - \tilde{\alpha} = 0.95$ and $\tilde{\beta} = 0.10$) when $p_{\tilde{\alpha}}$ and $p_{\tilde{\beta}}$ change

$1 - \tilde{\alpha}$	$\tilde{\beta}$	$p_{\tilde{\alpha}}$	$p_{\tilde{\beta}}$	k_1	V^*	n^*	TCost(A,α,c)
0.95	0.10	0.001	0.01	5.79028	51.67260	451	349,719
			0.05	4.86442	36.35710	159	308,733
			0.10	4.44703	30.35870	105	301,584
		0.01	0.05	3.85243	22.76600	409	343,699
			0.10	3.43503	18.12170	202	314,573
			0.15	3.17779	15.53740	137	305,777

9.10 SENSITIVITY ANALYSIS

To formulate an optimum test plan, one needs information about acceleration factor, A, and estimates of the design parameters α and c. These estimates sometimes may significantly affect the values of the resulting decision variables. Therefore, it is important to conduct sensitivity analysis to evaluate the robustness of the resulting PALTSP.

Table 9.3: Optimal ALTSP ($A = 2.17, \alpha = 0.24, c = 1.6, k = 1.2$ and $\eta = 1.9, 1 - \tilde{\alpha} = 0.99$ and $\tilde{\beta} = 0.01$) when $p_{\tilde{\alpha}}$ and $p_{\tilde{\beta}}$ change

$1-\tilde{\alpha}$	$\tilde{\beta}$	$p_{\tilde{\alpha}}$	$p_{\tilde{\beta}}$	k_1	V^*	n^*	$TCost(A, \alpha, c)$
0.99	0.01	0.001	0.01	5.93372	54.30150	564	363,407
			0.05	5.11011	40.15740	205	315,005
			0.10	4.73881	34.50050	140	306,204
		0.01	0.05	3.95467	23.99900	497	356,375
			0.10	3.58337	19.70620	256	322,058
			0.15	3.35454	17.29460	178	311,368

Table 9.4: Optimal ALTSP ($A = 2.17$, $\alpha = 0.24$, $c = 1.6$, $k = 1.2$) when censoring time changes

η	τ^*
1.2	0.55777
1.4	0.56735
1.6	0.57260
1.8	0.57559
2.0	0.57736
2.2	0.57846
2.4	0.57916

In this section, how the optimal solution changes when the associated parameters' change is discussed. The analysis is performed by changing only one parameter at a time and keeping others unchanged. The sensitivity of the optimal test plan against the uncertain model parameter estimates is investigated. Table 9.5 summarizes the optimal test plans for various deviations from the design parameter estimates. The PDs of the resulting optimal settings are measured by

$$\mathrm{PD} = \left(\frac{|Z^{**} - Z^*|}{Z^*}\right) \times 100,$$

Table 9.5: Optimal ALTSP (A = 2.17, α = 0.24, c = 1.6, k = 1.2, and η = 1.9) when design parameter estimates change

% Change	α	τ^*	PD	c	τ^*	PD
1	0.24240	0.58218	2.04009	1.61600	0.57193	0.05780
−1	0.23760	0.57099	2.01809	1.58400	0.58135	0.06263
2	0.24480	0.58777	4.10226	1.63200	0.56738	0.11113
−2	0.23520	0.56537	4.01425	1.56800	0.58622	0.13048
3	0.24720	0.59335	6.18660	1.64800	0.56293	0.16032
−3	0.23280	0.55976	5.98858	1.55200	0.59120	0.20398
4	0.24960	0.59892	8.29320	1.66400	0.55857	0.20569
−4	0.23040	0.55413	7.94114	1.53600	0.59629	0.28357
5	0.25200	0.60448	10.42210	1.68000	0.55432	0.24753
−5	0.22800	0.54850	9.87201	1.52000	0.60150	0.36973

where Z^* is the setting obtained with the given design parameters, and Z^{**} is the one obtained when the parameter is misspecified. Table 9.5 shows that irrespective of whether the incorrect variance is smaller or larger than the true variance, the optimum plan is robust as the PD in variance is small.

9.11 SUMMARY

The present chapter has delved upon optimum design of time-censored step-stress PALTSP based on hazard acceleration with Burr Type-XII life distribution. Variable single sampling plan for proportion nonconforming based on k-method has been used. Optimum sampling plan consists in finding optimum sample size and optimum stress change point by minimizing expected total cost per lot. The expected total cost comprises warranty costs with respect to acceptance or rejection of the lot, sampling cost, and testing cost such that the producer's and consumer's interest are safeguarded. Bilevel programming approach is used for this purpose. The procedure developed has been explained using a numerical example. The results of sensitivity analysis show that the optimal sampling plan is robust to small deviations in baseline parameter estimates.

CHAPTER 10

OPTIMUM TIME-CENSORED STEP-STRESS PALTSP WITH COMPETING CAUSES OF FAILURE USING TAMPERED FAILURE RATE MODEL

10.1 INTRODUCTION

In this chapter, design of optimum time-censored step-stress partially accelerated life test sampling plan with competing causes of failure proposed by Srivastava and Sharma (2015b) has been discussed. Variable repetitive group sampling plan (VRGSP) has been used. Prior to this, VRGSP under normal operating conditions have been proposed by Balamurali and Jun (2006) for normal distribution, and Jun *et al.* (2010) for Weibull distribution (see Section 1.4.2.4). Optimum stress change point and optimum sample size have been obtained by minimizing the average sample number of a lot such that producer's and consumer's interests are safeguarded. Bilevel programming approach is used for the purpose (see Appendix C.1). The method developed has been explained using an example. The effects of mis-specification of pre-estimates of the design parameters have also been investigated.

10.1.1 Notation

n number of test units

n_i number of failed units at stress x_i, $i = 1, 2$

n_c number of units censored $n_c = n - n_1 - n_2$

n_{1j} the number of units that fail before τ due to the failure cause j, $j = 1, 2$

n_{2j} the number of units that fail after τ due to the failure cause j, $j = 1, 2$

10.2 MODEL

10.2.1 Assumptions

1. Each unit has two statistically independent potential failure times corresponding to two causes of failure.
2. Failure time of a unit is the smallest of its two potential failure times.
3. The lifetime of an item tested at both use and at accelerated condition follows a Weibull distribution.
4. For each failure cause, a TFR model (Bhattacharyya and Soejoeti, 1989, see Section 1.12.2.1 also) is assumed.

10.2.2 Test Procedure

1. All n items are first tested at use condition.
2. If any item out of n items does not fail at use condition by pre-specified time τ, then it is put on accelerated condition, and run until censoring time η.
3. The test is continued until:
   (a) all test items fail, or
   (b) a prescribed censoring time η

   whichever occurs earlier, and the test conditions remain the same.
4. Failure times and failure causes of test units are jointly observed continuously.

10.2.3 Model Formulation

The cdf of Weibull distribution with shape parameter δ and scale parameter θ is given by:

$$G'(t) = 1 - \exp\left\{-\frac{t^{\delta}}{\theta}\right\}, \quad 0 \le t < \infty \tag{10.1}$$

Thus, the cdf of lifetime T_j due to failure cause j ($j = 1, 2$) under normal operating condition is

$$H_j(t) = 1 - \exp\left\{-\frac{t^\delta}{\theta_j}\right\}, \quad 0 \le t < \infty, \; j = 1, 2 \tag{10.2}$$

and the failure time of a unit is the smallest of its two potential failure times, i.e., $T = \min\{T_1, T_2\}$ with cdf given by:

$$H(t) = 1-(1-H_1(t))(1-H_2(t)) = 1-\exp\{-(\theta_1^{-1}+\theta_2^{-1})t^\delta\} \tag{10.3}$$

which is again Weibull with scale parameter $\theta_\bullet^{-1} = \theta_1^{-1} + \theta_2^{-1}$, and shape parameter δ.

It is assumed that there is a lower specification, L^*, so that an item having the lifetime, T, less than L^* is regarded as nonconforming. Using (10.1), the fraction nonconforming (or unreliability at time L) is obtained as:

$$p = P[T < L^*] = 1 - \exp\left\{-\frac{L^{*\delta}}{\theta_\bullet}\right\}, \quad 0 \le t < \infty. \tag{10.4}$$

(10.4) can be re-written as:

$$p = 1 - \exp\{-w'\}, \tag{10.5}$$

where $w' = L^{*\delta}/\theta_\bullet = -\ln(1-p)$. A part is considered as "good" if the unreliability is low. Also, the mean life is given by:

$$\mu = \int_0^\infty \exp\left\{-\frac{t^\delta}{\theta_\bullet}\right\} dt = \frac{\Gamma(1/\delta)\theta_\bullet^{1/\delta}}{\delta}. \tag{10.6}$$

Under partially accelerated environmental condition, using the TFR model and Weibull distributed life assumptions, the cdf of lifetime W_j due to failure cause $j = 1, 2$ is given by:

$$G_j(w) = G_j(w; \theta_j)$$
$$= \begin{cases} 1 - \exp\left\{-\dfrac{w^\delta}{\theta_j}\right\}, & \text{if } 0 < w < \tau \\ 1 - \exp\left\{-\dfrac{A(w^\delta - \tau^\delta)}{\theta_j}\right\}, & \text{if } \tau \le w < \infty \end{cases} \tag{10.7}$$

for $j = 1, 2$, and the corresponding pdf of W_j is given by

$$g_j(w) = g_j(w;\theta_j)$$
$$= \begin{cases} \dfrac{\delta w^{\delta-1}}{\theta_j} \exp\left\{-\dfrac{w^\delta}{\theta_j}\right\}, & \text{if } 0 < w < \tau \\ \dfrac{\delta A w^{\delta-1}}{\theta_j} \exp\left\{-\dfrac{\tau^\delta}{\theta_j}\right\} \exp\left\{-\dfrac{A(w^\delta - \tau^\delta)}{\theta_j}\right\}, & \text{if } \tau \le w < \infty \end{cases} \tag{10.8}$$

for $j = 1, 2$ and acceleration factor, $A > 1$. Let $W = \min\{W_1, W_2\}$ denote the overall failure time of a test unit. Then, its cdf and pdf are obtained to be

$$F(w;\theta_1,\theta_2)$$
$$= 1 - (1 - G_1(w))(1 - G_2(w))$$
$$= \begin{cases} 1 - \exp\{-(\theta_1^{-1} + \theta_2^{-1})w^\delta\}, & \text{if } 0 \le w < \tau \\ 1 - \exp\{-(\theta_1^{-1} + \theta_2^{-1})\tau^\delta\} & \\ \exp\{-A(\theta_1^{-1} + \theta_2^{-1})(w^\delta - \tau^\delta)\}, & \text{if } \tau \le w < \infty, \end{cases} \tag{10.9}$$

$$f(w;\theta_1,\theta_2)$$
$$= \begin{cases} \delta w^{\delta-1}(\theta_1^{-1} + \theta_2^{-1}) \exp\{-(\theta_1^{-1} + \theta_2^{-1})w^\delta\}, & \text{if } 0 \le w < \tau \\ \delta A w^{\delta-1}(\theta_1^{-1} + \theta_2^{-1}) \exp\{-(\theta_1^{-1} + \theta_2^{-1})w^\delta\} & \\ \exp\{-A((\theta_1^{-1} + \theta_2^{-1})(w^\delta - \tau^\delta)\}, & \text{if } \tau \le w < \infty, \end{cases} \tag{10.10}$$

$$= \begin{cases} \delta w^{\delta-1}\theta_\bullet^{-1} \exp\left\{-\dfrac{w^\delta}{\theta_\bullet}\right\}, & \text{if } 0 < w < \tau \\ \delta A w^{\delta-1}\theta_\bullet^{-1} \exp\left\{-\dfrac{\tau^\delta}{\theta_\bullet}\right\} \exp\left\{-\dfrac{A(w^\delta - \tau^\delta)}{\theta_\bullet}\right\}, & \text{if } \tau \le w < \infty, \end{cases} \tag{10.11}$$

respectively, where $\theta_\bullet^{-1} = \theta_1^{-1} + \theta_2^{-1}$. Furthermore, let j denote the indicator for the cause of failure. Then, under the assumptions, the joint pdf of (W, C) is given by

$$\begin{aligned} f_{w,c}(w,j) &= g_j(w)(1-G_{j'}(w)) \\ &= \begin{cases} \delta w^{\delta-1}\theta_j^{-1}\exp\Big\{-\dfrac{w^\delta}{\theta_\bullet}\Big\}, & \text{if } 0<w<\tau \\ \delta A w^{\delta-1}\theta_j^{-1}\exp\Big\{-\dfrac{\tau^\delta}{\theta_\bullet}\Big\}\exp\Big\{-\dfrac{A(w^\delta-\tau^\delta)}{\theta_\bullet}\Big\}, & \text{if } \tau\le w<\infty \end{cases} \end{aligned} \quad (10.12)$$

for $j, j' = 1, 2$ and $j' \neq j$.

Let the relative risk imposed on a test unit before τ due to failure cause j be

$$\pi_{1j} = \Pr[C = j | 0 < W < \tau] = \frac{\theta_j^{-1}}{\theta_1^{-1} + \theta_2^{-1}}, \quad j = 1, 2. \quad (10.13)$$

Similarly, let the relative risk after τ due to the cause j be denoted by

$$\pi_{2j} = \Pr[C = j | W \geq \tau] = \frac{\theta_j^{-1}}{\theta_1^{-1} + \theta_2^{-1}} = \pi_{1j}, \quad j = 1, 2. \quad (10.14)$$

These are simply the proportion of failure rates in the given time frame. It follows that W and C are independent given the time frame in which a failure has occurred.

10.3 LOT ACCEPTANCE PROCEDURE

In this section, procedure underlying single sampling plan (SSP) and VRGSP have been explained in Sections 10.3.1 and 10.3.2, respectively.

10.3.1 SSP

1. Draw a random sample of size n from a lot.
2. Calculate the quantity $\nu = \hat{\mu}$, where $\hat{\mu}$ is the estimate of mean lifetime, μ, given in (10.6). Accept the lot if $\nu \geq L^*k$ and reject the lot if $\nu < L^*k$.

10.3.2 VRGSP

The VRGSP for the Weibull distribution is as follows:

1. Take a random sample of size n from a lot.
2. Calculate the quantity $\nu = \hat{\mu}$, where $\hat{\mu}$ is the estimate of mean lifetime, μ, given in (10.6). Accept the lot if $\nu \geq L^* k_a$ and reject the lot if $\nu < L^* k_r$.
3. If $L^* k_r \leq \nu < L^* k_a$, then repeat above steps (1)–(2) through re-sampling.

10.4 LOG-LIKELIHOOD

Under the assumption of the TFR model, the likelihood function of θ based on the Type-I censored sample, using (10.9) and (10.12), is

$$\begin{aligned} L(\theta) &= \prod_{j=1}^{2}\prod_{i=1}^{n_{1j}} f(w_i, c_i) \prod_{j=1}^{2}\prod_{i=1}^{n_{2j}} f(w_i, c_i)(1 - F(T))^{n_c} \\ &= \prod_{i=1}^{n_{11}} \left(\frac{\delta w_i^{\delta-1}\theta_1^{-1}}{\exp\left\{\frac{w_i^\delta}{\theta_\bullet}\right\}} \right) \times \prod_{i=1}^{n_{12}} \left(\frac{\delta w_i^{\delta-1}\theta_2^{-1}}{\exp\left\{\frac{w_i^\delta}{\theta_\bullet}\right\}} \right) \\ &\quad \times \prod_{i=1}^{n_{21}} \left(\frac{\delta A w_i^{\delta-1}\theta_1^{-1}}{\exp\left\{\frac{\tau^\delta}{\theta_\bullet}\right\}\exp\left\{\frac{A(w_i^\delta - \tau^\delta)}{\theta_\bullet}\right\}} \right) \\ &\quad \times \prod_{i=1}^{n_{22}} \left(\frac{\delta A w_i^{\delta-1}\theta_2^{-1}}{\exp\left\{\frac{\tau^\delta}{\theta_\bullet}\right\}\exp\left\{\frac{A(w_i^\delta - \tau^\delta)}{\theta_\bullet}\right\}} \right) \\ &\quad \times \left(\frac{1}{\exp\left\{\frac{n_c\tau^\delta}{\theta_\bullet}\right\}\exp\left\{\frac{A n_c(T^\delta - \tau^\delta)}{\theta_\bullet}\right\}} \right), \end{aligned} \tag{10.15}$$

where $n_{1\bullet} = n_{11} + n_{12}$, $n_{2\bullet} = n_{21} + n_{22}$, and $n = n_{1\bullet} + n_{2\bullet} + n_c$. Thus, the log-likelihood L of a Type-I censored observation is

$$\begin{aligned}\log L = {} & (n_{11} + n_{12} + n_{21} + n_{22}) \log \delta - (n_{11} + n_{21}) \log \theta_1 \\ & - (n_{12} + n_{22}) \log \theta_2 + (n_{21} + n_{22}) \log A + (\delta - 1) \\ & \times \left(\sum_{i=1}^{n_{11}} \log w_i + \sum_{i=1}^{n_{12}} \log w_i + \sum_{i=1}^{n_{21}} \log w_i + \sum_{i=1}^{n_{22}} \log w_i \right) \\ & - \sum_{i=1}^{n_{11}} \frac{w_i^\delta}{\theta_\bullet} - \sum_{i=1}^{n_{12}} \frac{w_i^\delta}{\theta_\bullet} - \sum_{i=1}^{n_{21}} \frac{A(w_i^\delta - \tau^\delta)}{\theta_\bullet} - \sum_{i=1}^{n_{22}} \frac{A(w_i^\delta - \tau^\delta)}{\theta_\bullet} \\ & - (n_{21} + n_{22} + n_c) \frac{\tau^\delta}{\theta_\bullet} - n_c \frac{A(T^\delta - \tau^\delta)}{\theta_\bullet}. \end{aligned} \tag{10.16}$$

The maximum likelihood estimates are the parameter values that maximize $\log L$. The base of log is e.

10.5 FISHER INFORMATION MATRIX

For a single observation, the first partial derivatives of $\log L$ with respect to θ_1, θ_2, and A is,

$$\begin{aligned}\frac{\partial \log L}{\partial \theta_1} = {} & -\frac{(n_{11} + n_{21})}{\theta_1} + \frac{(n_{2\bullet} + n_c)\tau^\delta}{\theta_1^2} + \frac{n_c A(T^\delta - \tau^\delta)}{\theta_1^2} \\ & + \frac{1}{\theta_1^2} \cdot \left(\sum_{i=1}^{n_{11}} w_i^\delta + \sum_{i=1}^{n_{12}} w_i^\delta + A \sum_{i=1}^{n_{21}} (w_i^\delta - \tau^\delta) \right. \\ & \left. + A \sum_{i=1}^{n_{22}} (w_i^\delta - \tau^\delta) \right), \end{aligned} \tag{10.17}$$

$$\frac{\partial \log L}{\partial \theta_2} = -\frac{(n_{12} + n_{22})}{\theta_2} + \frac{(n_{2\bullet} + n_c)\tau^\delta}{\theta_2^2} + \frac{n_c A(T^\delta - \tau^\delta)}{\theta_2^2}$$

$$+\frac{1}{\theta_2^2}\cdot\left(\sum_{i=1}^{n_{11}} w_i^\delta+\sum_{i=1}^{n_{12}} w_i^\delta+A\sum_{i=1}^{n_{21}}(w_i^\delta-\tau^\delta)\right.$$
$$\left.+A\sum_{i=1}^{n_{22}}(w_i^\delta-\tau^\delta)\right), \tag{10.18}$$

$$\frac{\partial\log L}{\partial A}=\frac{n_{2\bullet}}{A}-\frac{1}{\theta_\bullet}\cdot\left(\sum_{i=1}^{n_{21}}(w_i^\delta-\tau^\delta)+\sum_{i=1}^{n_{22}}(w_i^\delta-\tau^\delta)+n_c(T^\delta-\tau^\delta)\right). \tag{10.19}$$

The elements of the Fisher information matrix F obtained by taking the negative expectations of the second partial and mixed partial derivatives of $\log L$ with respect to θ_1, θ_2, and A are given in Appendix B.5.

10.6 OPERATING CHARACTERISTIC (OC) FUNCTION

The VRGSP is characterized by the parameters, namely, n, k_a, k_r. It reduces to a SSP if $k_a = k_r$. Based on the asymptotic distribution theory, $\sqrt{n}(\nu-\mu)/\sqrt{\mathrm{Var}(\nu)} \sim N(0,1)$ as $n\to\infty$, therefore the lot acceptance probability based on a single sample (group) when the lot quality is p will be

$$P_a(p)=\Pr[\nu\ge L^*k_a|p]=1-\Phi\left(\sqrt{n}\,(L^*k_a-\mu)/\sqrt{\mathrm{Var}(\nu)}\right), \tag{10.20}$$

where, $\Phi(\cdot)$ is the cdf of a standard normal variate, and $w' = L^{*\delta}/\theta_\bullet = -\log(1-p)$ (see (10.5)) and on the other hand, the lot rejection probability is given by

$$P_r(p)=\Pr[\nu<L^*k_r|p]$$
$$=\Phi\left(\sqrt{n}\left(k_r-\Gamma(1/\delta)/\delta w'^{(1/\delta)}\right)\Big/\sqrt{\mathrm{Var}(\nu)/L^{*2}}\right) \tag{10.21}$$

where $\mathrm{Var}(\nu)$ is given in Appendix B.6.

The OC curve is obtained by plotting $L(p) \approx L_p$ against fraction nonconforming p, where the lot acceptance probability based on a single sample (group) when the lot quality is p is given by

$$\begin{aligned} L_{\mathrm{SSP}}(p) &= \Pr[\nu \geq L^* k|p] \\ &= \Phi\left(\sqrt{n}\left(k - \Gamma(1/\delta)/\delta w'^{(1/\delta)}\right) \Big/ \sqrt{\mathrm{Var}(\nu)/L^{*2}}\right), \end{aligned} \tag{10.22}$$

and the lot acceptance probability based on a repetitive group sample when the lot quality is p is given by

$$\begin{aligned} & L_{\mathrm{VRGSP}}(p) \\ & \quad = \frac{P_a(p)}{P_a(p) + P_r(p)} \\ & \quad = \frac{1 - \Phi\left(\frac{\sqrt{n}\left(k_a - \Gamma(1/\delta)/\delta w'^{1/\delta}\right)}{\sqrt{\mathrm{Var}(\nu)/L^{*2}}}\right)}{1 - \Phi\left(\frac{\sqrt{n}\left(k_a - \Gamma(1/\delta)/\delta w'^{(1/\delta)}\right)}{\sqrt{\mathrm{Var}(\nu)/L^{*2}}}\right) + \Phi\left(\frac{\sqrt{n}\left(k_r - \Gamma(1/\delta)/\delta w'^{1/\delta}\right)}{\sqrt{\mathrm{Var}(\nu)/L^{*2}}}\right)}. \end{aligned} \tag{10.23}$$

10.7 FORMULATION OF AN OPTIMIZATION PROBLEM

Bilevel programming is a mathematical optimization problem that contains an optimization problem in the constraints (see Appendix C.1).

In the first stage, optimum stress change point is obtained by using D-optimality criterion which consist in minimizing $n/|F|$; followed by the second stage in which optimal sample size has been obtained by minimizing the average sample number such that producer's and consumer's interest are safeguarded. In this section, bilevel programming has been devised for VRGSP and SSP in Sections 10.7.1 and 10.7.2, respectively.

10.7.1 Bilevel Programming: VRGSP

$$
\begin{cases}
\operatorname*{Min}_{\tau} n/|F| \\
\text{s.t. } 0 \leq \tau \leq \eta, \\
\quad \text{where } (\tau) \text{ solves} \\
\quad \begin{cases}
\operatorname*{Min}_{n,k} \mathrm{ASN}_{\mathrm{VRGSP}} = \mathrm{ASN}(w_1) + \mathrm{ASN}(w_2) \\
= \dfrac{n}{1-\Phi\left(\frac{\sqrt{n}\left(k_a - \Gamma(1/\delta)/\delta w_1'^{(1/\delta)}\right)}{\sqrt{\mathrm{Var}(\nu)/L^{*2}}}\right) + \Phi\left(\frac{\sqrt{n}\left(k_r - \Gamma(1/\delta)/\delta w_1'^{(1/\delta)}\right)}{\sqrt{\mathrm{Var}(\nu)/L^{*2}}}\right)} \\
+ \dfrac{n}{1-\Phi\left(\frac{\sqrt{n}\left(k_a - \Gamma(1/\delta)/\delta w_2'^{(1/\delta)}\right)}{\sqrt{\mathrm{Var}(\nu)/L^{*2}}}\right) + \Phi\left(\frac{\sqrt{n}\left(k_r - \Gamma(1/\delta)/\delta w_2'^{(1/\delta)}\right)}{\sqrt{\mathrm{Var}(\nu)/L^{*2}}}\right)} \\
\text{s.t.} \\
A_1(w_1) = \dfrac{1-\Phi\left(\frac{\sqrt{n}\left(k_a - \Gamma(1/\delta)/\delta w_1'^{(1/\delta)}\right)}{\sqrt{\mathrm{Var}(\nu)/L^{*2}}}\right)}{1-\Phi\left(\frac{\sqrt{n}\left(k_a - \Gamma(1/\delta)/\delta w_1'^{1/\delta}\right)}{\sqrt{\mathrm{Var}(\nu)/L^{*2}}}\right) + \Phi\left(\frac{\sqrt{n}\left(k_r - \Gamma(1/\delta)/\delta w_1'^{(1/\delta)}\right)}{\sqrt{\mathrm{Var}(\nu)/L^{*2}}}\right)} \\
\quad \geq 1 - \tilde{\alpha}, \\
A_2(w_2) = \dfrac{1-\Phi\left(\frac{\sqrt{n}\left(k_a - \Gamma(1/\delta)/\delta w_2'^{(1/\delta)}\right)}{\sqrt{\mathrm{Var}(\nu)/L^{*2}}}\right)}{1-\Phi\left(\frac{\sqrt{n}\left(k_a - \Gamma(1/\delta)/\delta w_2'^{(1/\delta)}\right)}{\sqrt{\mathrm{Var}(\nu)/L^{*2}}}\right) + \Phi\left(\frac{\sqrt{n}\left(k_r - \Gamma(1/\delta)/\delta w_2'^{(1/\delta)}\right)}{\sqrt{\mathrm{Var}(\nu)/L^{*2}}}\right)} \\
\quad \leq \tilde{\beta}, \\
n > 2, k_a > k_r, k_a > 0, k_r > 0,
\end{cases}
\end{cases}
\tag{10.24}
$$

Let $\Psi_3 = \{(\tau) : 0 \leq \tau \leq \eta\}, \Psi_4 = \{(\mathrm{ASN}_{\mathrm{VRGSP}}) : A_1(w_2') \geq 1 - \tilde{\alpha}, n > 2, A_2(w_2') \leq \tilde{\beta}, k_a > k_r, k_a > 0, k_r > 0\}$ and $(\tau^*) = \operatorname*{Argmin}_{\tau}\{\frac{n}{|\mathrm{F}|} : (\tau) \in \Psi_1\}$. Then the optimization problem reduces to:

$$
\operatorname*{Min}_{n}\{(\mathrm{ASN}_{\mathrm{VRGSP}}) : n \in \Omega_2, (\tau) = (\tau^*)\}. \tag{10.25}
$$

In the above, w_1' and w_2' are the values of w' in (10.5) evaluated at $p = p_1$ and $p = p_2$, respectively.

10.7.2 Bilevel Programming: SSP

The optimal design problem can be formulated as a nonlinear optimization problem: The optimal design problem can be formulated as:

$$\begin{cases} \underset{\tau}{\text{Min}}\, n/|F| \\ \text{s.t. } 0 \leq \tau \leq \eta, \\ \text{where } (\tau) \text{ solves} \\ \begin{cases} \underset{n,k}{\text{Min}}\ \text{ASN}_{\text{SSP}} = \text{ASN}(w_1') + \text{ASN}(w_2') = 2n \\ \text{s.t. } A_3(w_1') = 1 - \Phi\left(\dfrac{\sqrt{n}\left(k - \Gamma(1/\delta)/\delta w_1'^{(1/\delta)}\right)}{\sqrt{\text{Var}(\nu)/L^{*2}}}\right) \geq 1 - \tilde{\alpha}, \\ A_4(w_2') = 1 - \Phi\left(\dfrac{\sqrt{n}\left(k - \Gamma(1/\delta)/\delta w_2'^{(1/\delta)}\right)}{\sqrt{\text{Var}(\nu)/L^{*2}}}\right) \leq \tilde{\beta}, \\ n \geq 1, n \in Z^+. \end{cases} \end{cases} \tag{10.26}$$

Let

$$\Psi_1 = \{(\tau) : 0 \leq \tau \leq \eta\},$$
$$\Psi_2 = \{(\text{ASN}_{\text{SSP}}) : A_3(w_1') \geq 1 - \tilde{\alpha}, A_4(w_2') \leq \tilde{\beta}, n \geq 1, n \in Z^+\},$$

and $(\tau^*) = \underset{\tau}{\text{Argmin}}\left\{\frac{n}{|F|} : (\tau) \in \Psi_1\right\}$. Then the optimization problem reduces to:

$$\underset{n}{\text{Min}}\{(\text{ASN}_{\text{SSP}}) : n \in \Omega_2, (\tau) = (\tau^*)\}. \tag{10.27}$$

In the above, w_1' and w_2' are the values of w' in (10.5) evaluated at $p = p_1$ and $p = p_2$, respectively.

Since, the optimum ramp-test depends on $\theta_1, \theta_2, A, \delta,$ and η; one must obtain their values from experience, similar data, or a preliminary test to achieve the minimum of $n/|\text{F}|$ in the ALTSP for the optimum values of τ. The optimal value of sample size is obtained by minimizing the average sample number.

10.8 ILLUSTRATIVE EXAMPLE

The method developed has been explained by an example in this section. It has been demonstrated that the VRGS plan has smaller average sample number than single sampling variable plan. This implies that the VRGSP will give desired protection with minimum inspection so that the cost of inspection will greatly be reduced.

Consider the hypothetical dataset as

$$\theta_1 = 25, \quad \theta_2 = 35, \quad A = 1.1, \quad \delta = 1.5, \quad \text{and} \quad \eta = 2.5.$$

The optimal value of τ is $\tau^* = 1.485$ obtained using (10.24) for VRGSP, and (10.26) for SSP. The estimate of relative risks are $\hat{\pi}_{11} = 0.583$, and $\hat{\pi}_{12} = 0.416$. This shows that there is a 58% chance for a test unit to fail by the failure cause 1 and 42% chance for a test unit to fail by the failure cause 2. The optimal values of sample size and acceptability constant with respect to producer's and consumer's nonconforming proportions are $n^*_{\text{VRGSP}} = 56$, $k^*_a = 8.932$ and $k^*_r = 7.575$ under VRGSP, and $n^*_{\text{SSP}} = 66$, $k^* = 8.552$ under SSP. With the optimal values so obtained, the decision about disposition of a lot is made using Section 10.4.1 for SSP and Section 10.4.2 for VRGSP.

Figure 10.1 depicts OC curve under VRGSP and SSP by plotting L_p against different values of p using the two points $(p_1, 1 - \tilde{\alpha})$ and $(p_2, \tilde{\beta})$ as (0.01, 0.99) and (0.10, 0.05).

Table 10.1 presents optimum VRGSP and SSP for various values of p_1 and p_2 when $(1 - \tilde{\alpha}, \tilde{\beta}) = (0.99, 0.10)$. It can be seen that the $\text{ASN}_{\text{VRGSP}}$ is smaller than the ASN_{SSP}. It is also observed that for given p_2 as p_1 decreases, n^* decreases.

10.9 SENSITIVITY ANALYSES

To formulate an optimum test plan, one needs information about the values of θ_1, θ_2, and A which are usually unknown. Therefore, they have to be approximated from experience, similar data, or preliminary test. Incorrect choice of pre-estimates gives a nonoptimal test plan and results in poor estimates of the parameters of life distribution at design stress. The effect of incorrect choice of decision

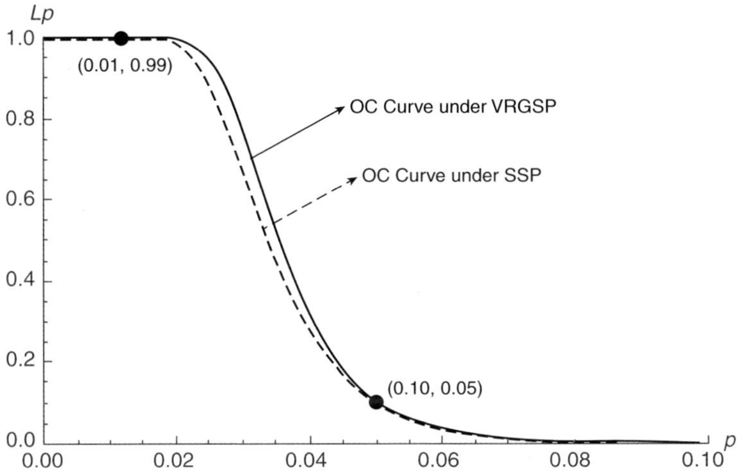

Figure 10.1: OC Curve

Table 10.1: Effects of various values of $(p_1, 1-\tilde{\alpha})$ and $(p_2, \tilde{\beta})$ in optimal ALT plans ($\theta_1 = 25, \theta_2 = 35, A = 1.1, \delta = 1.5$, and $\eta = 2.5$)

p_2	p_1	n^*_{VRGSP}	k_a	k_r	$\text{ASN}_{\text{VRGSP}}$	n^*_{SSP}	k
0.500	0.200	85	1.512	1.238	207	116	1.420
	0.100	48	1.598	1.390	104	54	1.545
	0.050	35	1.647	1.554	73	37	1.629
	0.010	25	1.731	1.731	50	25	1.731
	0.005	23	1.751	23.173	46	23	1.751
0.300	0.100	77	2.375	1.952	186	103	2.237
	0.050	47	2.496	2.184	101	52	2.418
	0.030	37	2.552	2.369	77	39	2.515
	0.010	28	2.649	2.649	55	28	2.649
	0.005	25	2.697	2.697	50	25	2.697
0.100	0.050	156	5.021	4.137	417	255	4.680
	0.020	55	5.535	4.704	123	65	5.304
	0.010	39	5.730	5.252	81	41	5.629
	0.005	31	5.861	5.813	63	31	5.853
	0.001	24	6.125	6.125	47	24	6.125
0.050	0.020	109	8.378	6.841	277	162	7.825
	0.010	56	8.932	7.575	125	66	8.552
	0.005	39	9.251	8.455	82	41	9.080
	0.001	26	9.752	9.752	52	26	9.752
	0.0005	24	9.893	9.893	48	24	9.893

Table 10.2: Effects of incorrect pre-estimates of θ_1, θ_2, A ($\%\Delta\tau^*$ due to $\pm1\%$ to $\pm5\%$ change in θ_1, θ_2, A)

% Change	A	τ^0	$\%\Delta\tau^*$	θ_1	τ^0	$\%\Delta\tau^*$	θ_2	τ^0	$\%\Delta\tau^*$
1	1.111	1.483	0.125	25.250	1.485	0.026	35.350	1.485	0.018
−1	1.089	1.487	0.127	24.750	1.484	0.026	34.650	1.485	0.019
2	1.122	1.481	0.127	25.500	1.486	0.050	35.700	1.485	0.036
−2	1.078	1.489	0.256	24.500	1.484	0.053	34.300	1.484	0.038
3	1.133	1.479	0.369	25.750	1.486	0.075	36.050	1.486	0.054
−3	1.067	1.491	0.388	24.250	1.484	0.080	33.950	1.484	0.057
4	1.144	1.478	0.488	26.000	1.486	0.099	36.400	1.486	0.071
−4	1.056	1.493	0.521	24.000	1.483	0.107	33.600	1.484	0.077
5	1.155	1.476	0.605	26.250	1.487	0.123	36.750	1.486	0.088
−5	1.045	1.495	0.657	23.750	1.483	0.136	33.250	1.483	0.097

parameter estimates for some selected values of θ_1, θ_2, and A in terms of relative stress change point, τ^* are presented in Table 10.2 using

$$\%\Delta\tau^* = (|(\tau^* - \tau^0)/\tau^*|) \times 100,$$

where τ^* is the optimal stress change point for the plans obtained with the correctly specified values, and τ^0 is the stress change point for the plans obtained with mis-specified values. The results show that the optimal plans are robust to small deviations from the baseline parameters.

10.10 SUMMARY

This chapter deals with the design of optimum time-censored step-stress PALTSP with competing causes of failure using TFR model. The VRGSP has been used with Weibull life distribution. The optimal plan consists in finding optimal stress change point, sample size, and acceptability constants. The method developed has been explained using an example. The VRGSP is compared with single sampling variable plan, and it is found that the optimum sample size is reduced by using the VRGSP as compared to the SSP for variables.

Because of the minimum inspection involved in VRGSP, the cost of inspection is also reduced with desired amount of protection. Sensitivity analysis has also been carried out, and it is found that the optimal plans described are robust to small deviations from the baseline parameters.

APPENDIX A

A.0 TRUNCATED DISTRIBUTIONS AND ORDER STATISTICS

A.01 Truncated Distributions

A **truncated distribution** is a conditional distribution resulting when the domain of the parent distribution is restricted to a smaller region. Let X be a continuous random variable with cdf $F(.)$, and pdf $f(.)$.

Definition A.1: The distribution of X is said to be truncated at the point $X = a$ if all the values of $X \leq a$ are discarded.

Thus, the pdf $g(.)$ of the distribution truncated at $X = a$ is given by

$$g(x) = \begin{cases} \dfrac{f(x)}{P[X > a]}; & x > a \\ 0; & \text{otherwise} \end{cases}$$

Definition A.2: The distribution of X is said to be truncated at both ends with relevant range of variation as (a, b) if its pdf is given by

$$g(x) = \begin{cases} \dfrac{f(x)}{P[a < X < b]}; & a < x < b \\ 0; & \text{otherwise.} \end{cases}$$

$$= \begin{cases} \dfrac{f(x)}{F(b) - F(a)}; & a < x < b \\ 0; & \text{otherwise} \end{cases}$$

Example A.1: The pdf of truncated standard normal distribution truncated at both ends with relevant range of variation as (a, b) is:

$$g(x) = \begin{cases} \dfrac{\phi(x)}{\Phi(b) - \Phi(a)}; & a < x < b \\ 0; & \text{otherwise,} \end{cases}$$

where $\phi(.)$ and $\Phi(.)$ are respectively pdf and cdf of standard normal distribution.

Example A.2: The pdf of truncated logistic distribution truncated at point zero is:

$$g(x) = \frac{\exp\left\{-\frac{x-\mu}{\sigma}\right\}}{A\sigma\left(1 + \exp\left\{-\frac{x-\mu}{\sigma}\right\}\right)^2}, \quad x,\ \sigma > 0,\ -\infty < \mu < \infty,$$

where

$$A = \frac{1}{\left(1 + \exp\left\{-\frac{\mu}{\sigma}\right\}\right)}.$$

A.02 Order Statistics

Let $X_1, X_2, \ldots, X_n$ be independent and identically distributed random variables with absolutely continuous distribution function $F(.)$, and let them be arranged in increasing order of magnitude as $X_{(1)} < X_{(2)} < \cdots < X_{(n)}$. Then, $X_{(r)}$ is called the rth order statistic, $r = 1, 2, \ldots, n$.

The cdf and pdf respectively of rth order statistic is given by:

$$F_r(x) = \sum_{j=r}^{n} \binom{n}{j} (F(x))^j (1 - F(x))^{n-j}, \quad \text{and}$$

$$f_r(x) = \frac{n!}{(r-1)!\,1!\,(n-r)!} \cdot (F(x))^{(r-1)} \cdot f(x) \cdot (1 - F(x))^{(n-r)}.$$

The joint pdf of rth and s^{th} order statistics $(1 \leq r < s \leq n)$ denoted by $f_{rs}(x, y)$ is

$$\begin{aligned} f_{rs}(x,y) = {} & \frac{n!}{(r-1)!\,1!\,(s-r-1)!\,1!\,(n-s)!} \cdot (F(x))^{(r-1)} f(x) \\ & \times (F(y) - F(x))^{(s-r-1)} f(y)(1 - F(y))^{(n-s)}. \end{aligned}$$

In this work, the concept of order statistics is used when dealing with failure censored data.

A.1 EXPONENTIAL DISTRIBUTION AS A LIMITING CASE OF THE BURR TYPE-XII DISTRIBUTION

Consider the cdf of the Burr Type-XII distribution as

$$G(t) = 1 - \left(1 + \left(\frac{t}{\alpha}\right)^c\right)^{-k}; \quad t \geq 0; \;\; c, k, \alpha > 0.$$

Let $\alpha = k^{1/c}$ and $c = 1$. Therefore,

$$\begin{aligned} G(t) &= 1 - \left(1 + \frac{t}{k}\right)^{-k}; \quad t \geq 0; k > 0 \\ &= 1 - \exp\left\{-k \ln\left(1 + \frac{t}{k}\right)\right\}; \quad t \geq 0; \;\; k > 0 \\ &= 1 - \exp\left\{-k\left(\frac{t}{k} - \frac{1}{2}\left(\frac{t}{k}\right)^2 + \cdots\right)\right\}; \quad t \geq 0; \;\; k > 0 \\ &= 1 - \exp\left\{-t + \frac{t^2}{2k} - \cdots\right\}; \quad t \geq 0; \;\; k > 0. \end{aligned}$$

As $k \to \infty$, $G(t) = 1 - \exp(-t)$; $t \geq 0$, which is the cdf of exponential distribution.

A.2 WEIBULL DISTRIBUTION AS A LIMITING CASE OF THE BURR TYPE-XII DISTRIBUTION

Consider the cdf of the Burr Type-XII distribution as

$$G(t) = 1 - \left(1 + \left(\frac{t}{\alpha}\right)^c\right)^{-k}; \quad t \geq 0; \;\; c, k, \alpha > 0.$$

Let $\alpha = k^{1/c}$. Therefore,

$$\begin{aligned}
G(t) &= 1 - \left(1 + \left(\frac{t}{k^{1/c}}\right)^c\right)^{-k}; \quad t \geq 0; \;\; c, k > 0 \\
&= 1 - \left(1 + \frac{t^c}{k}\right)^{-k}; \quad t \geq 0; \;\; c, k > 0 \\
&= 1 - \exp\left\{-k \ln\left(1 + \frac{t^c}{k}\right)\right\}; \quad t \geq 0; \;\; c, k > 0 \\
&= 1 - \exp\left\{-k\left(\frac{t^c}{k} - \frac{1}{2}\left(\frac{t^c}{k}\right)^2 + \cdots\right)\right\}; \quad t \geq 0; \;\; c, k > 0 \\
&= 1 - \exp\left\{-t^c + \frac{(t^c)^2}{2k} - \cdots\right\}; \quad t \geq 0; \;\; c, k > 0.
\end{aligned}$$

As $k \to \infty$, $G(t) = 1 - \exp(-t^c)$; $t \geq 0$; $c > 0$, which is the cdf of the Weibull distribution.

A.3 CALCULATIONS OF THE ELEMENTS OF THE FISHER INFORMATION MATRIX IN 3.2.4

The first-order partial derivatives are

$$\frac{\partial \ln L(\beta, \mu, \sigma)}{\partial \beta} = \frac{n_a}{\beta} - \sum_{i=1}^{n} \delta_{2i}\left(\frac{y_i - \tau}{\sigma}\right) - \sum_{i=1}^{n} \bar{\delta}_{1i}\bar{\delta}_{2i}\left(\frac{\eta - \tau}{\sigma}\right)$$

$$+2\sum_{i=1}^{n}\delta_{2i}\left(\frac{y_i-\tau}{\sigma}\right)\left(\frac{\exp\left\{-\frac{(y_i-\tau)\beta+\tau-\mu}{\sigma}\right\}}{1+\exp\left\{-\frac{(y_i-\tau)\beta+\tau-\mu}{\sigma}\right\}}\right)$$

$$+\sum_{i=1}^{n}\bar{\delta}_{1i}\bar{\delta}_{2i}\left(\frac{\eta-\tau}{\sigma}\right)\left(\frac{\exp\left\{-\frac{(\eta-\tau)\beta+\tau-\mu}{\sigma}\right\}}{1+\exp\left\{-\frac{(\eta-\tau)\beta+\tau-\mu}{\sigma}\right\}}\right),$$

$$\frac{\partial \ln L(\beta,\mu,\sigma)}{\partial\mu}$$

$$=-\frac{n\exp\left\{-\frac{\mu}{\sigma}\right\}}{\sigma\left(1+\exp\left\{-\frac{\mu}{\sigma}\right\}\right)}+\sum_{i=1}^{n}\frac{\delta_{1i}}{\sigma}$$

$$-2\sum_{i=1}^{n}\frac{\delta_{1i}}{\sigma}\left(\frac{\exp\left\{-\frac{(y_i-\mu)}{\sigma}\right\}}{1+\exp\left\{-\frac{(y_i-\mu)}{\sigma}\right\}}\right)$$

$$+\sum_{i=1}^{n}\frac{\delta_{2i}}{\sigma}-\frac{2}{\sigma}\sum_{i=1}^{n}\delta_{2i}\left(\frac{\exp\left\{-\frac{(y_i-\tau)\beta+\tau-\mu}{\sigma}\right\}}{1+\exp\left\{-\frac{(y_i-\tau)\beta+\tau-\mu}{\sigma}\right\}}\right)$$

$$+\sum_{i=1}^{n}\frac{\bar{\delta}_{1i}\bar{\delta}_{2i}}{\sigma}-\frac{1}{\sigma}\sum_{i=1}^{n}\bar{\delta}_{1i}\bar{\delta}_{2i}\left(\frac{\exp\left\{-\frac{(\eta-\tau)\beta+\tau-\mu}{\sigma}\right\}}{1+\exp\left\{-\frac{(\eta-\tau)\beta+\tau-\mu}{\sigma}\right\}}\right)$$

and

$$\frac{\partial \ln L(\beta,\mu,\sigma)}{\partial\sigma}$$

$$=\frac{n\mu\exp\left\{-\frac{\mu}{\sigma}\right\}}{\sigma^2\left(1+\exp\left\{-\frac{\mu}{\sigma}\right\}\right)}-\left(\frac{n_u+n_a}{\sigma}\right)$$

$$+\sum_{i=1}^{n}\delta_{2i}\left(\frac{(y_i-\tau)\beta+\tau-\mu}{\sigma^2}\right)$$

$$+\sum_{i=1}^{n}\delta_{1i}\left(\frac{y_i-\mu}{\sigma^2}\right)-2\sum_{i=1}^{n}\delta_{1i}\left(\frac{y_i-\mu}{\sigma^2}\right)\left(\frac{\exp\left\{-\frac{(y_i-\mu)}{\sigma}\right\}}{1+\exp\left\{-\frac{(y_i-\mu)}{\sigma}\right\}}\right)$$

$$- 2\sum_{i=1}^{n} \delta_{2i} \left(\frac{(y_i - \tau)\beta + \tau - \mu}{\sigma^2} \right) \left(\frac{\exp\left\{ -\frac{(y_i-\tau)\beta+\tau-\mu}{\sigma} \right\}}{1 + \exp\left\{ -\frac{(y_i-\tau)\beta+\tau-\mu}{\sigma} \right\}} \right)$$

$$- \sum_{i=1}^{n} \bar{\delta}_{1i}\bar{\delta}_{2i} \left(\frac{(\eta - \tau)\beta + \tau - \mu}{\sigma^2} \right) \left(\frac{\exp\left\{ -\frac{(\eta-\tau)\beta+\tau-\mu}{\sigma} \right\}}{1 + \exp\left\{ -\frac{(\eta-\tau)\beta+\tau-\mu}{\sigma} \right\}} \right)$$

$$+ \sum_{i=1}^{n} \bar{\delta}_{1i}\bar{\delta}_{2i} \left(\frac{(\eta - \tau)\beta + \tau - \mu}{\sigma^2} \right).$$

The second-order partial derivatives are

$$\frac{\partial^2 \ln L(\beta, \mu, \sigma)}{\partial \beta^2}$$

$$= -2\sum_{i=1}^{n} \delta_{2i} \left(\frac{y_i - \tau}{\sigma} \right)^2 \left(\frac{\exp\left\{ -\frac{(y_i-\tau)\beta+\tau-\mu}{\sigma} \right\}}{\left(1 + \exp\left\{ -\frac{(y_i-\tau)\beta+\tau-\mu}{\sigma} \right\}\right)^2} \right)$$

$$- \frac{n_a}{\beta^2} - \sum_{i=1}^{n} \bar{\delta}_{1i}\bar{\delta}_{2i} \left(\frac{\eta - \tau}{\sigma} \right)^2 \left(\frac{\exp\left\{ -\frac{(\eta-\tau)\beta+\tau-\mu}{\sigma} \right\}}{\left(1 + \exp\left\{ -\frac{(\eta-\tau)\beta+\tau-\mu}{\sigma} \right\}\right)^2} \right),$$

$$\frac{\partial^2 \ln L(\beta, \mu, \sigma)}{\partial \mu \partial \beta}$$

$$= \frac{2}{\sigma}\sum_{i=1}^{n} \delta_{2i} \left(\frac{y_i - \tau}{\sigma} \right) \left(\frac{\exp\left\{ -\frac{(y_i-\tau)\beta+\tau-\mu}{\sigma} \right\}}{\left(1 + \exp\left\{ -\frac{(y_i-\tau)\beta+\tau-\mu}{\sigma} \right\}\right)^2} \right)$$

$$+ \frac{1}{\sigma}\sum_{i=1}^{n} \bar{\delta}_{1i}\bar{\delta}_{2i} \left(\frac{\eta - \tau}{\sigma} \right) \left(\frac{\exp\left\{ -\frac{(\eta-\tau)\beta+\tau-\mu}{\sigma} \right\}}{\left(1 + \exp\left\{ -\frac{(\eta-\tau)\beta+\tau-\mu}{\sigma} \right\}\right)^2} \right),$$

$$\frac{\partial^2 \ln L(\beta,\mu,\sigma)}{\partial\sigma\partial\beta}$$

$$= -2\sum_{i=1}^{n}\delta_{2i}\left(\frac{y_i-\tau}{\sigma^2}\right)\left(\frac{\exp\left\{-\frac{(y_i-\tau)\beta+\tau-\mu}{\sigma}\right\}}{1+\exp\left\{-\frac{(y_i-\tau)\beta+\tau-\mu}{\sigma}\right\}}\right)$$

$$+\sum_{i=1}^{n}\delta_{2i}\left(\frac{y_i-\tau}{\sigma^2}\right)+\sum_{i=1}^{n}\bar{\delta}_{1i}\bar{\delta}_{2i}\left(\frac{\eta-\tau}{\sigma^2}\right)$$

$$-\sum_{i=1}^{n}\bar{\delta}_{1i}\bar{\delta}_{2i}\left(\frac{\eta-\tau}{\sigma^2}\right)\left(\frac{\exp\left\{-\frac{(\eta-\tau)\beta+\tau-\mu}{\sigma}\right\}}{1+\exp\left\{-\frac{(\eta-\tau)\beta+\tau-\mu}{\sigma}\right\}}\right)$$

$$+2\sum_{i=1}^{n}\left\{\delta_{2i}\left(\frac{y_i-\tau}{\sigma}\right)\left(\frac{(y_i-\tau)\beta+\tau-\mu}{\sigma^2}\right)\right.$$

$$\left.\times\left(\frac{\exp\left\{-\frac{(y_i-\tau)\beta+\tau-\mu}{\sigma}\right\}}{\left(1+\exp\left\{-\frac{(y_i-\tau)\beta+\tau-\mu}{\sigma}\right\}\right)^2}\right)\right\}$$

$$+\sum_{i=1}^{n}\left\{\bar{\delta}_{1i}\bar{\delta}_{2i}\left(\frac{\eta-\tau}{\sigma}\right)\left(\frac{(\eta-\tau)\beta+\tau-\mu}{\sigma^2}\right)\right.$$

$$\left.\times\left(\frac{\exp\left\{-\frac{(\eta-\tau)\beta+\tau-\mu}{\sigma}\right\}}{\left(1+\exp\left\{-\frac{(\eta-\tau)\beta+\tau-\mu}{\sigma}\right\}\right)^2}\right)\right\},$$

$$\frac{\partial^2 \ln L(\beta,\mu,\sigma)}{\partial\mu^2}$$

$$= \frac{n\exp\left\{-\frac{\mu}{\sigma}\right\}}{\sigma^2\left(1+\exp\left\{-\frac{\mu}{\sigma}\right\}\right)^2}$$

$$
\begin{aligned}
&- 2\sum_{i=1}^{n}\frac{\delta_{1i}}{\sigma^2}\left(\frac{\exp\left\{-\frac{(y_i-\mu)}{\sigma}\right\}}{\left(1+\exp\left\{-\frac{(y_i-\mu)}{\sigma}\right\}\right)^2}\right)\\
&- 2\sum_{i=1}^{n}\frac{\delta_{2i}}{\sigma^2}\left(\frac{\exp\left\{-\frac{(y_i-\tau)\beta+\tau-\mu}{\sigma}\right\}}{\left(1+\exp\left\{-\frac{(y_i-\tau)\beta+\tau-\mu}{\sigma}\right\}\right)^2}\right)\\
&- \frac{1}{\sigma^2}\sum_{i=1}^{n}\bar{\delta}_{1i}\bar{\delta}_{2i}\left(\frac{\exp\left\{-\frac{(\eta-\tau)\beta+\tau-\mu}{\sigma}\right\}}{\left(1+\exp\left\{-\frac{(\eta-\tau)\beta+\tau-\mu}{\sigma}\right\}\right)^2}\right),
\end{aligned}
$$

$$
\begin{aligned}
&\frac{\partial^2 \ln L(\beta,\mu,\sigma)}{\partial\sigma\partial\mu}\\
&= 2\sum_{i=1}^{n}\frac{\delta_{1i}}{\sigma^2}\left(\frac{\exp\left\{-\frac{(y_i-\mu)}{\sigma}\right\}}{1+\exp\left\{-\frac{(y_i-\mu)}{\sigma}\right\}}\right) - \frac{n\mu\exp\left\{-\frac{\mu}{\sigma}\right\}}{\sigma^3\left(1+\exp\left\{-\frac{\mu}{\sigma}\right\}\right)^2}\\
&- 2\sum_{i=1}^{n}\frac{\delta_{1i}}{\sigma}\left(\frac{y_i-\mu}{\sigma^2}\right)\left(\frac{\exp\left\{-\frac{(y_i-\mu)}{\sigma}\right\}}{\left(1+\exp\left\{-\frac{(y_i-\mu)}{\sigma}\right\}\right)^2}\right) - \sum_{i=1}^{n}\frac{\delta_{1i}}{\sigma^2}\\
&- \sum_{i=1}^{n}\frac{\delta_{2i}}{\sigma^2} + \frac{2}{\sigma^2}\sum_{i=1}^{n}\delta_{2i}\left(\frac{\exp\left\{-\frac{(y_i-\tau)\beta+\tau-\mu}{\sigma}\right\}}{1+\exp\left\{-\frac{(y_i-\tau)\beta+\tau-\mu}{\sigma}\right\}}\right)\\
&+ \frac{n\exp\left\{-\frac{\mu}{\sigma}\right\}}{\sigma^2\left(1+\exp\left\{-\frac{\mu}{\sigma}\right\}\right)} - \frac{2}{\sigma}\sum_{i=1}^{n}\delta_{2i}\left(\frac{(y_i-\tau)\beta+\tau-\mu}{\sigma^2}\right)\\
&\times\left(\frac{\exp\left\{-\frac{(y_i-\tau)\beta+\tau-\mu}{\sigma}\right\}}{\left(1+\exp\left\{-\frac{(y_i-\tau)\beta+\tau-\mu}{\sigma}\right\}\right)^2}\right)\\
&- \frac{1}{\sigma}\sum_{i=1}^{n}\bar{\delta}_{1i}\bar{\delta}_{2i}\left(\frac{(\eta-\tau)\beta+\tau-\mu}{\sigma^2}\right)
\end{aligned}
$$

$$\times\left(\frac{\exp\left\{-\frac{(\eta-\tau)\beta+\tau-\mu}{\sigma}\right\}}{\left(1+\exp\left\{-\frac{(\eta-\tau)\beta+\tau-\mu}{\sigma}\right\}\right)^2}\right)$$

$$-\sum_{i=1}^{n}\frac{\bar{\delta}_{1i}\bar{\delta}_{2i}}{\sigma^2}+\frac{1}{\sigma^2}\sum_{i=1}^{n}\bar{\delta}_{1i}\bar{\delta}_{2i}\left(\frac{\exp\left\{-\frac{(\eta-\tau)\beta+\tau-\mu}{\sigma}\right\}}{1+\exp\left\{-\frac{(\eta-\tau)\beta+\tau-\mu}{\sigma}\right\}}\right)$$

and

$$\frac{\partial^2 \ln L(\beta,\mu,\sigma)}{\partial\sigma^2}$$

$$=\frac{n\mu^2\exp\{-\frac{\mu}{\sigma}\}}{\sigma^4\left(1+\exp\{-\frac{\mu}{\sigma}\}\right)^2}-\frac{2n\mu\exp\{-\frac{\mu}{\sigma}\}}{\sigma^3\left(1+\exp\{-\frac{\mu}{\sigma}\}\right)}+\left(\frac{n_u+n_a}{\sigma^2}\right)$$

$$+4\sum_{i=1}^{n}\delta_{1i}\left(\frac{y_i-\mu}{\sigma^3}\right)\left(\frac{\exp\left\{-\frac{(y_i-\mu)}{\sigma}\right\}}{1+\exp\left\{-\frac{(y_i-\mu)}{\sigma}\right\}}\right)$$

$$-2\sum_{i=1}^{n}\delta_{1i}\left(\frac{y_i-\mu}{\sigma^3}\right)-2\sum_{i=1}^{n}\delta_{1i}\left(\frac{y_i-\mu}{\sigma^2}\right)^2$$

$$\times\left(\frac{\exp\left\{-\frac{(y_i-\mu)}{\sigma}\right\}}{\left(1+\exp\left\{-\frac{(y_i-\mu)}{\sigma}\right\}\right)^2}\right)$$

$$-2\sum_{i=1}^{n}\delta_{2i}\left(\frac{(y_i-\tau)\beta+\tau-\mu}{\sigma^2}\right)^2$$

$$\times\left(\frac{\exp\left\{-\frac{(y_i-\tau)\beta+\tau-\mu}{\sigma}\right\}}{\left(1+\exp\left\{-\frac{(y_i-\tau)\beta+\tau-\mu}{\sigma}\right\}\right)^2}\right)$$

$$+4\sum_{i=1}^{n}\delta_{2i}\left(\frac{(y_i-\tau)\beta+\tau-\mu}{\sigma^3}\right)\left(\frac{\exp\left\{-\frac{(y_i-\tau)\beta+\tau-\mu}{\sigma}\right\}}{1+\exp\left\{-\frac{(y_i-\tau)\beta+\tau-\mu}{\sigma}\right\}}\right)$$

$$-2\sum_{i=1}^{n}\bar{\delta}_{1i}\bar{\delta}_{2i}\left(\frac{(\eta-\tau)\beta+\tau-\mu}{\sigma^3}\right)\left(\frac{\exp\left\{-\frac{(\eta-\tau)\beta+\tau-\mu}{\sigma}\right\}}{1+\exp\left\{-\frac{(\eta-\tau)\beta+\tau-\mu}{\sigma}\right\}}\right)$$

$$-\sum_{i=1}^{n}\bar{\delta}_{1i}\bar{\delta}_{2i}\left(\frac{(\eta-\tau)\beta+\tau-\mu}{\sigma^2}\right)^2$$

$$\times\left(\frac{\exp\left\{-\frac{(\eta-\tau)\beta+\tau-\mu}{\sigma}\right\}}{\left(1+\exp\left\{-\frac{(\eta-\tau)\beta+\tau-\mu}{\sigma}\right\}\right)^2}\right)$$

$$-2\sum_{i=1}^{n}\bar{\delta}_{1i}\bar{\delta}_{2i}\left(\frac{(\eta-\tau)\beta+\tau-\mu}{\sigma^3}\right)$$

$$-2\sum_{i=1}^{n}\delta_{2i}\left(\frac{(y_i-\tau)\beta+\tau-\mu}{\sigma^3}\right).$$

The elements of the Fisher information matrix for an observation are the negative expectations of these second-order partial derivatives. These are as follows:

$$E\left[-\frac{\partial^2\ln L(\beta,\mu,\sigma)}{\partial\beta^2}\right]=\frac{nP_a}{\beta^2}+\frac{2n\beta}{A\sigma^3}\int_{\tau}^{\eta}(y-\tau)^2(K(y))^2dy$$
$$+\frac{n(1-P_a-P_u)}{\sigma^2}(\eta-\tau)^2K(\eta),$$

$$E\left[-\frac{\partial^2\ln L(\beta,\mu,\sigma)}{\partial\mu\partial\beta}\right]=\frac{-n(1-P_a-P_u)(\eta-\tau)K(\eta)}{\sigma^2}$$
$$-\frac{2n\beta}{A\sigma^3}\int_{\tau}^{\eta}(y-\tau)(K(y))^2dy,$$

$$E\left[-\frac{\partial^2\ln L(\beta,\mu,\sigma)}{\partial\sigma\partial\beta}\right]=-\frac{n\beta}{A\sigma^3}\int_{\tau}^{\eta}(y-\tau)K(y)(1-2B(y))dy$$
$$-\frac{n(1-P_a-P_u)}{\sigma^2}(\eta-\tau)(1-B(\eta)),$$

$$E\left[-\frac{\partial^2 \ln L(\beta,\mu,\sigma)}{\partial\mu^2}\right] = \frac{-ng(2\mu)}{\sigma^2} + \frac{n(1-P_a-P_u)K(\eta)}{\sigma^2}$$
$$+ \frac{2n}{A\sigma^3}\int_0^{\tau}(g(y))^2 dy + \frac{2n\beta}{A\sigma^3}\int_{\tau}^{\eta}(K(y))^2\, dy,$$

$$E\left[-\frac{\partial^2 \ln L(\beta,\mu,\sigma)}{\partial\sigma\partial\mu}\right] = \frac{n(1-P_a-P_u)(1-B(\eta))}{\sigma^2}$$
$$+ \frac{n\beta}{A\sigma^3}\int_{\tau}^{\eta}K(y)(1-2B(y))dy$$
$$+ \frac{n}{A\sigma^3}\int_0^{\tau}g(y)(1-2Q(y))dy - \frac{nQ(2\mu)}{\sigma^2}$$

and

$$E\left[-\frac{\partial^2 \ln L(\beta,\mu,\sigma)}{\partial\sigma^2}\right]$$
$$= -\frac{2n(1-P_a-P_u)(\tau+\beta(\eta-\tau)-\mu)(1-H(\eta))}{\sigma^3}$$
$$+ \frac{2n\mu}{\sigma^3}M(2\mu) + \frac{2n\beta}{A\sigma^4}\int_{\tau}^{\eta}(\tau+\beta(y-\tau)-\mu)K(y)(1-2H(y))dy$$
$$- \frac{n(P_a+P_u)}{\sigma^2} + \frac{2n}{A\sigma^4}\int_0^{\tau}(y-\mu)g(y)(1-2M(y))dy,$$

where

$$K(y) = \frac{\exp\left\{-\frac{(y-\tau)\beta+\tau-\mu}{\sigma}\right\}}{\left(1+\exp\left\{-\frac{(y-\tau)\beta+\tau-\mu}{\sigma}\right\}\right)^2},$$

$$g(y) = \frac{\exp\left\{-\frac{(y-\mu)}{\sigma}\right\}}{\left(1+\exp\left\{-\frac{(y-\mu)}{\sigma}\right\}\right)^2},$$

$$M(y) = \left(1+\exp\left\{-\frac{(y-\mu)}{\sigma}\right\} - \frac{(y-\mu)}{2\sigma}\right)g(y),$$

$$Q(y) = \left(1 + \exp\left\{-\frac{(y-\mu)}{\sigma}\right\} - \frac{(y-\mu)}{\sigma}\right) g(y),$$

$$H(y) = \left(1 + \exp\left\{-\frac{(y-\tau)\beta + \tau - \mu}{\sigma}\right\} - \frac{(y-\tau)\beta + \tau - \mu}{2\sigma}\right) K(y)$$

and

$$B(y) = \left(1 + \exp\left\{-\frac{(y-\tau)\beta + \tau - \mu}{\sigma}\right\} - \frac{(y-\tau)\beta + \tau - \mu}{\sigma}\right) K(y).$$

A.4 CALCULATIONS OF ELEMENTS OF THE FISHER INFORMATION MATRIX GIVEN IN 3.3.1.4

The first-order partial derivatives are

$$\begin{aligned}
\frac{\partial \ln L(\beta,\mu,\sigma)}{\partial \beta} &= \frac{n_a}{\beta} - \sum_{i=1}^{n} \delta_{2i}\left(\frac{y_i-\tau}{\sigma}\right) - \sum_{i=1}^{n} \bar{\delta}_{1i}\bar{\delta}_{2i}\left(\frac{y_{(r)}-\tau}{\sigma}\right) \\
&\quad + 2\sum_{i=1}^{n} \delta_{2i}\left(\frac{y_i-\tau}{\sigma}\right)\left(\frac{\exp\left\{-\frac{(y_i-\tau)\beta+\tau-\mu}{\sigma}\right\}}{1+\exp\left\{-\frac{(y_i-\tau)\beta+\tau-\mu}{\sigma}\right\}}\right) \\
&\quad + \sum_{i=1}^{n} \bar{\delta}_{1i}\bar{\delta}_{2i}\left(\frac{y_{(r)}-\tau}{\sigma}\right)\left(\frac{\exp\left\{-\frac{(y_{(r)}-\tau)\beta+\tau-\mu}{\sigma}\right\}}{1+\exp\left\{-\frac{(y_{(r)}-\tau)\beta+\tau-\mu}{\sigma}\right\}}\right),
\end{aligned}$$

$$\begin{aligned}
\frac{\partial \ln L(\beta,\mu,\sigma)}{\partial \mu} &= -\frac{n\exp\{-\frac{\mu}{\sigma}\}}{\sigma\left(1+\exp\{-\frac{\mu}{\sigma}\}\right)} + \sum_{i=1}^{n}\frac{\delta_{1i}}{\sigma} \\
&\quad - 2\sum_{i=1}^{n}\frac{\delta_{1i}}{\sigma}\left(\frac{\exp\left\{-\frac{(y_i-\mu)}{\sigma}\right\}}{1+\exp\left\{-\frac{(y_i-\mu)}{\sigma}\right\}}\right) + \sum_{i=1}^{n}\frac{\delta_{2i}}{\sigma}
\end{aligned}$$

$$-\frac{2}{\sigma}\sum_{i=1}^{n}\delta_{2i}\left(\frac{\exp\left\{-\frac{(y_i-\tau)\beta+\tau-\mu}{\sigma}\right\}}{1+\exp\left\{-\frac{(y_i-\tau)\beta+\tau-\mu}{\sigma}\right\}}\right)$$

$$+\sum_{i=1}^{n}\frac{\bar{\delta}_{1i}\bar{\delta}_{2i}}{\sigma}-\sum_{i=1}^{n}\bar{\delta}_{1i}\bar{\delta}_{2i}\left(\frac{\exp\left\{-\frac{(y_{(r)}-\tau)\beta+\tau-\mu}{\sigma}\right\}}{1+\exp\left\{-\frac{(y_{(r)}-\tau)\beta+\tau-\mu}{\sigma}\right\}}\right)$$

and

$$\frac{\partial \ln L(\beta,\mu,\sigma)}{\partial\sigma}$$

$$=\frac{n\mu\exp\{-\frac{\mu}{\sigma}\}}{\sigma^2\left(1+\exp\{-\frac{\mu}{\sigma}\}\right)}-\left(\frac{n_u+n_a}{\sigma}\right)+\sum_{i=1}^{n}\delta_{2i}\left(\frac{(y_i-\tau)\beta+\tau-\mu}{\sigma^2}\right)$$

$$+\sum_{i=1}^{n}\delta_{1i}\left(\frac{y_i-\mu}{\sigma^2}\right)-2\sum_{i=1}^{n}\delta_{1i}\left(\frac{y_i-\mu}{\sigma^2}\right)\left(\frac{\exp\left\{-\frac{(y_i-\mu)}{\sigma}\right\}}{1+\exp\left\{-\frac{(y_i-\mu)}{\sigma}\right\}}\right)$$

$$-2+\sum_{i=1}^{n}\delta_{2i}\left(\frac{(y_i-\tau)\beta+\tau-\mu}{\sigma^2}\right)\left(\frac{\exp\left\{-\frac{(y_i-\tau)\beta+\tau-\mu}{\sigma}\right\}}{1+\exp\left\{-\frac{(y_i-\tau)\beta+\tau-\mu}{\sigma}\right\}}\right)$$

$$-\sum_{i=1}^{n}\bar{\delta}_{1i}\bar{\delta}_{2i}\left(\frac{(y_{(r)}-\tau)\beta+\tau-\mu}{\sigma^2}\right)\left(\frac{\exp\left\{-\frac{(y_{(r)}-\tau)\beta+\tau-\mu}{\sigma}\right\}}{1+\exp\left\{-\frac{(y_{(r)}-\tau)\beta+\tau-\mu}{\sigma}\right\}}\right)$$

$$+\sum_{i=1}^{n}\bar{\delta}_{1i}\bar{\delta}_{2i}\left(\frac{(y_{(r)}-\tau)\beta+\tau-\mu}{\sigma^2}\right).$$

The second-order partial derivatives are

$$\frac{\partial^2 \ln L(\beta,\mu,\sigma)}{\partial\beta^2}$$

$$=-2\sum_{i=1}^{n}\delta_{2i}\left(\frac{y_i-\tau}{\sigma}\right)^2\left(\frac{\exp\left\{-\frac{(y_i-\tau)\beta+\tau-\mu}{\sigma}\right\}}{\left(1+\exp\left\{-\frac{(y_i-\tau)\beta+\tau-\mu}{\sigma}\right\}\right)^2}\right)-\frac{n_a}{\beta^2}$$

$$-\sum_{i=1}^{n}\bar{\delta}_{1i}\bar{\delta}_{2i}\left(\frac{y_{(r)}-\tau}{\sigma}\right)^2\left(\frac{\exp\left\{-\frac{(y_{(r)}-\tau)\beta+\tau-\mu}{\sigma}\right\}}{\left(1+\exp\left\{-\frac{(y_{(r)}-\tau)\beta+\tau-\mu}{\sigma}\right\}\right)^2}\right),$$

$$\frac{\partial^2 \ln L(\beta,\mu,\sigma)}{\partial\mu\partial\beta}$$

$$=\frac{2}{\sigma}\sum_{i=1}^{n}\delta_{2i}\left(\frac{y_i-\tau}{\sigma}\right)\left(\frac{\exp\left\{-\frac{(y_i-\tau)\beta+\tau-\mu}{\sigma}\right\}}{\left(1+\exp\left\{-\frac{(y_i-\tau)\beta+\tau-\mu}{\sigma}\right\}\right)^2}\right)$$

$$+\frac{1}{\sigma}\sum_{i=1}^{n}\bar{\delta}_{1i}\bar{\delta}_{2i}\left(\frac{y_{(r)}-\tau}{\sigma}\right)\left(\frac{\exp\left\{-\frac{(y_{(r)}-\tau)\beta+\tau-\mu}{\sigma}\right\}}{\left(1+\exp\left\{-\frac{(y_{(r)}-\tau)\beta+\tau-\mu}{\sigma}\right\}\right)^2}\right),$$

$$\frac{\partial^2 \ln L(\beta,\mu,\sigma)}{\partial\sigma\partial\beta}$$

$$=-2\sum_{i=1}^{n}\delta_{2i}\left(\frac{y_i-\tau}{\sigma^2}\right)\left(\frac{\exp\left\{-\frac{(y_i-\tau)\beta+\tau-\mu}{\sigma}\right\}}{1+\exp\left\{-\frac{(y_i-\tau)\beta+\tau-\mu}{\sigma}\right\}}\right)$$

$$+\sum_{i=1}^{n}\delta_{2i}\left(\frac{y_i-\tau}{\sigma^2}\right)+\sum_{i=1}^{n}\bar{\delta}_{1i}\bar{\delta}_{2i}\left(\frac{y_{(r)}-\tau}{\sigma^2}\right)$$

$$-\sum_{i=1}^{n}\bar{\delta}_{1i}\bar{\delta}_{2i}\left(\frac{y_{(r)}-\tau}{\sigma^2}\right)\left(\frac{\exp\left\{-\frac{(y_{(r)}-\tau)\beta+\tau-\mu}{\sigma}\right\}}{1+\exp\left\{-\frac{(y_{(r)}-\tau)\beta+\tau-\mu}{\sigma}\right\}}\right)$$

$$+2\sum_{i=1}^{n}\left\{\delta_{2i}\left(\frac{(y_i-\tau)\beta+\tau-\mu}{\sigma^2}\right)\right.$$

$$\left.\times\left(\frac{\left(\frac{y_i-\tau}{\sigma}\right)\exp\left\{-\frac{(y_i-\tau)\beta+\tau-\mu}{\sigma}\right\}}{\left(1+\exp\left\{-\frac{(y_i-\tau)\beta+\tau-\mu}{\sigma}\right\}\right)^2}\right)\right\}$$

$$+\sum_{i=1}^{n}\left\{\bar{\delta}_{1i}\bar{\delta}_{2i}\left(\frac{y_{(r)}-\tau}{\sigma}\right)\left(\frac{(y_{(r)}-\tau)\beta+\tau-\mu}{\sigma^2}\right)\right.$$

$$\left.\times\left(\frac{\exp\left\{-\frac{(y_{(r)}-\tau)\beta+\tau-\mu}{\sigma}\right\}}{\left(1+\exp\left\{-\frac{(y_{(r)}-\tau)\beta+\tau-\mu}{\sigma}\right\}\right)^2}\right)\right\},$$

$$\frac{\partial^2 \ln L(\beta,\mu,\sigma)}{\partial\mu^2}$$

$$=\frac{n\exp\{-\frac{\mu}{\sigma}\}}{\sigma^2\left(1+\exp\{-\frac{\mu}{\sigma}\}\right)^2}-2\sum_{i=1}^{n}\frac{\delta_{1i}}{\sigma^2}\left(\frac{\exp\left\{-\frac{(y_i-\mu)}{\sigma}\right\}}{\left(1+\exp\left\{-\frac{(y_i-\mu)}{\sigma}\right\}\right)^2}\right)$$

$$-2\sum_{i=1}^{n}\frac{\delta_{2i}}{\sigma^2}\left(\frac{\exp\left\{-\frac{(y_i-\tau)\beta+\tau-\mu}{\sigma}\right\}}{\left(1+\exp\left\{-\frac{(y_i-\tau)\beta+\tau-\mu}{\sigma}\right\}\right)^2}\right)$$

$$-\frac{1}{\sigma^2}\sum_{i=1}^{n}\bar{\delta}_{1i}\bar{\delta}_{2i}\left(\frac{\exp\left\{-\frac{(y_{(r)}-\tau)\beta+\tau-\mu}{\sigma}\right\}}{\left(1+\exp\left\{-\frac{(y_{(r)}-\tau)\beta+\tau-\mu}{\sigma}\right\}\right)^2}\right),$$

$$\frac{\partial^2 \ln L(\beta,\mu,\sigma)}{\partial\sigma\partial\mu}$$

$$=2\sum_{i=1}^{n}\frac{\delta_{1i}}{\sigma^2}\left(\frac{\exp\left\{-\frac{(y_i-\mu)}{\sigma}\right\}}{1+\exp\left\{-\frac{(y_i-\mu)}{\sigma}\right\}}\right)-\frac{n\mu\exp\{-\frac{\mu}{\sigma}\}}{\sigma^3\left(1+\exp\{-\frac{\mu}{\sigma}\}\right)^2}$$

$$-2\sum_{i=1}^{n}\frac{\delta_{1i}}{\sigma}\left(\frac{y_i-\mu}{\sigma^2}\right)\left(\frac{\exp\left\{-\frac{(y_i-\mu)}{\sigma}\right\}}{\left(1+\exp\left\{-\frac{(y_i-\mu)}{\sigma}\right\}\right)^2}\right)-\sum_{i=1}^{n}\frac{\delta_{1i}}{\sigma^2}$$

$$
\begin{aligned}
&-\sum_{i=1}^{n}\frac{\delta_{2i}}{\sigma^2}+\frac{2}{\sigma^2}\sum_{i=1}^{n}\delta_{2i}\left(\frac{\exp\left\{-\frac{(y_i-\tau)\beta+\tau-\mu}{\sigma}\right\}}{1+\exp\left\{-\frac{(y_i-\tau)\beta+\tau-\mu}{\sigma}\right\}}\right)\\
&+\frac{n\exp\{-\frac{\mu}{\sigma}\}}{\sigma^2\left(1+\exp\{-\frac{\mu}{\sigma}\}\right)}-\frac{1}{\sigma}\sum_{i=1}^{n}\left\{\bar{\delta}_{1i}\bar{\delta}_{2i}\left(\frac{(y_{(r)}-\tau)\beta+\tau-\mu}{\sigma^2}\right)\right.\\
&\left.\times\left(\frac{\exp\left\{-\frac{(y_{(r)}-\tau)\beta+\tau-\mu}{\sigma}\right\}}{\left(1+\exp\left\{-\frac{(y_{(r)}-\tau)\beta+\tau-\mu}{\sigma}\right\}\right)^2}\right)\right\}\\
&-\frac{2}{\sigma}\sum_{i=1}^{n}\delta_{2i}\left(\frac{(y_i-\tau)\beta+\tau-\mu}{\sigma^2}\right)\\
&\times\left(\frac{\exp\left\{-\frac{(y_i-\tau)\beta+\tau-\mu}{\sigma}\right\}}{\left(1+\exp\left\{-\frac{(y_i-\tau)\beta+\tau-\mu}{\sigma}\right\}\right)^2}\right)\\
&-\sum_{i=1}^{n}\frac{\bar{\delta}_{1i}\bar{\delta}_{2i}}{\sigma^2}+\frac{1}{\sigma^2}\sum_{i=1}^{n}\bar{\delta}_{1i}\bar{\delta}_{2i}\left(\frac{\exp\left\{-\frac{(y_{(r)}-\tau)\beta+\tau-\mu}{\sigma}\right\}}{1+\exp\left\{-\frac{(y_{(r)}-\tau)\beta+\tau-\mu}{\sigma}\right\}}\right)
\end{aligned}
$$

and

$$
\begin{aligned}
&\frac{\partial^2\ln L(\beta,\mu,\sigma)}{\partial\sigma^2}\\
&\quad=\frac{n\mu^2\exp\{-\frac{\mu}{\sigma}\}}{\sigma^4\left(1+\exp\{-\frac{\mu}{\sigma}\}\right)^2}-\frac{2n\mu\exp\{-\frac{\mu}{\sigma}\}}{\sigma^3\left(1+\exp\{-\frac{\mu}{\sigma}\}\right)}+\left(\frac{n_u+n_a}{\sigma^2}\right)\\
&\quad+4\sum_{i=1}^{n}\delta_{1i}\left(\frac{y_i-\mu}{\sigma^3}\right)\left(\frac{\exp\left\{-\frac{(y_i-\mu)}{\sigma}\right\}}{1+\exp\left\{-\frac{(y_i-\mu)}{\sigma}\right\}}\right)\\
&\quad-2\sum_{i=1}^{n}\delta_{1i}\left(\frac{y_i-\mu}{\sigma^3}\right)-2\sum_{i=1}^{n}\delta_{1i}\left(\frac{y_i-\mu}{\sigma^2}\right)^2
\end{aligned}
$$

$$
\begin{aligned}
&\times \left(\frac{\exp\left\{-\frac{(y_i-\mu)}{\sigma}\right\}}{\left(1+\exp\left\{-\frac{(y_i-\mu)}{\sigma}\right\}\right)^2} \right) \\
&- 2\sum_{i=1}^{n} \bar{\delta}_{1i}\bar{\delta}_{2i} \left(\frac{(y_{(r)}-\tau)\beta+\tau-\mu}{\sigma^3} \right) \\
&- 2\sum_{i=1}^{n} \delta_{2i} \left(\frac{(y_i-\tau)\beta+\tau-\mu}{\sigma^3} \right) \\
&- 2\sum_{i=1}^{n} \delta_{2i} \left(\frac{(y_i-\tau)\beta+\tau-\mu}{\sigma^2} \right)^2 \\
&\times \left(\frac{\exp\left\{-\frac{(y_i-\tau)\beta+\tau-\mu}{\sigma}\right\}}{\left(1+\exp\left\{-\frac{(y_i-\tau)\beta+\tau-\mu}{\sigma}\right\}\right)^2} \right) \\
&+ 4\sum_{i=1}^{n} \delta_{2i} \left(\frac{(y_i-\tau)\beta+\tau-\mu}{\sigma^3} \right) \left(\frac{\exp\left\{-\frac{(y_i-\tau)\beta+\tau-\mu}{\sigma}\right\}}{1+\exp\left\{-\frac{(y_i-\tau)\beta+\tau-\mu}{\sigma}\right\}} \right) \\
&- 2\sum_{i=1}^{n} \bar{\delta}_{1i}\bar{\delta}_{2i} \left(\frac{(y_{(r)}-\tau)\beta+\tau-\mu}{\sigma^3} \right) \\
&\times \left(\frac{\exp\left\{-\frac{(y_{(r)}-\tau)\beta+\tau-\mu}{\sigma}\right\}}{1+\exp\left\{-\frac{(y_{(r)}-\tau)\beta+\tau-\mu}{\sigma}\right\}} \right) \\
&- \sum_{i=1}^{n} \bar{\delta}_{1i}\bar{\delta}_{2i} \left(\frac{(y_{(r)}-\tau)\beta+\tau-\mu}{\sigma^2} \right)^2 \\
&\times \left(\frac{\exp\left\{-\frac{(y_{(r)}-\tau)\beta+\tau-\mu}{\sigma}\right\}}{\left(1+\exp\left\{-\frac{(y_{(r)}-\tau)\beta+\tau-\mu}{\sigma}\right\}\right)^2} \right).
\end{aligned}
$$

The elements of the Fisher information matrix for an observation are the negative expectations of these second-order partial derivatives.

These are as follows:

$$E\left[-\frac{\partial^2 \ln L(\beta,\mu,\sigma)}{\partial\beta^2}\right] = \frac{nP_a}{\beta^2} + \frac{2n\beta}{A\sigma^3}\int_\tau^{y_{(r)}} (y-\tau)^2 \left(K(y)\right)^2 dy$$
$$+ \frac{n(1-P_a-P_u)}{\sigma^2}(y_{(r)}-\tau)^2 K(y_{(r)}),$$

$$E\left[-\frac{\partial^2 \ln L(\beta,\mu,\sigma)}{\partial\mu\partial\beta}\right] = \frac{-n(1-P_a-P_u)\left(y_{(r)}-\tau\right)K(y_{(r)})}{\sigma^2}$$
$$- \frac{2n\beta}{A\sigma^3}\int_\tau^{y_{(r)}} (y-\tau)(K(y))^2 dy,$$

$$E\left[-\frac{\partial^2 \ln L(\beta,\mu,\sigma)}{\partial\sigma\partial\beta}\right] = -\frac{n\beta}{A\sigma^3}\int_\tau^{y_{(r)}} (y-\tau)K(y)(1-2B(y))dy$$
$$- \frac{n(1-P_a-P_u)}{\sigma^2}(y_{(r)}-\tau)(1-B(y_{(r)})),$$

$$E\left[-\frac{\partial^2 \ln L(\beta,\mu,\sigma)}{\partial\mu^2}\right] = \frac{-ng(2\mu)}{\sigma^2} + \frac{n(1-P_a-P_u)K(y_{(r)})}{\sigma^2}$$
$$+ \frac{2n}{A\sigma^3}\int_0^\tau (g(y))^2 dy + \frac{2n\beta}{A\sigma^3}\int_\tau^{y_{(r)}} (K(y))^2 dy,$$

$$E\left[-\frac{\partial^2 \ln L(\beta,\mu,\sigma)}{\partial\sigma\partial\mu}\right] = \frac{n(1-P_a-P_u)(1-B(y_{(r)}))}{\sigma^2}$$
$$+ \int_\tau^{y_{(r)}} K(y)(1-2B(y))dy\left(\frac{n\beta}{A\sigma^3}\right)$$
$$+ \frac{n}{A\sigma^3}\int_0^\tau g(y)(1-2Q(y))dy - \frac{nQ(2\mu)}{\sigma^2}$$

and

$$E\left[-\frac{\partial^2 \ln L(\beta,\mu,\sigma)}{\partial\sigma^2}\right]$$
$$= -\frac{2n(1-P_a-P_u)(\tau+\beta(y_{(r)}-\tau)-\mu)(1-H(y_{(r)}))}{\sigma^3}$$
$$+ \frac{2n\mu}{\sigma^3}M(2\mu) + \frac{2n\beta}{A\sigma^4}\int_\tau^{y_{(r)}} (\tau+\beta(y-\tau)-\mu)K(y)(1-2H(y))dy$$
$$- \frac{n(P_a+P_u)}{\sigma^2} + \frac{2n}{A\sigma^4}\int_0^\tau (y-\mu)g(y)(1-2M(y))dy,$$

where

$$K(y) = \frac{\exp\left\{-\frac{(y-\tau)\beta+\tau-\mu}{\sigma}\right\}}{\left(1+\exp\left\{-\frac{(y-\tau)\beta+\tau-\mu}{\sigma}\right\}\right)^2},$$

$$g(y) = \frac{\exp\left\{-\frac{(y-\mu)}{\sigma}\right\}}{\left(1+\exp\left\{-\frac{(y-\mu)}{\sigma}\right\}\right)^2},$$

$$M(y) = \left(1+\exp\left\{-\frac{(y-\mu)}{\sigma}\right\} - \frac{(y-\mu)}{2\sigma}\right) g(y),$$

$$Q(y) = \left(1+\exp\left\{-\frac{(y-\mu)}{\sigma}\right\} - \frac{(y-\mu)}{\sigma}\right) g(y),$$

$$H(y) = \left(1+\exp\left\{-\frac{(y-\tau)\beta+\tau-\mu}{\sigma}\right\} - \frac{(y-\tau)\beta+\tau-\mu}{2\sigma}\right) K(y)$$

and

$$B(y) = \left(1+\exp\left\{-\frac{(y-\tau)\beta+\tau-\mu}{\sigma}\right\} - \frac{(y-\tau)\beta+\tau-\mu}{\sigma}\right) K(y).$$

A.5 CALCULATIONS OF THE ELEMENTS OF THE FISHER INFORMATION MATRIX GIVEN IN SECTION 3.3.2.4

The first-order partial derivatives are

$$\begin{aligned}\frac{\partial \ln L(\beta,\mu,\sigma)}{\partial\beta} \\ &= \frac{nq_r - nq_l}{\beta} - \sum_{i=1}^{n} \delta_{2i}\left(\frac{u_{(i)} - u_{(nq_l)}}{\sigma}\right) \\ &\quad - \sum_{i=1}^{n} \bar{\delta}_{1i}\bar{\delta}_{2i}\left(\frac{u_{(nq_r)} - u_{(nq_l)}}{\sigma}\right)\end{aligned}$$

$$+ 2\sum_{i=1}^{n} \delta_{2i} \left(\frac{u_{(i)} - u_{(nq_l)}}{\sigma}\right)\left(\frac{\exp\left\{-\frac{u_{(nq_l)}+(u_{(i)}-u_{(nq_l)})\beta-\mu}{\sigma}\right\}}{1+\exp\left\{-\frac{u_{(nq_l)}+(u_{(i)}-u_{(nq_l)})\beta-\mu}{\sigma}\right\}}\right)$$

$$+ \sum_{i=1}^{n} \bar{\delta}_{1i}\bar{\delta}_{2i} \left(\frac{\exp\left\{-\frac{u_{(nq_l)}+(u_{(nq_r)}-u_{(nq_l)})\beta-\mu}{\sigma}\right\}}{1+\exp\left\{-\frac{u_{(nq_l)}+(u_{(nq_r)}-u_{(nq_l)})\beta-\mu}{\sigma}\right\}}\right)$$

$$\times \left(\frac{u_{(nq_r)} - u_{(nq_l)}}{\sigma}\right),$$

$$\frac{\partial \ln L(\beta, \mu, \sigma)}{\partial \mu}$$

$$= \sum_{i=1}^{n} \delta_{1i} \left(\frac{1}{\sigma}\right) + \sum_{i=1}^{n} \delta_{2i} \left(\frac{1}{\sigma}\right) + \sum_{i=1}^{n} \bar{\delta}_{1i}\bar{\delta}_{2i} \left(\frac{1}{\sigma}\right)$$

$$- \frac{n\exp\{-\frac{\mu}{\sigma}\}}{\sigma\left(1+\exp\{-\frac{\mu}{\sigma}\}\right)} - \frac{2}{\sigma}\sum_{i=1}^{n} \delta_{1i} \left(\frac{\exp\left\{-\frac{u_{(i)}-\mu}{\sigma}\right\}}{1+\exp\left\{-\frac{u_{(i)}-\mu}{\sigma}\right\}}\right)$$

$$- \frac{2}{\sigma}\sum_{i=1}^{n} \delta_{2i} \left(\frac{\exp\left\{-\frac{u_{(nq_l)}+(u_{(i)}-u_{(nq_l)})\beta-\mu}{\sigma}\right\}}{1+\exp\left\{-\frac{u_{(nq_l)}+(u_{(i)}-u_{(nq_l)})\beta-\mu}{\sigma}\right\}}\right)$$

$$- \frac{1}{\sigma}\sum_{i=1}^{n} \bar{\delta}_{1i}\bar{\delta}_{2i} \left(\frac{\exp\left\{-\frac{u_{(nq_l)}+(u_{(nq_r)}-u_{(nq_l)})\beta-\mu}{\sigma}\right\}}{1+\exp\left\{-\frac{u_{(nq_l)}+(u_{(nq_r)}-u_{(nq_l)})\beta-\mu}{\sigma}\right\}}\right)$$

and

$$\frac{\partial \ln L(\beta, \mu, \sigma)}{\partial \sigma}$$

$$= \frac{n\mu\exp\{-\frac{\mu}{\sigma}\}}{\sigma^2\left(1+\exp\{-\frac{\mu}{\sigma}\}\right)} + \sum_{i=1}^{n} \delta_{2i} \left(\frac{u_{(nq_l)} + (u_{(i)} - u_{(nq_l)})\beta - \mu}{\sigma^2}\right)$$

$$-\frac{nq_r}{\sigma}+\sum_{i=1}^{n}\delta_{1i}\left(\frac{u_{(i)}-\mu}{\sigma^2}\right)$$

$$-2\sum_{i=1}^{n}\delta_{1i}\left(\frac{u_{(i)}-\mu}{\sigma^2}\right)\left(\frac{\exp\left\{-\frac{u_{(i)}-\mu}{\sigma}\right\}}{1+\exp\left\{-\frac{u_{(i)}-\mu}{\sigma}\right\}}\right)$$

$$+\sum_{i=1}^{n}\bar{\delta}_{1i}\bar{\delta}_{2i}\left(\frac{u_{(nq_l)}+(u_{(nq_r)}-u_{(nq_l)})\beta-\mu}{\sigma^2}\right)$$

$$-2\sum_{i=1}^{n}\delta_{2i}\left(-\frac{u_{(nq_l)}+(u_{(i)}-u_{(nq_l)})\beta-\mu}{\sigma^2}\right)$$

$$\times\left(\frac{\exp\left\{-\frac{u_{(nq_l)}+(u_{(i)}-u_{(nq_l)})\beta-\mu}{\sigma}\right\}}{1+\exp\left\{-\frac{u_{(nq_l)}+(u_{(i)}-u_{(nq_l)})\beta-\mu}{\sigma}\right\}}\right)$$

$$-\left(\frac{u_{(nq_l)}+(u_{(nq_r)}-u_{(nq_l)})\beta-\mu}{\sigma^2}\right)$$

$$\times\sum_{i=1}^{n}\bar{\delta}_{1i}\bar{\delta}_{2i}\left(\frac{\exp\left\{-\frac{u_{(nq_l)}+(u_{(nq_r)}-u_{(nq_l)})\beta-\mu}{\sigma}\right\}}{1+\exp\left\{-\frac{u_{(nq_l)}+(u_{(nq_r)}-u_{(nq_l)})\beta-\mu}{\sigma}\right\}}\right).$$

The second-order partial derivatives are

$$\frac{\partial^2\ln L(\beta,\mu,\sigma)}{\partial\beta^2}$$

$$=-2\sum_{i=1}^{n}\left\{\delta_{2i}\left(\frac{u_{(i)}-u_{(nq_l)}}{\sigma}\right)^2\right.$$

$$\left.\times\left(\frac{\exp\left\{-\frac{(u_{(i)}-u_{(nq_l)})\beta+u_{(nq_l)}-\mu}{\sigma}\right\}}{\left(1+\exp\left\{-\frac{(u_{(i)}-u_{(nq_l)})\beta+u_{(nq_l)}-\mu}{\sigma}\right\}\right)^2}\right)\right\}$$

$$- \frac{(nq_r - nq_l)}{\beta^2} - \sum_{i=1}^{n} \left\{ \bar{\delta}_{1i} \bar{\delta}_{2i} \left(\frac{u_{(nq_r)} - u_{(nq_l)}}{\sigma} \right)^2 \right.$$

$$\left. \times \left(\frac{\exp\left\{ -\frac{(u_{(nq_r)} - u_{(nq_l)})\beta + u_{(nq_l)} - \mu}{\sigma} \right\}}{\left(1 + \exp\left\{ -\frac{(u_{(nq_r)} - u_{(nq_l)})\beta + u_{(nq_l)} - \mu}{\sigma} \right\}\right)^2} \right) \right\},$$

$$\frac{\partial^2 \ln L(\beta, \mu, \sigma)}{\partial \mu \partial \beta}$$

$$= \frac{2}{\sigma} \sum_{i=1}^{n} \left\{ \delta_{2i} \left(\frac{u_{(i)} - u_{(nq_l)}}{\sigma} \right) \right.$$

$$\left. \times \left(\frac{\exp\left\{ -\frac{(u_{(i)} - u_{(nq_l)})\beta + u_{(nq_l)} - \mu}{\sigma} \right\}}{\left(1 + \exp\left\{ -\frac{(u_{(i)} - u_{(nq_l)})\beta + u_{(nq_l)} - \mu}{\sigma} \right\}\right)^2} \right) \right\}$$

$$+ \frac{1}{\sigma} \sum_{i=1}^{n} \left\{ \bar{\delta}_{1i} \bar{\delta}_{2i} \left(\frac{u_{(nq_r)} - u_{(nq_l)}}{\sigma} \right) \right.$$

$$\left. \times \left(\frac{\exp\left\{ -\frac{(u_{(nq_r)} - u_{(nq_l)})\beta + u_{(nq_l)} - \mu}{\sigma} \right\}}{\left(1 + \exp\left\{ -\frac{(u_{(nq_r)} - u_{(nq_l)})\beta + u_{(nq_l)} - \mu}{\sigma} \right\}\right)^2} \right) \right\},$$

$$\frac{\partial^2 \ln L(\beta, \mu, \sigma)}{\partial \sigma \partial \beta}$$

$$= \sum_{i=1}^{n} \delta_{2i} \left(\frac{u_{(i)} - u_{(nq_l)}}{\sigma^2} \right) - 2 \sum_{i=1}^{n} \left\{ \delta_{2i} \left(\frac{u_{(i)} - u_{(nq_l)}}{\sigma^2} \right) \right.$$

$$\left. \times \left(\frac{\exp\left\{ -\frac{(u_{(i)} - u_{(nq_l)})\beta + u_{(nq_l)} - \mu}{\sigma} \right\}}{1 + \exp\left\{ -\frac{(u_{(i)} - u_{(nq_l)})\beta + u_{(nq_l)} - \mu}{\sigma} \right\}} \right) \right\}$$

$$+\sum_{i=1}^{n}\bar{\delta}_{1i}\bar{\delta}_{2i}\left(\frac{u_{(nq_r)}-u_{(nq_l)}}{\sigma^2}\right)-\sum_{i=1}^{n}\left\{\bar{\delta}_{1i}\bar{\delta}_{2i}\left(\frac{u_{(nq_r)}-u_{(nq_l)}}{\sigma^2}\right)\right.$$

$$\left.\times\left(\frac{\exp\left\{-\frac{(u_{(nq_r)}-u_{(nq_l)})\beta+u_{(nq_l)}-\mu}{\sigma}\right\}}{1+\exp\left\{-\frac{(u_{(nq_r)}-u_{(nq_l)})\beta+u_{(nq_l)}-\mu}{\sigma}\right\}}\right)\right\}$$

$$+2\sum_{i=1}^{n}\left\{\delta_{2i}\left(\frac{u_{(i)}-u_{(nq_l)}}{\sigma}\right)\left(\frac{(u_{(i)}-u_{(nq_l)})\beta+u_{(nq_l)}-\mu}{\sigma^2}\right)\right.$$

$$\left.\times\left(\frac{\exp\left\{-\frac{(u_{(i)}-u_{(nq_l)})\beta+u_{(nq_l)}-\mu}{\sigma}\right\}}{\left(1+\exp\left\{-\frac{(u_{(i)}-u_{(nq_l)})\beta+u_{(nq_l)}-\mu}{\sigma}\right\}\right)^2}\right)\right\}$$

$$+\sum_{i=1}^{n}\left\{\bar{\delta}_{1i}\bar{\delta}_{2i}\left(\frac{u_{(nq_r)}-u_{(nq_l)}}{\sigma}\right)\left(\frac{(u_{(nq_r)}-u_{(nq_l)})\beta+u_{(nq_l)}-\mu}{\sigma^2}\right)\right.$$

$$\left.\times\left(\frac{\exp\left\{-\frac{(u_{(nq_r)}-u_{(nq_l)})\beta+u_{(nq_l)}-\mu}{\sigma}\right\}}{\left(1+\exp\left\{-\frac{(u_{(nq_r)}-u_{(nq_l)})\beta+u_{(nq_l)}-\mu}{\sigma}\right\}\right)^2}\right)\right\},$$

$$\frac{\partial^2\ln L(\beta,\mu,\sigma)}{\partial\mu^2}$$

$$=\frac{n\exp\left\{-\frac{\mu}{\sigma}\right\}}{\sigma^2\left(1+\exp\left\{-\frac{\mu}{\sigma}\right\}\right)^2}-2\sum_{i=1}^{n}\frac{\delta_{1i}}{\sigma^2}\left(\frac{\exp\left\{-\frac{(u_{(i)}-\mu)}{\sigma}\right\}}{\left(1+\exp\left\{-\frac{(u_{(i)}-\mu)}{\sigma}\right\}\right)^2}\right)$$

$$-2\sum_{i=1}^{n}\frac{\delta_{2i}}{\sigma^2}\left(\frac{\exp\left\{-\frac{(u_{(i)}-u_{(nq_l)})\beta+u_{(nq_l)}-\mu}{\sigma}\right\}}{\left(1+\exp\left\{-\frac{(u_{(i)}-u_{(nq_l)})\beta+u_{(nq_l)}-\mu}{\sigma}\right\}\right)^2}\right)$$

$$-\frac{1}{\sigma^2}\sum_{i=1}^{n}\bar{\delta}_{1i}\bar{\delta}_{2i}\left(\frac{\exp\left\{-\frac{(u_{(nq_r)}-u_{(nq_l)})\beta+u_{(nq_l)}-\mu}{\sigma}\right\}}{\left(1+\exp\left\{-\frac{(u_{(nq_r)}-u_{(nq_l)})\beta+u_{(nq_l)}-\mu}{\sigma}\right\}\right)^2}\right),$$

$$\frac{\partial^2 \ln L(\beta,\mu,\sigma)}{\partial\sigma\partial\mu}$$

$$= 2\sum_{i=1}^{n}\frac{\delta_{1i}}{\sigma^2}\left(\frac{\exp\left\{-\frac{(u_{(i)}-\mu)}{\sigma}\right\}}{1+\exp\left\{-\frac{(u_{(i)}-\mu)}{\sigma}\right\}}\right) - \frac{n\mu\exp\{-\frac{\mu}{\sigma}\}}{\sigma^3\left(1+\exp\{-\frac{\mu}{\sigma}\}\right)^2}$$

$$-2\sum_{i=1}^{n}\frac{\delta_{1i}}{\sigma}\left(\frac{u_{(i)}-\mu}{\sigma^2}\right)\left(\frac{\exp\left\{-\frac{(u_{(i)}-\mu)}{\sigma}\right\}}{\left(1+\exp\left\{-\frac{(u_{(i)}-\mu)}{\sigma}\right\}\right)^2}\right) - \sum_{i=1}^{n}\frac{\delta_{1i}}{\sigma^2}$$

$$-\sum_{i=1}^{n}\frac{\delta_{2i}}{\sigma^2} + \frac{2}{\sigma^2}\sum_{i=1}^{n}\delta_{2i}\left(\frac{\exp\left\{-\frac{(u_{(i)}-u_{(nq_l)})\beta+u_{(nq_l)}-\mu}{\sigma}\right\}}{1+\exp\left\{-\frac{(u_{(i)}-u_{(nq_l)})\beta+u_{(nq_l)}-\mu}{\sigma}\right\}}\right)$$

$$+\frac{n\exp\{-\frac{\mu}{\sigma}\}}{\sigma^2\left(1+\exp\{-\frac{\mu}{\sigma}\}\right)}$$

$$-\frac{2}{\sigma}\sum_{i=1}^{n}\left\{\delta_{2i}\left(\frac{(u_{(i)}-u_{(nq_l)})\beta+u_{(nq_l)}-\mu}{\sigma^2}\right)\right.$$

$$\left.\times\left(\frac{\exp\left\{-\frac{(u_{(i)}-u_{(nq_l)})\beta+u_{(nq_l)}-\mu}{\sigma}\right\}}{\left(1+\exp\left\{-\frac{(u_{(i)}-u_{(nq_l)})\beta+u_{(nq_l)}-\mu}{\sigma}\right\}\right)^2}\right)\right\}$$

$$+\frac{1}{\sigma^2}\sum_{i=1}^{n}\bar{\delta}_{1i}\bar{\delta}_{2i}\left(\frac{\exp\left\{-\frac{(u_{(nq_r)}-u_{(nq_l)})\beta+u_{(nq_l)}-\mu}{\sigma}\right\}}{1+\exp\left\{-\frac{(u_{(nq_r)}-u_{(nq_l)})\beta+u_{(nq_l)}-\mu}{\sigma}\right\}}\right)$$

$$-\sum_{i=1}^{n}\frac{\bar{\delta}_{1i}\bar{\delta}_{2i}}{\sigma^2} - \frac{1}{\sigma}\sum_{i=1}^{n}\left\{\bar{\delta}_{1i}\bar{\delta}_{2i}\left(\frac{(u_{(nq_r)}-u_{(nq_l)})\beta+u_{(nq_l)}-\mu}{\sigma^2}\right)\right.$$

$$\left.\times\left(\frac{\exp\left\{-\frac{(u_{(nq_r)}-u_{(nq_l)})\beta+u_{(nq_l)}-\mu}{\sigma}\right\}}{\left(1+\exp\left\{-\frac{(u_{(nq_r)}-u_{(nq_l)})\beta+u_{(nq_l)}-\mu}{\sigma}\right\}\right)^2}\right)\right\}$$

and

$$\begin{aligned}
&\frac{\partial^2 \ln L(\beta,\mu,\sigma)}{\partial\sigma^2} \\
&\quad= \frac{n\mu^2\exp\{-\frac{\mu}{\sigma}\}}{\sigma^4\left(1+\exp\{-\frac{\mu}{\sigma}\}\right)^2} - \frac{2n\mu\exp\{-\frac{\mu}{\sigma}\}}{\sigma^3\left(1+\exp\{-\frac{\mu}{\sigma}\}\right)} + \frac{nq_r}{\sigma^2} \\
&\qquad + 4\sum_{i=1}^{n}\delta_{1i}\left(\frac{u_{(i)}-\mu}{\sigma^3}\right)\left(\frac{\exp\left\{-\frac{(u_{(i)}-\mu)}{\sigma}\right\}}{1+\exp\left\{-\frac{(u_{(i)}-\mu)}{\sigma}\right\}}\right) \\
&\qquad - 2\sum_{i=1}^{n}\delta_{1i}\left(\frac{u_{(i)}-\mu}{\sigma^3}\right) \\
&\qquad - 2\sum_{i=1}^{n}\delta_{1i}\left(\frac{u_{(i)}-\mu}{\sigma^2}\right)^2\left(\frac{\exp\left\{-\frac{(u_{(i)}-\mu)}{\sigma}\right\}}{\left(1+\exp\left\{-\frac{(u_{(i)}-\mu)}{\sigma}\right\}\right)^2}\right) \\
&\qquad - 2\sum_{i=1}^{n}\left\{\delta_{2i}\left(\frac{(u_{(i)}-u_{(nq_l)})\beta+u_{(nq_l)}-\mu}{\sigma^2}\right)^2\right. \\
&\qquad \times\left.\left(\frac{\exp\left\{-\frac{(u_{(i)}-u_{(nq_l)})\beta+u_{(nq_l)}-\mu}{\sigma}\right\}}{\left(1+\exp\left\{-\frac{(u_{(i)}-u_{(nq_l)})\beta+u_{(nq_l)}-\mu}{\sigma}\right\}\right)^2}\right)\right\} \\
&\qquad - 2\sum_{i=1}^{n}\bar{\delta}_{1i}\bar{\delta}_{2i}\left(\frac{(u_{(nq_r)}-u_{(nq_l)})\beta+u_{(nq_l)}-\mu}{\sigma^3}\right) \\
&\qquad -2\sum_{i=1}^{n}\delta_{2i}\left(\frac{(u_{(i)}-u_{(nq_l)})\beta+u_{(nq_l)}-\mu}{\sigma^3}\right) \\
&\qquad +4\sum_{i=1}^{n}\left\{\delta_{2i}\left(\frac{(u_{(i)}-u_{(nq_l)})\beta+u_{(nq_l)}-\mu}{\sigma^3}\right)\right. \\
&\qquad \times\left.\left(\frac{\exp\left\{-\frac{(u_{(i)}-u_{(nq_l)})\beta+u_{(nq_l)}-\mu}{\sigma}\right\}}{1+\exp\left\{-\frac{(u_{(i)}-u_{(nq_l)})\beta+u_{(nq_l)}-\mu}{\sigma}\right\}}\right)\right\}
\end{aligned}$$

$$-2\sum_{i=1}^{n}\left\{\bar{\delta}_{1i}\bar{\delta}_{2i}\left(\frac{(u_{(nq_r)}-u_{(nq_l)})\beta+u_{(nq_l)}-\mu}{\sigma^3}\right)\right.$$

$$\left.\times\left(\frac{\exp\left\{-\frac{(u_{(nq_r)}-u_{(nq_l)})\beta+u_{(nq_l)}-\mu}{\sigma}\right\}}{1+\exp\left\{-\frac{(u_{(nq_r)}-u_{(nq_l)})\beta+u_{(nq_l)}-\mu}{\sigma}\right\}}\right)\right\}$$

$$-\sum_{i=1}^{n}\left\{\bar{\delta}_{1i}\bar{\delta}_{2i}\left(\frac{(u_{(nq_r)}-u_{(nq_l)})\beta+u_{(nq_l)}-\mu}{\sigma^2}\right)^2\right.$$

$$\left.\times\left(\frac{\exp\left\{-\frac{(u_{(nq_r)}-u_{(nq_l)})\beta+u_{(nq_l)}-\mu}{\sigma}\right\}}{\left(1+\exp\left\{-\frac{(u_{(nq_r)}-u_{(nq_l)})\beta+u_{(nq_l)}-\mu}{\sigma}\right\}\right)^2}\right)\right\}.$$

The elements of the Fisher information matrix for an observation are the negative expectations of these second-order partial derivatives.

The conditional expectations of negative of second-order derivatives given $u_{(nq_r)} = y$ and $u_{(nq_l)} = x$ are

$$E\left[-\frac{\partial^2 \ln L(\beta,\mu,\sigma)}{\partial\beta^2}\,|u_{(nq_r)}=y, u_{(nq_l)}=x\right]$$

$$=\frac{nP_a}{\beta^2}+\frac{n(1-P_u-P_a)(y-x)^2K(y)}{\sigma^2}$$

$$+\frac{2n\beta}{A\sigma^3}\int_x^y (u-x)^2(K(u))^2du, \qquad \text{(A.1)}$$

$$E\left[-\frac{\partial^2 \ln L(\beta,\mu,\sigma)}{\partial\mu\partial\beta}\,|u_{(nq_r)}=y, u_{(nq_l)}=x\right]$$

$$=-\frac{n(1-P_u-P_a)(y-x)K(y)}{\sigma^2}$$

$$-\frac{2n\beta}{A\sigma^3}\int_x^y (u-x)(K(u))^2du, \qquad \text{(A.2)}$$

$$E\left[-\frac{\partial^2 \ln L(\beta,\mu,\sigma)}{\partial\sigma\partial\beta}\,\middle|u_{(nq_r)}=y,u_{(nq_l)}=x\right]$$
$$=-\frac{n(1-P_u-P_a)(y-x)(1-B(y))}{\sigma^2}$$
$$-\frac{n\beta}{A\sigma^3}\int_x^y (u-x)K(u)(1-2B(u))du, \tag{A.3}$$

$$E\left[-\frac{\partial^2 \ln L(\beta,\mu,\sigma)}{\partial\mu^2}\,\middle|u_{(nq_r)}=y,u_{(nq_l)}=x\right]$$
$$=-\frac{ng(2\mu)}{\sigma^2}+\frac{n(1-P_u-P_a)K(y)}{\sigma^2}$$
$$+\frac{2n}{A\sigma^3}\int_0^x (g(u))^2du+\frac{2n\beta}{A\sigma^3}\int_x^y (K(u))^2du, \tag{A.4}$$

$$E\left[-\frac{\partial^2 \ln L(\beta,\mu,\sigma)}{\partial\sigma\partial\mu}\,\middle|u_{(nq_r)}=y,u_{(nq_l)}=x\right]$$
$$=-\frac{nQ(2\mu)}{\sigma^2}+\frac{n(1-P_u-P_a)(1-B(y))}{\sigma^2}$$
$$+\frac{n}{A\sigma^3}\int_0^x g(u)(1-2Q(u))du+\frac{n\beta}{A\sigma^3}\int_x^y K(u)(1-2B(u))du, \tag{A.5}$$

and

$$E\left[-\frac{\partial^2 \ln L(\beta,\mu,\sigma)}{\partial\sigma^2}\,\middle|u_{(nq_r)}=y,u_{(nq_l)}=x\right]$$
$$=\frac{2n}{A\sigma^4}\int_0^x (u-\mu)g(u)(1-2M(u))du$$
$$+\frac{2n(1-P_u-P_a)(x+(y-x)\beta-\mu)(1-H(y))}{\sigma^3}$$
$$+\frac{2n\mu M(2\mu)}{\sigma^3}$$
$$+\frac{2n\beta}{A\sigma^4}\int_x^y (x+(u-x)\beta-\mu)K(u)(1-2H(u))du$$
$$-\frac{n(P_u+P_a)}{\sigma^2}, \tag{A.6}$$

where

$$K(u) = \frac{\exp\left\{-\frac{x+(u-x)\beta-\mu}{\sigma}\right\}}{\left(1+\exp\left\{-\frac{x+(u-x)\beta-\mu}{\sigma}\right\}\right)^2},$$

$$B(y) = \left(1+\exp\left\{-\frac{x+(u-x)\beta-\mu}{\sigma}\right\} - \frac{x+(u-x)\beta-\mu}{\sigma}\right) K(u),$$

$$g(u) = \frac{\exp\{-\frac{u-\mu}{\sigma}\}}{(1+\exp\{-\frac{u-\mu}{\sigma}\})^2},$$

$$Q(u) = \left(1+\exp\left\{-\frac{u-\mu}{\sigma}\right\} - \frac{u-\mu}{\sigma}\right) g(u),$$

$$M(u) = \left(1+\exp\left\{-\frac{u-\mu}{\sigma}\right\} - \frac{u-\mu}{2\sigma}\right) g(u),$$

and

$$H(y) = \left(1+\exp\left\{-\frac{x+(u-x)\beta-\mu}{\sigma}\right\} - \frac{x+(u-x)\beta-\mu}{2\sigma}\right) K(u).$$

The elements of the Fisher Information matrix are obtained by taking the expectation of the conditional expectations obtained in (A.1)–(A.6), and using the fact that if $Y_1, Y_2, \ldots, Y_n$ is a random sample of size n from a population with absolutely continuous distribution function $F(x)$, satisfying $F(0) = 0$, and $Y_{(1)} < Y_{(2)} < \cdots < Y_{(r)} \cdots < Y_{(n)}$ is the associated ordered sample, then the expectation of some function $t(x, y), 1 \leq r < s \leq n$ of rth, and sth order statistic (see Section A.02) is

$$E\left[t(Y_{(r)}, Y_{(s)})\right] = a \int_0^\infty \int_0^y t(x, y) z dx dy,$$

where

$$a = \frac{n!}{(r-1)!(s-r-1)!(n-s)!}$$

and

$$
\begin{aligned}
z &= F(x)^{r-1} f(x)(F(y) - F(x))^{s-r-1} f(y)(1 - F(y))^{n-s} \\
&= \left(\frac{1}{A}\left(\frac{1}{1+\exp\{-\frac{x-\mu}{\sigma}\}} - \frac{1}{1+\exp\{\frac{\mu}{\sigma}\}}\right)\right)^{r-1} \\
&\quad \times \frac{\exp\{-\frac{x-\mu}{\sigma}\}}{A\sigma\left(1+\exp\{-\frac{x-\mu}{\sigma}\}\right)^2} \\
&\quad \times \left(\frac{1}{A}\left(\frac{1}{1+\exp\left\{-\frac{u_{(nq_l)}+\beta(y-u_{(nq_l)})-\mu}{\sigma}\right\}}\right.\right. \\
&\quad \left.\left. - \frac{1}{1+\exp\{-\frac{x-\mu}{\sigma}\}}\right)\right)^{s-r-1} \\
&\quad \frac{\beta\exp\left\{-\frac{u_{(nq_l)}+\beta(y-u_{(nq_l)})-\mu}{\sigma}\right\}}{A\sigma\left(1+\exp\left\{-\frac{u_{(nq_l)}+\beta(y-u_{(nq_l)})-\mu}{\sigma}\right\}\right)^2} \\
&\quad \times \left(1 - \frac{1}{A}\left(\frac{1}{1+\exp\left\{-\frac{u_{(nq_l)}+\beta(y-u_{(nq_l)})-\mu}{\sigma}\right\}}\right.\right. \\
&\quad \left.\left. - \frac{1}{1+\exp\{\frac{\mu}{\sigma}\}}\right)\right)^{n-s}.
\end{aligned}
$$

These are as follows

$$
\begin{aligned}
&E\left[-\frac{\partial^2 \ln L(\beta,\mu,\sigma)}{\partial\beta^2}\right] \\
&\quad = E\left[E\left[-\frac{\partial^2 \ln L(\beta,\mu,\sigma)}{\partial\beta^2}\,\middle|\, u_{(nq_r)} = y, u_{(nq_l)} = x\right]\right]
\end{aligned}
$$

$$= a\int_0^\infty \int_0^y \left(\frac{nP_a}{\beta^2} + \frac{n(1-P_u-P_a)(y-x)^2K(y)}{\sigma^2}\right.$$

$$\left. + \int_x^y \left(\frac{2n\beta}{A\sigma^3}\right)(u-x)^2(K(u))^2du\right) zdxdy,$$

$$E\left[-\frac{\partial^2 \ln L(\beta,\mu,\sigma)}{\partial\mu\partial\beta}\right]$$

$$= E\left[E\left[-\frac{\partial^2 \ln L(\beta,\mu,\sigma)}{\partial\mu\partial\beta}\,|u_{(nq_r)}=y, u_{(nq_l)}=x\right]\right]$$

$$= a\int_0^\infty \int_0^y \left(-\frac{n(1-P_u-P_a)(y-x)K(y)}{\sigma^2}\right.$$

$$\left. - \int_x^y \left(\frac{2n\beta}{A\sigma^3}\right)(u-x)(K(u))^2du\right) zdxdy,$$

$$E\left[-\frac{\partial^2 \ln L(\beta,\mu,\sigma)}{\partial\sigma\partial\beta}\right]$$

$$= E\left[E\left[-\frac{\partial^2 \ln L(\beta,\mu,\sigma)}{\partial\sigma\partial\beta}\,|u_{(nq_r)}=y, u_{(nq_l)}=x\right]\right]$$

$$= a\int_0^\infty \int_0^y \left(-\frac{n(1-P_u-P_a)(y-x)(1-B(y))}{\sigma^2}\right.$$

$$\left. - \int_x^y \left(\frac{n\beta}{A\sigma^3}\right)(u-x)K(u)(1-2B(u))du\right) zdxdy,$$

$$E\left[-\frac{\partial^2 \ln L(\beta,\mu,\sigma)}{\partial\mu^2}\right]$$

$$= E\left[E\left[-\frac{\partial^2 \ln L(\beta,\mu,\sigma)}{\partial\mu^2}\,|u_{(nq_r)}=y, u_{(nq_l)}=x\right]\right]$$

$$= a\int_0^\infty \int_0^y \left(-\frac{ng(2\mu)}{\sigma^2} + \frac{n(1-P_u-P_a)K(y)}{\sigma^2}\right.$$

$$\left. + \int_0^x \left(\frac{2n}{A\sigma^3}\right)(g(u))^2 du + \int_x^y \left(\frac{2n\beta}{A\sigma^3}\right)(K(u))^2 du\right) zdxdy,$$

$$E\left[-\frac{\partial^2 \ln L(\beta,\mu,\sigma)}{\partial\sigma\partial\mu}\right]$$

$$= E\left[E\left[-\frac{\partial^2 \ln L(\beta,\mu,\sigma)}{\partial\sigma\partial\mu} | u_{(nq_r)} = y, u_{(nq_l)} = x\right]\right]$$

$$= a\int_0^\infty \int_0^y \left(-\frac{nQ(2\mu)}{\sigma^2} + \frac{n(1-P_u-P_a)(1-B(y))}{\sigma^2}\right.$$

$$+ \int_0^x \left(\frac{n}{A\sigma^3}\right) g(u)(1-2Q(u))du$$

$$\left. + \int_x^y \left(\frac{n\beta}{A\sigma^3}\right) K(u)(1-2B(u))du\right) zdxdy$$

and

$$E\left[-\frac{\partial^2 \ln L(\beta,\mu,\sigma)}{\partial\sigma^2}\right]$$

$$= E\left[E\left[-\frac{\partial^2 \ln L(\beta,\mu,\sigma)}{\partial\sigma^2} | u_{(nq_r)} = y, u_{(nq_l)} = x\right]\right]$$

$$= a\int_0^\infty \int_0^y \left(\frac{2n(1-P_u-P_a)(x+(y-x)\beta-\mu)(1-H(y))}{\sigma^3}\right.$$

$$+ \frac{2n\mu}{\sigma^3}$$

$$M(2\mu) - \frac{n(P_u+P_a)}{\sigma^2} + \int_0^x \left(\frac{2n}{A\sigma^4}\right)(u-\mu)g(u)(1-2M(u))du$$

$$\left. + \int_x^y \left(\frac{2n\beta}{A\sigma^4}\right)(x+(u-x)\beta-\mu)K(u)(1-2H(u))du\right) zdxdy.$$

A.6 JUSTIFICATION FOR THE USE OF LINEAR CUMULATIVE EXPOSURE MODEL IN THE BURR TYPE-XII DISTRIBUTION IN SECTION 4.2.3

Consider the cdf of the Burr Type-XII distribution as

$$\begin{aligned} &G(t; c, k, \alpha) \\ &\quad = 1 - \left(1 + \left(\frac{t}{\alpha}\right)^{c}\right)^{-k}; \quad t \geq 0;\ c, k, \alpha > 0 \\ &\quad \Rightarrow G(\varepsilon(t); c, k, \alpha) = 1 - \left(1 + \left(\frac{\varepsilon(t)}{\alpha}\right)^{c}\right)^{-k}; \quad t \geq 0;\ c, k, \alpha > 0 \end{aligned}$$

For $\alpha = 1$,

$$\begin{aligned} G(\varepsilon(t)) &\equiv G(\varepsilon(t); c, k, 1) \\ &= 1 - (1 + (\varepsilon(t))^{c})^{-k}; \quad t \geq 0;\ c, k > 0 \\ &= 1 - \left\{1 + \left[\int_0^t \left(\frac{1}{\alpha(s(u))}\right) du\right]^{c}\right\}^{-k}; \\ &\qquad t \geq 0;\ c, k > 0 \quad (\text{using } (4.7)) \\ &= 1 - \left\{1 + \left[\int_0^t \left(\frac{1}{e^{\gamma_0}(s_0/s(u))^{\gamma_1}}\right) du\right]^{c}\right\}^{-k}; \\ &\qquad t \geq 0;\ c, k > 0 \quad (\text{using } (4.5)) \\ &= F(t) \end{aligned}$$

It is necessary to set $\alpha = 1$, as then only $F(t)$ will be equal to $G(\varepsilon(t))$, and a linear cumulative exposure model will hold.

A.7 JUSTIFICATION FOR THE USE OF LINEAR CUMULATIVE EXPOSURE MODEL IN THE BURR TYPE-XII DISTRIBUTION IN SECTION 4.3.3

Consider the cdf of the Burr Type-XII distribution as

$$\begin{aligned} &G(t; c, k, \alpha) \\ &\quad = 1 - \left(1 + \left(\frac{t}{\alpha}\right)^{c}\right)^{-k}; \quad t \geq 0;\ c, k, \alpha > 0 \end{aligned}$$

$$\Rightarrow G(\varepsilon(t); c, k, \alpha) = 1 - \left(1 + \left(\frac{\varepsilon(t)}{\alpha}\right)^c\right)^{-k}; \quad t \geq 0;\ c, k, \alpha > 0$$

For $\alpha = 1$,

$$\begin{aligned}
G(\varepsilon(t)) &\equiv G(\varepsilon(t); c, k, 1) \\
&= 1 - (1 + (\varepsilon(t))^c)^{-k}; \quad t \geq 0;\ c, k > 0 \\
&= 1 - \left\{1 + \left[\int_0^t \left(\frac{1}{\alpha(s(u))}\right) du\right]^c\right\}^{-k}; \\
&\qquad t \geq 0;\ c, k > 0 \quad \text{(using (4.25))} \\
&= 1 - \left\{1 + \left[\int_0^t \left(\frac{1}{e^{\gamma_0} (s_h/s(u))^{\gamma_1}}\right) du\right]^c\right\}^{-k}; \\
&\qquad t \geq 0;\ c, k > 0 \quad \text{(using (4.23))} \\
&= F(t).
\end{aligned}$$

It is necessary to set $\alpha = 1$, as then only the $F(t)$ will be equal to $G(\varepsilon(t))$, and a linear cumulative exposure model will hold.

A.8 CALCULATIONS OF THE ELEMENTS OF THE FISHER INFORMATION MATRIX GIVEN IN SECTION 4.2.6

The second-order partial derivatives are

$$\frac{\partial^2 L}{\partial \gamma_0^2} = -c^2 \left[I(z) \left(\frac{(k+1)e^{-z}}{(1+e^{-z})^2} \right) + \bar{I}(z) \left(\frac{ke^{-\eta'}}{(1+e^{-\eta'})^2} \right) \right], \tag{A.7}$$

$$\begin{aligned}
\frac{\partial^2 L}{\partial \gamma_0 \partial \gamma_1} &= -\left(\frac{1}{1+\gamma_1}\right)\left(\frac{\partial L}{\partial \gamma_0}\right) + \left(\frac{c}{1+\gamma_1}\right) \left[\frac{T_1}{c} \left(\frac{\partial^2 L}{\partial \gamma_0^2}\right) \right. \\
&\quad \left. + I(z) \left(\frac{(k+1)e^{-z} z}{(1+e^{-z})^2} \right) + \bar{I}(z) \left(\frac{ke^{-\eta'} \eta'}{(1+e^{-\eta'})^2} \right) \right],
\end{aligned} \tag{A.8}$$

$$\frac{\partial^2 L}{\partial \gamma_0 \partial c} = \frac{2}{c} \left(\frac{\partial L}{\partial \gamma_0}\right) - \frac{T_1}{c} \left(\frac{\partial^2 L}{\partial \gamma_0^2}\right) + \left(\frac{1+\gamma_1}{c}\right)\left(\frac{\partial^2 L}{\partial \gamma_0 \partial \gamma_1}\right), \tag{A.9}$$

$$\frac{\partial^2 L}{\partial \gamma_1^2} = -\frac{1}{(1+\gamma_1)}\left(\frac{\partial L}{\partial \gamma_1}\right) - \frac{1}{(1+\gamma_1)^2}\left(\frac{\partial L}{\partial \gamma_0}\right)$$
$$- \frac{c}{(1+\gamma_1)^2}\left(\frac{\partial L}{\partial c}\right) + \frac{c}{(1+\gamma_1)}\left(\frac{\partial^2 L}{\partial c \partial \gamma_1}\right)$$
$$+ \frac{T_1}{(1+\gamma_1)}\left(\frac{\partial^2 L}{\partial \gamma_1 \partial \gamma_0}\right), \tag{A.10}$$

$$\frac{\partial^2 L}{\partial \gamma_1 \partial c} = \frac{1}{(1+\gamma_1)}\left(\frac{\partial L}{\partial c}\right) + \frac{1}{c}\left(\frac{\partial L}{\partial \gamma_1}\right) + \frac{T_1}{c(1+\gamma_1)}\left(\frac{\partial L}{\partial \gamma_0}\right)$$
$$+ \frac{c}{(1+\gamma_1)}\left(\frac{\partial^2 L}{\partial c^2}\right) + \frac{T_1}{c}\left(\frac{\partial^2 L}{\partial \gamma_1 \partial \gamma_0}\right) - \frac{T_1^2}{c(1+\gamma_1)}\left(\frac{\partial^2 L}{\partial \gamma_0^2}\right), \tag{A.11}$$

and

$$\frac{\partial^2 L}{\partial c^2} = -\frac{1}{c^2}\left[I(z)\left(1 + \frac{(k+1)z^2 e^{-z}}{(1+e^{-z})^2}\right) + \bar{I}(z)\left(\frac{k e^{-\eta'} \eta'^2}{(1+e^{-\eta'})^2}\right)\right]. \tag{A.12}$$

The elements of the Fisher information matrix for an observation at $\xi_i, i = 1, 2, \ldots, m$ are the negative expectations of these second-order partial derivatives. Using (A.7)–(A.12), these are as follows:

$$E\left[-\frac{\partial^2 L}{\partial \gamma_0^2}\right] = c^2(k+1)k \int_{-\infty}^{\eta'} \left(\frac{e^{-z(k+1)}}{(1+e^{-z})^{(k+3)}}\right) dz$$
$$+ \frac{k c^2 e^{-\eta'(k+1)}}{(1+e^{-\eta'})^{(k+2)}}, \tag{A.13}$$

$$E\left[-\frac{\partial^2 L}{\partial \gamma_0 \partial \gamma_1}\right] = -\frac{c}{(1+\gamma_1)}\left[(k+1)k \int_{-\infty}^{\eta'} \left(\frac{z e^{-z(k+1)}}{(1+e^{-z})^{(k+3)}}\right) dz\right.$$
$$\left. + \frac{k \eta' e^{-\eta'(k+1)}}{(1+e^{-\eta'})^{(k+2)}}\right] + \frac{T_1}{(1+\gamma_1)} E\left[-\frac{\partial^2 L}{\partial \gamma_0^2}\right], \tag{A.14}$$

$$E\left[-\frac{\partial^2 L}{\partial \gamma_0 \partial c}\right] = -\frac{T_1}{c} E\left[-\frac{\partial^2 L}{\partial \gamma_0^2}\right] + \frac{(1+\gamma_1)}{c} E\left[-\frac{\partial^2 L}{\partial \gamma_0 \partial \gamma_1}\right], \tag{A.15}$$

$$E\left[-\frac{\partial^2 L}{\partial \gamma_1^2}\right] = \frac{c}{(1+\gamma_1)}E\left[-\frac{\partial^2 L}{\partial \gamma_1 \partial c}\right] + \frac{T_1}{(1+\gamma_1)}E\left[-\frac{\partial^2 L}{\partial \gamma_0 \partial \gamma_1}\right], \tag{A.16}$$

$$E\left[-\frac{\partial^2 L}{\partial \gamma_1 \partial c}\right] = \frac{c}{(1+\gamma_1)}E\left[-\frac{\partial^2 L}{\partial c^2}\right] + \frac{T_1}{c}E\left[-\frac{\partial^2 L}{\partial \gamma_0 \partial \gamma_1}\right] - \frac{T_1^2}{c(1+\gamma_1)}E\left[-\frac{\partial^2 L}{\partial \gamma_0^2}\right] \tag{A.17}$$

and

$$E\left[-\frac{\partial^2 L}{\partial c^2}\right] = \frac{k}{c^2}\int_{-\infty}^{\eta'}\left(1+\frac{(k+1)z^2e^{-z}}{(1+e^{-z})^2}\right)\left(\frac{e^{-zk}}{(1+e^{-z})^{(k+1)}}\right)dz + \frac{k\eta'^2 e^{-\eta'(k+1)}}{c^2(1+e^{-\eta'})^{(k+2)}}, \tag{A.18}$$

where

$$E[I(z)] = 1-(1+e^{\eta'})^{-k},$$
$$E[\bar{I}(z)] = (1+e^{\eta'})^{-k},$$
$$E\left[\frac{\partial L}{\partial \gamma_{i'}}\right] = 0, \quad \text{for } i'=0,1, \quad \text{and}$$
$$E\left[\frac{\partial L}{\partial c}\right] = 0.$$

Thus, (A.13)–(A.18) give the elements of Fisher Information Matrix.

A.9 CALCULATIONS OF THE ELEMENTS OF THE FISHER INFORMATION MATRIX IN SECTION 4.3.6

The first-order partial derivatives are

$$\frac{\partial L}{\partial \gamma_0} = \sum_{i=1}^{2} u_i(t)\left[\frac{c(k(e^{-\gamma_0}c_i(t))^c-1)}{(e^{-\gamma_0}c_i(t))^c+1}\right] + u_3(t)\left[\frac{ck(e^{-\gamma_0}c_2(\eta))^c}{(e^{-\gamma_0}c_2(\eta))^c+1}\right], \tag{A.19}$$

$$\begin{aligned}\frac{\partial L}{\partial \gamma_1} &= u_1(t)\left[c\ln\left(\frac{t}{\tau}\right)\left\{\frac{1-k(e^{-\gamma_0}c_1(t))^c}{(e^{-\gamma_0}c_1(t))^c+1}\right\}\right.\\ &\quad \left.+\left\{\frac{(ck+1)(e^{-\gamma_0}c_1(t))^c+(1-c)}{(1+\gamma_1)((e^{-\gamma_0}c_1(t))^c+1)}\right\}\right]\\ &\quad +u_2(t)\left[\frac{\tau((ck+1)(e^{-\gamma_0}c_2(t))^c+(1-c))}{(1+\gamma_1)^2c_2(t)((e^{-\gamma_0}c_2(t))^c+1)}\right]\\ &\quad +u_3(t)\left[\frac{\tau ck(e^{-\gamma_0}c_2(\eta))^c}{(1+\gamma_1)^2c_2(\eta)((e^{-\gamma_0}c_2(\eta))^c+1)}\right],\end{aligned} \tag{A.20}$$

and

$$\begin{aligned}\frac{\partial L}{\partial c} &= \left(\frac{1}{c}-\gamma_0\right)\sum_{i=1}^{2}u_i(t)\\ &\quad -(k+1)\sum_{i=1}^{2}u_i(t)\left[\frac{(e^{-\gamma_0}c_i(t))^c\ln(e^{-\gamma_0}c_i(t))}{(e^{-\gamma_0}c_i(t))^c+1}\right]\\ &\quad -u_3(t)\left[\frac{k(e^{-\gamma_0}c_2(\eta))^c\ln(e^{-\gamma_0}c_2(\eta))}{(e^{-\gamma_0}c_2(\eta))^c+1}\right]\\ &\quad +\sum_{i=1}^{2}u_i(t)\ln(c_i(t)).\end{aligned} \tag{A.21}$$

The second order partial derivatives are

$$\begin{aligned}\frac{\partial^2 L}{\partial \gamma_0^2} &= -c^2\left\{\sum_{i=1}^{2}u_i(t)\left[\frac{(k+1)(e^{-\gamma_0}c_i(t))^c}{((e^{-\gamma_0}c_i(t))^c+1)^2}\right]\right.\\ &\quad \left.+u_3(t)\left[\frac{k(e^{-\gamma_0}c_2(\eta))^c}{((e^{-\gamma_0}c_2(\eta))^c+1)^2}\right]\right\},\end{aligned} \tag{A.22}$$

$$\begin{aligned}\frac{\partial^2 L}{\partial \gamma_0 \partial \gamma_1} &= c^2\left\{\left(\frac{(k+1)u_1(t)(e^{-\gamma_0}c_1(t))^c}{((e^{-\gamma_0}c_1(t))^c+1)^2}\right)\left(\ln\left(\frac{t}{\tau}\right)-\frac{1}{(1+\gamma_1)}\right)\right.\\ &\quad -\left(\frac{u_2(t)}{(1+\gamma_1)^2}\right)\left(\frac{(k+1)\tau(e^{-\gamma_0}c_2(t))^c}{c_2(t)((e^{-\gamma_0}c_2(t))^c+1)^2}\right)\end{aligned}$$

$$- \frac{u_3(t)\tau k(e^{-\gamma_0}c_2(\eta))^c}{(1+\gamma_1)^2 c_2(\eta)((e^{-\gamma_0}c_2(\eta))^c+1)^2}\Bigg\}, \tag{A.23}$$

$$\begin{aligned}
\frac{\partial^2 L}{\partial\gamma_0\partial c} &= \sum_{i=1}^{2} u_i(t)\left[\frac{k(e^{-\gamma_0}c_i(t))^c-1}{(e^{-\gamma_0}c_i(t))^c+1}\right.\\
&\quad \left.+\frac{c(e^{-\gamma_0}c_i(t))^c(k+1)\ln(e^{-\gamma_0}c_i(t))}{((e^{-\gamma_0}c_i(t))^c+1)^2}\right]\\
&\quad + u_3(t)k\left[\frac{(e^{-\gamma_0}c_2(\eta))^c}{(e^{-\gamma_0}c_2(\eta))^c+1}\right.\\
&\quad \left.+\frac{c(e^{-\gamma_0}c_2(\eta))^c\ln(e^{-\gamma_0}c_2(\eta))}{(1+(e^{-\gamma_0}c_2(\eta))^c)^2}\right],
\end{aligned} \tag{A.24}$$

$$\begin{aligned}
\frac{\partial^2 L}{\partial\gamma_1^2} &= \frac{u_3(t)ck\tau^2(e^{-\gamma_0}c_2(\eta))^c}{((e^{-\gamma_0}c_2(\eta))^c+1)(1+\gamma_1)^4 c_2(\eta)^2}\\
&\quad - \frac{u_2(t)c^2(k+1)\tau^2(e^{-\gamma_0}c_2(t))^c}{((e^{-\gamma_0}c_2(t))^c+1)^2(1+\gamma_1)^4 c_2(t)^2}\\
&\quad - \frac{u_3(t)c^2k\tau^2(e^{-\gamma_0}c_2(\eta))^c}{((e^{-\gamma_0}c_2(\eta))^c+1)^2(1+\gamma_1)^4 c_2(\eta)^2}\\
&\quad - \left(\frac{u_1(t)c^2(k+1)(e^{-\gamma_0}c_1(t))^c}{((e^{-\gamma_0}c_1(t))^c+1)^2}\right)\\
&\quad \times \left(\ln\left(\frac{t}{\tau}\right)-\frac{1}{(1+\gamma_1)}\right)^2\\
&\quad + \frac{u_2(t)\tau^2\left((ck+1)(e^{-\gamma_0}c_2(t))^c+(1-c)\right)}{(1+\gamma_1)^4 c_2(t)^2((e^{-\gamma_0}c_2(t))^c+1)},
\end{aligned} \tag{A.25}$$

$$\begin{aligned}
\frac{\partial^2 L}{\partial\gamma_1\partial c} &= u_1(t)\left[\ln\left(\frac{t}{\tau}\right)-\frac{1}{(1+\gamma_1)}\right]\\
&\quad + \frac{u_3(t)k\tau(e^{-\gamma_0}c_2(\eta))^c}{((e^{-\gamma_0}c_2(\eta))^c+1)(1+\gamma_1)^2 c_2(\eta)}\\
&\quad + \frac{u_3(t)ck\tau(e^{-\gamma_0}c_2(\eta))^c\ln(e^{-\gamma_0}c_2(\eta))}{((e^{-\gamma_0}c_2(\eta))^c+1)^2(1+\gamma_1)^2 c_2(\eta)}
\end{aligned}$$

$$
\begin{aligned}
&- u_2(t)\left[\frac{\tau}{(c_2(t)(1+\gamma_1)^2)}\right]\\
&+ \frac{u_2(t)c(k+1)\tau(e^{-\gamma_0}c_2(t))^c \ln(e^{-\gamma_0}c_2(t))}{((e^{-\gamma_0}c_2(t))^c+1)^2(1+\gamma_1)^2 c_2(t)}\\
&- \left[\ln\left(\frac{t}{\tau}\right) - \frac{1}{(1+\gamma_1)}\right]\left(\frac{u_1(t)(k+1)(e^{-\gamma_0}c_1(t))^c}{((e^{-\gamma_0}c_1(t))^c+1)}\right)\\
&- \left(\frac{u_1(t)(k+1)(e^{-\gamma_0}c_1(t))^c c\ln(e^{-\gamma_0}c_1(t))}{((e^{-\gamma_0}c_1(t))^c+1)^2}\right)\\
&\times \left[\ln\left(\frac{t}{\tau}\right) - \frac{1}{(1+\gamma_1)}\right] + \frac{u_2(t)(k+1)\tau(e^{-\gamma_0}c_2(t))^c}{((e^{-\gamma_0}c_2(t))^c+1)(1+\gamma_1)^2 c_2(t)},
\end{aligned}
\tag{A.26}
$$

and

$$
\begin{aligned}
\frac{\partial^2 L}{\partial c^2} &= -\frac{1}{c^2}\sum_{i=1}^{2} u_i(t) - (k+1)\sum_{i=1}^{2} u_i(t)\\
&\quad \times \left(\frac{(\ln(e^{-\gamma_0}c_i(t)))^2(e^{-\gamma_0}c_i(t))^c}{((e^{-\gamma_0}c_i(t))^c+1)^2}\right)\\
&\quad - u_3(t)\left(\frac{k(\ln(e^{-\gamma_0}c_2(\eta)))^2(e^{-\gamma_0}c_2(\eta))^c}{((e^{-\gamma_0}c_2(\eta))^c+1)^2}\right).
\end{aligned}
\tag{A.27}
$$

The elements of the Fisher information matrix for an observation at τ are the negative expectations of these second-order partial derivatives. Using (A.19)–(A.27) these are obtained as:

$$
\begin{aligned}
E\left[-\frac{\partial^2 L}{\partial \gamma_0^2}\right] &= c^2(k+1)\int_0^{\tau}\left(\frac{(e^{-\gamma_0}c_1(t))^c}{(1+(e^{-\gamma_0}c_1(t))^c)^2}\right) f_1(t)dt\\
&\quad + c^2(k+1)\int_{\tau}^{\eta}\left(\frac{(e^{-\gamma_0}c_2(t))^c}{(1+(e^{-\gamma_0}c_2(t))^c)^2}\right) f_2(t)dt\\
&\quad + \frac{c^2 k(e^{-\gamma_0}c_2(\eta))^c(1-F_2(\eta))}{(1+(e^{-\gamma_0}c_2(\eta))^c)^2},
\end{aligned}
\tag{A.28}
$$

$$
\begin{aligned}
E\left[-\frac{\partial^2 L}{\partial\gamma_0\partial\gamma_1}\right] &= -c^2(k+1)\int_0^\tau \left(\frac{(e^{-\gamma_0}c_1(t))^c}{(1+(e^{-\gamma_0}c_1(t))^c)^2}\right) \\
&\quad \times \left[\ln\left(\frac{t}{\tau}\right) - \frac{1}{(1+\gamma_1)}\right] f_1(t)dt + \frac{c^2\tau(k+1)}{(1+\gamma_1)^2} \\
&\quad \times \int_\tau^\eta \left(\frac{(e^{-\gamma_0}c_2(t))^c}{c_2(t)(1+(e^{-\gamma_0}c_2(t))^c)^2}\right) f_2(t)dt \\
&\quad + \frac{c^2\tau k(1-F_2(\eta))(e^{-\gamma_0}c_2(\eta))^c}{(1+(e^{-\gamma_0}c_2(\eta))^c)^2(1+\gamma_1)^2 c_2(\eta)}, \qquad \text{(A.29)}
\end{aligned}
$$

$$
\begin{aligned}
E\left[-\frac{\partial^2 L}{\partial\gamma_0\partial c}\right] &= -\int_0^\tau \left(\frac{k(e^{-\gamma_0}c_1(t))^c - 1}{1+(e^{-\gamma_0}c_1(t))^c}\right) f_1(t)dt \\
&\quad - \int_\tau^\eta \left(\frac{k(e^{-\gamma_0}c_2(t))^c - 1}{(1+(e^{-\gamma_0}c_2(t))^c)}\right) f_2(t)dt \\
&\quad - c(k+1)\int_0^\tau \left(\frac{(e^{-\gamma_0}c_1(t))^c \ln(e^{-\gamma_0}c_1(t))}{(1+(e^{-\gamma_0}c_1(t))^c)^2}\right) f_1(t)dt \\
&\quad -c(k+1)\int_\tau^\eta \left(\frac{(e^{-\gamma_0}c_2(t))^c \ln(e^{-\gamma_0}c_2(t))}{(1+(e^{-\gamma_0}c_2(t))^c)^2}\right) f_2(t)dt \\
&\quad - k(1-F_2(\eta))(e^{-\gamma_0}c_2(\eta))^c \\
&\quad \left[\frac{1}{1+(e^{-\gamma_0}c_2(\eta))^c} + \frac{c\ln(e^{-\gamma_0}c_2(\eta))}{(1+(e^{-\gamma_0}c_2(\eta))^c)^2}\right], \qquad \text{(A.30)}
\end{aligned}
$$

$$
\begin{aligned}
E\left[-\frac{\partial^2 L}{\partial\gamma_1^2}\right] &= c^2(k+1)\int_0^\tau \left(\frac{(e^{-\gamma_0}c_1(t))^c}{(1+(e^{-\gamma_0}c_1(t))^c)^2}\right) \\
&\quad \times \left[\ln\left(\frac{t}{\tau}\right) - \frac{1}{(1+\gamma_1)}\right]^2 f_1(t)dt - \frac{\tau^2}{(1+\gamma_1)^4} \\
&\quad \times \int_\tau^\eta \left(\frac{(ck+1)(e^{-\gamma_0}c_2(t))^c + 1 - c}{(1+(e^{-\gamma_0}c_2(t))^c)c_2(t)^2}\right) f_2(t)dt \\
&\quad - \left(\frac{\tau^2(1-F_2(\eta))}{(1+\gamma_1)^4}\right)\left(\frac{kc(e^{-\gamma_0}c_2(\eta))^c}{c_2(\eta)^2(1+(e^{-\gamma_0}c_2(\eta))^c)}\right)
\end{aligned}
$$

$$+\int_{\tau}^{\eta}\left(\frac{(e^{-\gamma_0}c_2(t))^c}{(1+(e^{-\gamma_0}c_2(t))^c)^2c_2(t)^2}\right)f_2(t)dt$$
$$\times\left(\frac{c^2(k+1)\tau^2}{(1+\gamma_1)^4}\right)$$
$$+\frac{k\tau^2c^2(1-F_2(\eta))(e^{-\gamma_0}c_2(\eta))^c}{(1+(e^{-\gamma_0}c_2(\eta))^c)^2(1+\gamma_1)^4c_2(\eta)^2}, \tag{A.31}$$

$$E\left[-\frac{\partial^2 L}{\partial\gamma_1\partial c}\right]=(k+1)\int_0^{\tau}\left(\frac{(e^{-\gamma_0}c_1(t))^c}{1+(e^{-\gamma_0}c_1(t))^c}\right)$$
$$\times\left[\ln\left(\frac{t}{\tau}\right)-\frac{1}{(1+\gamma_1)}\right]f_1(t)dt$$
$$+(k+1)\int_0^{\tau}\left(\frac{(e^{-\gamma_0}c_1(t))^c}{(1+(e^{-\gamma_0}c_1(t))^c)^2}\right)$$
$$\times\left[\ln\left(\frac{t}{\tau}\right)-\frac{1}{(1+\gamma_1)}\right]f_1(t)dt-\left(\frac{\tau}{(1+\gamma_1)^2}\right)$$
$$\times\int_{\tau}^{\eta}\left(\frac{(k+1)(e^{-\gamma_0}c_2(t))^c}{(1+(e^{-\gamma_0}c_2(t))^c)c_2(t)}\right)f_2(t)dt$$
$$-\int_0^{\tau}\left[\ln\left(\frac{t}{\tau}\right)-\frac{1}{(1+\gamma_1)}\right]f_1(t)dt$$
$$-\frac{k\tau(1-F_2(\eta))(e^{-\gamma_0}c_2(\eta))^c}{(1+(e^{-\gamma_0}c_2(\eta))^c)(1+\gamma_1)^2c_2(\eta)}$$
$$+\int_{\tau}^{\eta}\left(\frac{\tau}{(1+\gamma_1)^2c_2(t)}\right)f_2(t)dt-\left(\frac{c\tau(k+1)}{(1+\gamma_1)^2}\right)$$
$$\times\int_{\tau}^{\eta}\left(\frac{(e^{-\gamma_0}c_2(t))^c\ln(e^{-\gamma_0}c_2(t))}{(1+(e^{-\gamma_0}c_2(t))^c)^2c_2(t)}\right)f_2(t)dt$$
$$-\frac{k\tau c(1-F_2(\eta))(e^{-\gamma_0}c_2(\eta))^c\ln(e^{-\gamma_0}c_2(\eta))}{(1+(e^{-\gamma_0}c_2(\eta))^c)^2(1+\gamma_1)^2c_2(\eta)}, \tag{A.32}$$

and

$$E\left[-\frac{\partial^2 L}{\partial c^2}\right] = \frac{F_2(\eta)}{c^2}$$
$$+ (k+1)\int_0^{\tau}\left(\frac{(e^{-\gamma_0}c_1(t))^c[\ln(e^{-\gamma_0}c_1(t))]^2}{(1+(e^{-\gamma_0}c_1(t))^c)^2}\right) f_1(t)dt$$
$$+ \frac{k[\ln(e^{-\gamma_0}c_2(\eta))]^2(e^{-\gamma_0}c_2(\eta))^c(1-F_2(\eta))}{(1+(e^{-\gamma_0}c_2(\eta))^c)^2}$$
$$+ (k+1)\int_{\tau}^{\eta}\left(\frac{(e^{-\gamma_0}c_2(t))^c[\ln(e^{-\gamma_0}c_2(t))]^2}{(1+(e^{-\gamma_0}c_2(t))^c)^2}\right) f_2(t)dt, \tag{A.33}$$

where

$$E[u_1(t)] = \{1+[e^{-\gamma_0}c_1(\eta)]^c\}^{-k},$$
$$E[u_2(t)] = \{1+[e^{-\gamma_0}c_2(\tau)]^c\}^{-k} - \{1+[e^{-\gamma_0}c_2(\eta)]^c\}^{-k},$$
$$E[u_3(t)] = \{1+[e^{-\gamma_0}c_2(\eta)]^c\}^{-k}, E\left[\frac{\partial L}{\partial \gamma_i}\right] = 0, \quad \text{for } i = 0, 1, \quad \text{and}$$
$$E\left[\frac{\partial L}{\partial c}\right] = 0.$$

A.10 JUSTIFICATION FOR THE USE OF LINEAR CUMULATIVE EXPOSURE MODEL IN THE BURR TYPE-XII DISTRIBUTION IN SECTION 5.2.3

Consider the cdf of the Burr Type-XII distribution as

$$G(t; c, k, \alpha) = 1 - \left(1 + \left(\frac{t}{\alpha}\right)^c\right)^{-k}; \quad t \geq 0;\ c, k, \alpha > 0$$
$$\Rightarrow G(\varepsilon(t); c, k, \alpha)$$
$$= 1 - \left(1 + \left(\frac{\varepsilon(t)}{\alpha}\right)^c\right)^{-k}; \quad t \geq 0;\ c, k, \alpha > 0$$

For $\alpha = 1$,

$$\begin{aligned}
G(\varepsilon(t)) &\equiv G(\varepsilon(t); c, k, 1) \\
&= 1 - (1 + (\varepsilon(t))^c)^{-k}; \quad t \geq 0;\ c, k > 0 \\
&= \begin{cases}
1 - (1 + (G_1(t))^c)^{-k}; & 0 < t \leq t_1;\ c, k > 0 \\
& \text{(using (5.6))} \\
1 - (1 + (\varepsilon(t_1) + G_2(t))^c)^{-k}; & t_1 < t \leq t_2;\ c, k > 0 \\
& \text{(using (5.7))} \\
1 - (1 + (\varepsilon(t_2) + G_3(t))^c)^{-k}; & t_2 < t \leq t_3;\ c, k > 0 \\
& \text{(using (5.8))} \\
1 - (1 + (\varepsilon(t_3) + G_4(t))^c)^{-k}; & t_3 < t \leq \eta;\ c, k > 0 \\
& \text{(using (5.9))}
\end{cases} \\
&= \begin{cases}
F_1(t); 0 < t \leq t_1 \\
F_2(t); t_1 < t \leq t_2 \\
F_3(t); t_2 < t \leq t_3 \\
F_4(t); t_3 < t \leq \eta
\end{cases} \\
&= F(t).
\end{aligned}$$

It is necessary to set $\alpha = 1$, as then only the $F(t)$ will be equal to $G(\varepsilon(t))$, and a linear cumulative exposure model will hold.

A.11 CALCULATIONS OF THE ELEMENTS OF THE FISHER INFORMATION MATRIX IN SECTION 5.2.5

The first-order partial derivatives are

$$\begin{aligned}
\frac{\partial L}{\partial \gamma_0} &= -c\sum_{j=0}^{3}\delta_j + (k+1)c\left(\frac{\delta_0(G_1(t))^c}{1+(G_1(t))^c} + \frac{\delta_1(a_1(t))^c}{1+(a_1(t))^c}\right. \\
&\quad \left. + \frac{\delta_2(a_2(t))^c}{1+(a_2(t))^c} + \frac{\delta_3(a_3(t))^c}{1+(a_3(t))^c}\right) + \frac{\delta_4 kc(a_4(\eta))^c}{(1+(a_4(\eta))^c)}, \qquad \text{(A.34)}
\end{aligned}$$

$$
\begin{aligned}
\frac{\partial L}{\partial \gamma_1} &= \delta_0 \left[-c \ln s_0 + \ln(\beta t + s_0) - (c-1)\left(\frac{1}{1+\gamma_1} + R_1(t) \right) \right. \\
&\quad \left. + \left(\frac{(k+1)c(G_1(t))^c}{1+(G_1(t))^c} \right) \left(\ln s_0 + \frac{1}{1+\gamma_1} + R_1(t) \right) \right] \\
&\quad + \delta_1 \left[-c \ln s_0 + \ln s_1 + \frac{(c-1)\ln s_1 G_2(t)}{a_1(t)} - (c-1) \right. \\
&\quad \times \left(\frac{1}{1+\gamma_1} + R_1(t_1) \right) \frac{G_1(t_1)}{a_1(t)} + \left(\ln s_0 + \left(\frac{1}{1+\gamma_1} + R_1(t_1) \right) \right. \\
&\quad \left. \left. \times \frac{G_1(t_1)}{a_1(t)} - \frac{G_2(t)\ln s_1}{a_1(t)} \right) \left(\frac{(k+1)c(a_1(t))^c}{1+(a_1(t))^c} \right) \right] \\
&\quad + \delta_2 \left[\ln(\beta t - \beta t_2 + s_1) + \frac{(c-1)\ln s_1 G_2(t_2)}{a_2(t)} - c \ln s_0 \right. \\
&\quad - \left(\frac{G_1(t_1) + G_3(t)}{a_2(t)} \right) \left(\frac{c-1}{1+\gamma_1} \right) \\
&\quad - \frac{(c-1)(R_1(t_1)G_1(t_1) + R_2(t)G_3(t))}{(a_2(t))} + \left(\frac{(k+1)c(a_2(t))^c}{1+(a_2(t))^c} \right) \\
&\quad \times \left(\frac{R_1(t_1)G_1(t_1) + R_2(t)G_3(t) - \ln s_1 G_2(t_2)}{a_2(t)} + \left(\frac{1}{1+\gamma_1} \right) \right. \\
&\quad \left. \left. \times \left(\frac{G_1(t_1) + G_3(t)}{a_2(t)} \right) + \ln s_0 \right) \right] \\
&\quad + \delta_3 \left[\frac{(c-1)(\ln s_1 G_2(t_2) + \ln s_2 G_4(t))}{a_3(t)} - \left(\frac{(c-1)}{1+\gamma_1} \right) \right. \\
&\quad \times \left(\frac{G_1(t_1) + G_3(t_3)}{a_3(t)} \right) - c \ln s_0 + \ln s_2 \\
&\quad - \frac{(c-1)(R_1(t_1)G_1(t_1) + R_2(t_3)G_3(t_3))}{a_3(t)}
\end{aligned}
$$

$$+\left(\frac{(k+1)c(a_3(t))^c}{1+(a_3(t))^c}\right)\Bigg(\frac{R_1(t_1)G_1(t_1)+R_2(t_3)G_3(t_3)}{a_3(t)}$$

$$+\ln s_0+\left(\frac{1}{1+\gamma_1}\right)\left(\frac{G_1(t_1)+G_3(t_3)}{a_3(t)}\right)$$

$$\times-\frac{(G_2(t_2)\ln s_1+G_4(t)\ln s_2)}{a_3(t)}\Bigg)\Bigg]+\left(\frac{\delta_4 kc(a_4(\eta))^c}{1+(a_4(\eta))^c}\right)$$

$$\times\left[\left(\frac{1}{1+\gamma_1}\right)\left(\frac{G_1(t_1)+G_3(t_3)}{a_4(\eta)}\right)+\ln s_0\right.$$

$$\left.+\frac{R_1(t_1)G_1(t_1)+R_2(t_3)G_3(t_3)}{a_4(\eta)}-\frac{(G_2(t_2)\ln s_1+G_4(\eta)\ln s_2)}{a_4(\eta)}\right], \tag{A.35}$$

and

$$\frac{\partial L}{\partial c}=\sum_{j=0}^{3}\frac{\delta_j}{c}+\delta_0\ln G_1(t)+\delta_1\ln a_1(t)+\delta_2\ln a_2(t)+\delta_3\ln a_3(t)$$

$$-(k+1)\left(\frac{\delta_0(G_1(t))^c\ln G_1(t)}{1+(G_1(t))^c}+\frac{\delta_1(a_1(t))^c\ln a_1(t)}{1+(a_1(t))^c}\right.$$

$$\left.+\frac{\delta_2(a_2(t))^c\ln a_2(t)}{1+(a_2(t))^c}+\frac{\delta_3(a_3(t))^c\ln a_3(t)}{1+(a_3(t))^c}\right)$$

$$-\frac{\delta_4 k(a_4(\eta))^c\ln a_4(\eta)}{1+(a_4(\eta))^c}. \tag{A.36}$$

The second-order partial derivatives are

$$\frac{\partial^2 L}{\partial\gamma_0^2}$$

$$=-\left\{(k+1)c^2\left(\frac{\delta_0(G_1(t))^c}{(1+(G_1(t))^c)^2}+\frac{\delta_1(a_1(t))^c}{(1+(a_1(t))^c)^2}\right.\right.$$

$$\left.\left.+\frac{\delta_2(a_2(t))^c}{(1+(a_2(t))^c)^2}+\frac{\delta_3(a_3(t))^c}{(1+(a_3(t))^c)^2}\right)+\frac{\delta_4 kc^2(a_4(\eta))^c}{(1+(a_4(\eta))^c)^2}\right\}, \tag{A.37}$$

$$\frac{\partial^2 L}{\partial \gamma_1 \partial \gamma_0}$$

$$= -(k+1)c^2 \left\{ \left(\frac{\delta_0 (G_1(t))^c}{(1+(G_1(t))^c)^2} \right) \left[\ln s_0 + \frac{1}{1+\gamma_1} + R_1(t) \right] \right.$$

$$+ \left[\ln s_0 + \left(\frac{1}{1+\gamma_1} + R_1(t_1) \right) \left(\frac{G_1(t_1) - G_2(t) \ln s_1}{a_1(t)} \right) \right]$$

$$\times \left(\frac{\delta_1 (a_1(t))^c}{(1+(a_1(t))^c)^2} \right) + \left(\frac{\delta_2 (a_2(t))^c}{(1+(a_2(t))^c)^2} \right)$$

$$\times \left[\left(\frac{1}{1+\gamma_1} \right) \left(\frac{G_1(t_1) + G_3(t)}{a_2(t)} \right) + \ln s_0 \right.$$

$$\left. + \frac{R_1(t_1) G_1(t_1) + R_2(t) G_3(t)}{a_2(t)} - \frac{G_2(t_2) \ln s_1}{a_2(t)} \right]$$

$$+ \left(\frac{\delta_3 (a_3(t))^c}{(1+(a_3(t))^c)^2} \right) \left[\left(\frac{1}{1+\gamma_1} \right) \left(\frac{G_1(t_1) + G_3(t_3)}{a_3(t)} \right) + \ln s_0 \right.$$

$$\left. \left. + \frac{R_1(t_1) G_1(t_1) + R_2(t_3) G_3(t_3) - (G_2(t_2) \ln s_1 + G_4(t) \ln s_2)}{a_3(t)} \right] \right\}$$

$$- \left(\frac{\delta_4 k c^2 (a_4(\eta))^c}{(1+(a_4(\eta))^c)^2} \right) \left[\left(\frac{1}{1+\gamma_1} \right) \left(\frac{G_1(t_1) + G_3(t_3)}{a_4(\eta)} \right) + \ln s_0 \right.$$

$$\left. + \frac{R_1(t_1) G_1(t_1) + R_2(t_3) G_3(t_3) - (G_2(t_2) \ln s_1 + G_4(\eta) \ln s_2)}{a_4(\eta)} \right], \tag{A.38}$$

$$\frac{\partial^2 L}{\partial c \partial \gamma_0}$$

$$= (k+1)c \left(\frac{\delta_0 (G_1(t))^c \ln G_1(t)}{(1+(G_1(t))^c)^2} + \frac{\delta_1 (a_1(t))^c \ln a_1(t)}{(1+(a_1(t))^c)^2} \right.$$

$$\left. + \frac{\delta_2 (a_2(t))^c \ln a_2(t)}{(1+(a_2(t))^c)^2} + \frac{\delta_3 (a_3(t))^c \ln a_3(t)}{(1+(a_3(t))^c)^2} \right)$$

$$+ \frac{\delta_4 k c (a_4(\eta))^c \ln a_4(\eta)}{(1+(a_4(\eta))^c)^2} + \frac{1}{c} \left(\frac{\partial L}{\partial \gamma_0} \right), \tag{A.39}$$

$$
\begin{aligned}
&\frac{\partial^2 L}{\partial \gamma_1^2} \\
&= -\delta_0 \left\{ \left[\ln s_0 + \frac{1}{1+\gamma_1} + R_1(t) \right]^2 \left(\frac{(k+1)c^2 (G_1(t))^c}{(1+(G_1(t))^c)^2} \right) \right. \\
&\quad \times \left. \left[\left(\frac{(k+1)c(G_1(t))^c}{(1+(G_1(t))^c)} \right) - (c-1) \right] A(t) \right\} \\
&\quad -\delta_1 \left\{ \left[\ln s_0 + \left(\frac{1}{1+\gamma_1} + R_1(t_1) \right) \left(\frac{G_1(t_1)}{a_1(t)} \right) - \frac{G_2(t) \ln s_1}{G_1(t_1) + G_2(t)} \right]^2 \right. \\
&\quad \times \left(\frac{(k+1)c^2 (a_1(t))^c}{(1+(a_1(t))^c)^2} - \frac{(k+1)c(a_1(t))^c}{1+(a_1(t))^c} + (c-1) \right) + \left(\frac{B(t)}{a_1(t)} \right) \\
&\quad \times \left. \left(\frac{(k+1)c(a_1(t))^c}{(1+(a_1(t))^c)} - (c-1) \right) \right\} \\
&\quad - \delta_2 \left\{ \left[\left(\frac{1}{1+\gamma_1} \right) \left(1 - \frac{G_2(t_2)}{a_2(t)} \right) + \ln s_0 \right. \right. \\
&\quad + \left. \frac{R_1(t_1)G_1(t_1) + R_2(t)G_3(t) - \ln s_1 G_2(t_2)}{a_2(t)} \right]^2 \\
&\quad \times \left(\frac{(k+1)c^2 (a_2(t))^c}{(1+(a_2(t))^c)^2} + (c-1) - \frac{(k+1)c(a_2(t))^c}{1+(a_2(t))^c} \right) \\
&\quad + \left. \left(\frac{(k+1)c(a_2(t))^c}{1+(a_2(t))^c} - (c-1) \right) \left(\frac{D(t)}{a_2(t)} \right) \right\} \\
&\quad -\delta_3 \left\{ \left[\left(\frac{1}{1+\gamma_1} \right) \left(\frac{G_1(t_1) + G_3(t_3)}{a_3(t)} \right) + \ln s_0 \right. \right. \\
&\quad + \left. \frac{R_1(t_1)G_1(t_1) + R_2(t_3)G_3(t_3) - G_2(t_2) \ln s_1}{a_3(t)} - \frac{G_4(t) \ln s_2}{a_3(t)} \right]^2 \\
&\quad \times \left(\frac{(k+1)c^2 (a_3(t))^c}{(1+(a_3(t))^c)^2} - \frac{(k+1)c(a_3(t))^c}{(1+(a_3(t))^c)} + (c-1) \right) + \left(\frac{E(t)}{a_3(t)} \right) \\
&\quad \times \left. \left(\frac{(k+1)c(a_3(t))^c}{1+(a_3(t))^c} - (c-1) \right) \right\}
\end{aligned}
$$

$$-\delta_4 \left\{ \left[\left(1 - \frac{G_2(t_2) + G_4(\eta)}{a_4(\eta)}\right) \left(\frac{1}{1+\gamma_1}\right) + \ln s_0 \right.\right.$$

$$\left. + \frac{R_1(t_1)G_1(t_1) + R_2(t_3)G_3(t_3) - (G_2(t_2)\ln s_1 + G_4(\eta)\ln s_2)}{a_4(\eta)} \right]^2$$

$$\times \left(\frac{kc^2(a_4(\eta))^c}{(1+(a_4(\eta))^c)^2} - \frac{kc(a_4(\eta))^c}{1+(a_4(\eta))^c} \right)$$

$$\left. + \left(\frac{kc(a_4(\eta))^c}{1+(a_4(\eta))^c}\right) \left(\frac{E(\eta)}{a_4(\eta)}\right) \right\}, \tag{A.40}$$

$$\frac{\partial^2 L}{\partial \gamma_1 \partial c}$$

$$= \delta_0 \left[\ln s_0 + \frac{1}{1+\gamma_1} + R_1(t) \right] \left(\frac{(k+1)c(G_1(t))^c \ln G_1(t)}{(1+(G_1(t))^c)^2} - 1 \right.$$

$$\left. + \frac{(G_1(t))^c}{1+(G_1(t))^c}(k+1) \right) + \delta_1 \left[\ln s_0 + \left(\frac{1}{1+\gamma_1} + R_1(t_1) \right) \right.$$

$$\left. \times \left(\frac{G_1(t_1)}{(a_1(t))} \right) - \frac{G_2(t)\ln s_1}{(a_1(t))} \right] \left(\frac{(k+1)(a_1(t))^c}{1+(a_1(t))^c} - 1 \right.$$

$$\left. + \frac{(k+1)c(a_1(t))^c \ln a_1(t)}{(1+(a_1(t))^c)^2} \right) + \delta_2 \left[\ln s_0 + \left(\frac{1}{1+\gamma_1} \right) \right.$$

$$\left. \times \left(1 - \frac{G_2(t_2)}{a_2(t)}\right) + \frac{R_1(t_1)G_1(t_1) + R_2(t)G_3(t) - G_2(t_2)\ln s_1}{a_2(t)} \right]$$

$$\times \left(\frac{(k+1)(a_2(t))^c}{1+(a_2(t))^c} - 1 + \frac{(k+1)c(a_2(t))^c \ln a_2(t)}{(1+(a_2(t))^c)^2} \right)$$

$$+ \delta_3 \left[\ln s_0 + \left(\frac{1}{1+\gamma_1}\right) \left(1 - \frac{G_2(t_2) + G_4(t)}{a_3(t)}\right) \right.$$

$$\left. + \frac{R_1(t_1)G_1(t_1) + R_2(t_3)G_3(t_3) - \ln s_1 G_2(t_2)}{a_3(t)} - \frac{\ln s_2 G_4(t)}{a_3(t)} \right]$$

$$\times \left(\frac{(k+1)c(a_3(t))^c \ln a_3(t)}{(1+(a_3(t))^c)^2} - 1 + \frac{(k+1)(a_3(t))^c}{1+(a_3(t))^c} \right)$$

$$+ \delta_4 \left[\frac{R_1(t_1)G_1(t_1) + R_2(t_3)G_3(t_3) - (\ln s_1 G_2(t_2) + \ln s_2 G_4(\eta))}{a_4(\eta)} \right.$$

$$
+\left(\frac{G_1(t_1)+G_3(t_3)}{a_4(\eta)}\right)\left(\frac{1}{1+\gamma_1}\right)+\ln s_0\Bigg]\left(\frac{k(a_4(\eta))^c}{1+(a_4(\eta))^c}\right)
\times\left(1+\frac{c\ln a_4(\eta)}{1+(a_4(\eta))^c}\right), \tag{A.41}
$$

and

$$
\frac{\partial^2 L}{\partial c^2}=-\left(\frac{\delta_0(G_1(t))^c(\ln G_1(t))^2}{(1+(G_1(t))^c)^2}+\frac{\delta_1(a_1(t))^c(\ln a_1(t))^2}{(1+(a_1(t))^c)^2}
+\frac{\delta_2(a_2(t))^c(\ln a_2(t))^2}{(1+(a_2(t))^c)^2}+\frac{\delta_3(a_3(t))^c(\ln a_3(t))^2}{(1+(a_3(t))^c)^2}\right)(k+1)
-\frac{\delta_4 k(a_4(\eta))^c(\ln a_4(\eta))^2}{(1+(a_4(\eta))^c)^2}-\sum_{j=0}^{3}\frac{\delta_j}{c^2}. \tag{A.42}
$$

The elements of the Fisher information matrix for an observation are the negative expectations of these second order partial derivatives. Using (A.37)–(A.42), these are obtained as:

$$
E\left[-\frac{\partial^2 L}{\partial\gamma_0^2}\right]
=(k+1)c^2\left(\int_0^{t_1}\left(\frac{(G_1(t))^c}{(1+(G_1(t))^c)^2}\right)f_1(t)dt
+\int_{t_1}^{t_2}\left(\frac{(a_1(t))^c}{(1+(a_1(t))^c)^2}\right)f_2(t)dt
+\int_{t_2}^{t_3}\left(\frac{(a_2(t))^c}{(1+(a_2(t))^c)^2}\right)f_3(t)dt
+\int_{t_3}^{\eta}\left(\frac{(a_3(t))^c}{(1+(a_3(t))^c)^2}\right)f_4(t)dt\right)
+\frac{kc^2(a_4(\eta))^c}{(1+(a_4(\eta))^c)^{2+k}}, \tag{A.43}
$$

$$
\begin{aligned}
&E\left[-\frac{\partial^2 L}{\partial\gamma_1\partial\gamma_0}\right]\\
&= (k+1)c^2\left\{\int_0^{t_1}\left(\frac{(G_1(t))^c}{(1+(G_1(t))^c)^2}\right)\left[\ln s_0+\frac{1}{1+\gamma_1}+R_1(t)\right]\right.\\
&\quad\times f_1(t)dt+\int_{t_1}^{t_2}\left(\frac{(a_1(t))^c}{(1+(a_1(t))^c)^2}\right)\left[\ln s_0+\left(\frac{1}{1+\gamma_1}+R_1(t_1)\right)\right.\\
&\quad\times\left.\left(\frac{G_1(t_1)}{a_1(t)}\right)-\frac{G_2(t)\ln s_1}{a_1(t)}\right]f_2(t)dt+\int_{t_2}^{t_3}\left(\frac{(a_2(t))^c}{(1+(a_2(t))^c)^2}\right)\\
&\quad\times\left[\left(\frac{G_1(t_1)+G_3(t)}{a_2(t)}\right)\left(\frac{1}{1+\gamma_1}\right)\right.\\
&\quad+\left.\frac{R_1(t_1)G_1(t_1)+R_2(t)G_3(t)-G_2(t_2)\ln s_1}{a_2(t)}+\ln s_0\right]f_3(t)dt\\
&\quad+\int_{t_3}^{\eta}\left(\frac{(a_3(t))^c}{(1+(a_3(t))^c)^2}\right)\left[\left(\frac{G_1(t_1)+G_3(t_3)}{a_3(t)}\right)\left(\frac{1}{1+\gamma_1}\right)\right.\\
&\quad+\ln s_0+\frac{R_1(t_1)G_1(t_1)+R_2(t_3)G_3(t_3)-G_2(t_2)\ln s_1}{a_3(t)}\\
&\quad-\left.\left.\frac{G_4(t)\ln s_2}{a_3(t)}\right]f_4(t)dt\right\}+\left(\frac{kc^2(a_4(\eta))^c}{(1+(a_4(\eta))^c)^{2+k}}\right)\\
&\quad\times\left[\ln s_0+\left(\frac{1}{1+\gamma_1}\right)\left(\frac{G_1(t_1)+G_3(t_3)}{a_4(\eta)}\right)\right.\\
&\quad+\left.\frac{R_1(t_1)G_1(t_1)+R_2(t_3)G_3(t_3)}{a_4(\eta)}-\frac{(G_2(t_2)\ln s_1+G_4(\eta)\ln s_2)}{a_4(\eta)}\right],
\end{aligned}
\tag{A.44}
$$

$$
\begin{aligned}
&E\left[-\frac{\partial^2 L}{\partial c\partial\gamma_0}\right]\\
&= -\left(\int_0^{t_1}\left(\frac{(G_1(t))^c\ln G_1(t)}{(1+(G_1(t))^c)^2}\right)f_1(t)dt\right.\\
&\quad+\int_{t_1}^{t_2}\left(\frac{(a_1(t))^c\ln a_1(t)}{(1+(a_1(t))^c)^2}\right)f_2(t)dt
\end{aligned}
$$

$$
\begin{aligned}
&+\int_{t_2}^{t_3}\left(\frac{(a_2(t))^c\ln a_2(t)}{(1+(a_2(t))^c)^2}\right)f_3(t)dt\\
&+\int_{t_3}^{\eta}\left(\frac{(a_3(t))^c\ln a_3(t)}{(1+(a_3(t))^c)^2}\right)f_4(t)dt\Bigg)\\
&\times(k+1)c-\frac{kc(a_4(\eta))^c\ln a_4(\eta)}{(1+(a_4(\eta))^c)^{k+2}},
\end{aligned}
\tag{A.45}
$$

$$
\begin{aligned}
E&\left[-\frac{\partial^2 L}{\partial\gamma_1^2}\right]\\
&=\int_0^{t_1}\left\{\left[\ln s_0+\frac{1}{1+\gamma_1}+R_1(t)\right]^2\left(\frac{(k+1)c^2(G_1(t))^c}{(1+(G_1(t))^c)^2}\right)\right.\\
&\quad\left.+\left(\left(\frac{c(G_1(t))^c}{1+(G_1(t))^c}\right)(k+1)-(c-1)\right)A(t)\right\}f_1(t)dt\\
&\quad+\int_{t_1}^{t_2}\left\{\left[\ln s_0+\left(\frac{1}{1+\gamma_1}+R_1(t_1)\right)\left(\frac{G_1(t_1)}{a_1(t)}\right)-\frac{G_2(t)\ln s_1}{a_1(t)}\right]^2\right.\\
&\quad\times\left(\frac{(k+1)c^2(a_1(t))^c}{(1+(a_1(t))^c)^2}-\frac{(k+1)c(a_1(t))^c}{1+(a_1(t))^c}+(c-1)\right)\\
&\quad\left.+\left(\frac{(k+1)c(a_1(t))^c}{1+(a_1(t))^c}-(c-1)\right)\left(\frac{B(t)}{a_1(t)}\right)\right\}f_2(t)dt\\
&\quad+\int_{t_2}^{t_3}\left\{\left[\ln s_0+\left(\frac{1}{1+\gamma_1}\right)\left(\frac{G_1(t_1)+G_3(t)}{a_2(t)}\right)\right.\right.\\
&\quad\left.+\frac{R_1(t_1)G_1(t_1)+R_2(t)G_3(t)-G_2(t_2)\ln s_1}{a_2(t)}\right]^2\\
&\quad\times\left(\frac{(k+1)c^2(a_2(t))^c}{(1+(a_2(t))^c)^2}-\frac{(k+1)c(a_2(t))^c}{1+(a_2(t))^c}+(c-1)\right)\\
&\quad\left.+\left(\frac{(k+1)c(a_2(t))^c}{1+(a_2(t))^c}-(c-1)\right)\left(\frac{D(t)}{a_2(t)}\right)\right\}f_3(t)dt\\
&\quad+\int_{t_3}^{\eta}\left\{\left[\left(\frac{1}{1+\gamma_1}\right)\left(1-\frac{G_2(t_2)+G_4(t)}{a_3(t)}\right)+\ln s_0\right.\right.\\
&\quad\left.+\frac{R_1(t_1)G_1(t_1)+R_2(t_3)G_3(t_3)-(G_2(t_2)\ln s_1+G_4(t)\ln s_2)}{a_3(t)}\right]^2
\end{aligned}
$$

$$\times \left(\frac{(k+1)c^2(a_3(t))^c}{(1+(a_3(t))^c)^2} - \frac{(k+1)c(a_3(t))^c}{1+(a_3(t))^c} + (c-1) \right)$$

$$+ \left(\frac{(k+1)c(a_3(t))^c}{(1+(a_3(t))^c)} - (c-1) \right) \left(\frac{E(t)}{a_3(t)} \right) \Bigg\} f_4(t)dt$$

$$+ \Bigg\{ \Bigg[\left(1 - \frac{G_2(t_2) + G_4(\eta)}{(a_4(\eta))} \right) \left(\frac{1}{1+\gamma_1} \right) + \ln s_0$$

$$+ \frac{R_1(t_1)G_1(t_1) + R_2(t_3)G_3(t_3) - (G_2(t_2)\ln s_1 + G_4(\eta)\ln s_2)}{a_4(\eta)} \Bigg]^2$$

$$\times \left(\frac{kc^2(a_4(\eta))^c}{(1+(a_4(\eta))^c)^2} - \frac{kc(a_4(\eta))^c}{1+(a_4(\eta))^c} \right) + \left(\frac{kc(a_4(\eta))^c}{1+(a_4(\eta))^c} \right)$$

$$\times \left(\frac{E(\eta)}{a_4(\eta)} \right) \Bigg\} (1+(a_4(\eta))^c)^{-k}, \tag{A.46}$$

$$E\left[-\frac{\partial^2 L}{\partial \gamma_1 \partial c} \right]$$

$$= -\int_0^{t_1} \left[\ln s_0 + \frac{1}{1+\gamma_1} + R_1(t) \right] \left(\frac{(k+1)c(G_1(t))^c \ln G_1(t)}{(1+(G_1(t))^c)^2} \right.$$

$$\left. + (k+1) \left(\frac{(G_1(t))^c}{1+(G_1(t))^c} \right) - 1 \right) f_1(t)dt$$

$$- \int_{t_1}^{t_2} \left(\frac{(k+1)c(a_1(t))^c \ln(a_1(t))}{(1+(a_1(t))^c)^2} - 1 \frac{(k+1)(a_1(t))^c}{1+(a_1(t))^c} \right)$$

$$\times \left[\ln s_0 + \left(\frac{1}{1+\gamma_1} + R_1(t_1) \right) \left(\frac{G_1(t_1)}{a_1(t)} \right) - \frac{G_2(t)\ln s_1}{a_1(t)} \right] f_2(t)dt$$

$$- \int_{t_2}^{t_3} \Bigg[\left(\frac{1}{1+\gamma_1} \right) \left(1 - \frac{G_2(t_2)}{a_2(t)} \right) + \frac{R_1(t_1)G_1(t_1)}{a_2(t)}$$

$$+ \frac{R_2(t)G_3(t) - G_2(t_2)\ln s_1}{a_2(t)} + \ln s_0 \Bigg] \left(\frac{(k+1)(a_2(t))^c}{1+(a_2(t))^c} \right.$$

$$\left. - 1 + (k+1) \frac{c(a_2(t))^c \ln(a_2(t))}{(1+(a_2(t))^c)^2} \right) f_3(t)dt$$

$$- \int_{t_3}^{\eta} \Bigg[\left(\frac{1}{1+\gamma_1} \right) \left(1 - \frac{G_2(t_2) + G_4(t)}{a_3(t)} \right) + \ln s_0$$

$$+\frac{R_1(t_1)G_1(t_1)+R_2(t_3)G_3(t_3)-G_2(t_2)\ln s_1}{a_3(t)}-\frac{G_4(t)\ln s_2}{a_3(t)}\Bigg]$$
$$\times\left(\frac{(k+1)c(a_3(t))^c\ln a_3(t)}{(1+(a_3(t))^c)^2}+\frac{(k+1)(a_3(t))^c}{1+(a_3(t))^c}-1\right)f_4(t)dt$$
$$-\Bigg[\left(1-\frac{G_2(t_2)+G_4(\eta)}{a_4(\eta)}\right)\left(\frac{1}{1+\gamma_1}\right)+\ln s_0$$
$$+\frac{R_1(t_1)G_1(t_1)+R_2(t_3)G_3(t_3)-(G_2(t_2)\ln s_1+G_4(\eta)\ln s_2)}{a_4(\eta)}\Bigg]$$
$$\times\left(\frac{k(a_4(\eta))^c}{1+(a_4(\eta))^c}+\frac{kc(a_4(\eta))^c\ln a_4(\eta)}{(1+(a_4(\eta))^c)^2}\right)(1+(a_4(\eta))^c)^{-k}, \tag{A.47}$$

$$E\left[-\frac{\partial^2 L}{\partial c^2}\right]$$
$$=\int_0^{t_1}\left(\frac{f_1(t)}{c^2}\right)dt+\int_{t_1}^{t_2}\left(\frac{f_2(t)}{c^2}\right)dt+\int_{t_2}^{t_3}\left(\frac{f_3(t)}{c^2}\right)dt$$
$$+\int_{t_3}^{\eta}\left(\frac{f_4(t)}{c^2}\right)dt+(k+1)\left(\int_0^{t_1}\left(\frac{(G_1(t))^c(\ln G_1(t))^2}{(1+(G_1(t))^c)^2}\right)f_1(t)dt\right.$$
$$+\int_{t_1}^{t_2}\left(\frac{(a_1(t))^c(\ln a_1(t))^2}{(1+(a_1(t))^c)^2}\right)f_2(t)dt$$
$$+\int_{t_2}^{t_3}\left(\frac{(a_2(t))^c(\ln a_2(t))^2}{(1+(a_2(t))^c)^2}\right)f_3(t)dt$$
$$\left.+\int_{t_3}^{\eta}\left(\frac{(a_3(t))^c(\ln a_3(t))^2}{(1+(a_3(t))^c)^2}\right)f_4(t)dt\right)$$
$$+\frac{k(a_4(\eta))^c(\ln a_4(\eta))^2}{(1+(a_4(\eta))^c)^{2+k}}, \tag{A.48}$$

where

$$a_1(t)=G_1(t_1)+G_2(t),$$
$$a_2(t)=G_1(t_1)+G_2(t_2)+G_3(t),$$
$$a_3(t)=G_1(t_1)+G_2(t_2)+G_3(t_3)+G_4(t),$$

$$\begin{aligned}
a_4(\eta) &= G_1(t_1) + G_2(t_2) + G_3(t_3) + G_4(\eta),\\
E[I_4(t)] &= 1 - F_4(\eta),\\
E\left[\frac{\partial L}{\partial \gamma_i}\right] &= 0, \quad \text{for } i = 0, 1, \quad \text{and}\\
E\left[\frac{\partial L}{\partial c}\right] &= 0.
\end{aligned}$$

Thus, (A.43)–(A.48) give the elements of Fisher Information Matrix.

A.12 JUSTIFICATION FOR THE USE OF LINEAR CUMULATIVE EXPOSURE MODEL IN THE BURR TYPE-XII DISTRIBUTION IN SECTION 5.3.3

Consider the cdf of the Burr Type-XII distribution as

$$\begin{aligned}
&G(t; c, k, \alpha)\\
&\quad = 1 - \left(1 + \left(\frac{t}{\alpha}\right)^c\right)^{-k}; \quad t \geq 0;\ c, k, \alpha > 0\\
&\quad \Rightarrow G(\varepsilon(t); c, k, \alpha) = 1 - \left(1 + \left(\frac{\varepsilon(t)}{\alpha}\right)^c\right)^{-k}; \quad t \geq 0;\ c, k, \alpha > 0
\end{aligned}$$

For $\alpha = 1$,

$$\begin{aligned}
G(\varepsilon(t)) &\equiv G(\varepsilon(t); c, k, 1)\\
&= 1 - (1 + (\varepsilon(t))^c)^{-k}; \quad t \geq 0;\ c, k > 0\\
&= \begin{cases} 1 - (1 + (G_1(t))^c)^{-k}; & 0 < t \leq t_i;\ c, k > 0;\\ & i = 1, 2 \ \text{(using (5.19))}\\ 1 - (1 + (\varepsilon(t_i) + G_{i2}(t))^c)^{-k}; & t_i < t \leq \eta;\ c, k > 0;\\ & i = 1, 2 \ \text{(using (5.20))} \end{cases}\\
&= \begin{cases} F_1(t); & 0 < t \leq t_i;\ i = 1, 2\\ F_{i2}(t); & t_i < t \leq \eta;\ i = 1, 2 \end{cases}\\
&= F(t).
\end{aligned}$$

It is necessary to set $\alpha = 1$, as then only the $F(t)$ will be equal to $G(\varepsilon(t))$, and a linear cumulative exposure model will hold.

A.13 CALCULATION OF THE ELEMENTS OF THE FISHER INFORMATION MATRIX GIVEN IN SECTION 5.3.5

The first-order partial derivatives are

$$\begin{aligned}
\frac{\partial L}{\partial \gamma_0} &\\
&= -c\sum_{j=0}^{1} I_j + (k+1)c\left(\frac{I_0(G_1(x))^c}{1+(G_1(x))^c} + \frac{I_1(G_1(x_i)+G_{i2}(x))^c}{1+(G_1(x_i)+G_{i2}(x))^c}\right)\\
&\quad + \frac{I_2kc(G_1(x_i)+G_{i2}(\eta))^c}{1+(G_1(x_i)+G_{i2}(\eta))^c},
\end{aligned} \tag{A.49}$$

$$\begin{aligned}
\frac{\partial L}{\partial \gamma_1} &\\
&= I_0\left[-c\ln s_0 + \ln(\beta x + s_0) - (c-1)\left(\frac{1}{1+\gamma_1} + R_1(x)\right)\right.\\
&\quad \left. + \frac{(k+1)c(G_1(x))^c}{1+(G_1(x))^c}\left(\ln s_0 + \frac{1}{1+\gamma_1} + R_1(x)\right)\right]\\
&\quad + I_1\left[-c\ln s_0 + \ln s_i + \frac{(c-1)G_{i2}(x)\ln s_i}{G_1(x_i)+G_{i2}(x)} - (c-1)\right.\\
&\quad \times\left(\frac{1}{1+\gamma_1} + R_1(x_i)\right)\left(\frac{G_1(x_i)}{G_1(x_i)+G_{i2}(x)}\right)\\
&\quad + \left(\frac{(k+1)c(G_1(x_i)+G_{i2}(x))^c}{1+(G_1(x_i)+G_{i2}(x))^c}\right)\left(\ln s_0 + \left(\frac{1}{1+\gamma_1} + R_1(x_i)\right)\right.\\
&\quad \left.\left. \times\left(\frac{G_1(x_i)}{G_1(x_i)+G_{i2}(x)}\right) - \frac{G_{i2}(x)\ln s_i}{G_1(x_i)+G_{i2}(x)}\right)\right]
\end{aligned}$$

$$+ I_2\left(\frac{kc(G_1(x_i)+G_{i2}(\eta))^c}{1+(G_1(x_i)+G_{i2}(\eta))^c}\right)\left[\left(\frac{1}{1+\gamma_1}\right)\left(\frac{G_1(x_i)}{G_1(x_i)+G_{i2}(\eta)}\right)\right.$$

$$\left.+\ln s_0+\frac{R_1(x_i)G_1(x_i)-G_{i2}(\eta)\ln s_i}{G_1(x_i)+G_{i2}(\eta)}\right], \tag{A.50}$$

and

$$\frac{\partial L}{\partial c}=\sum_{j=0}^{1}\frac{I_j}{c}+I_0\ln G_1(x)+I_1\ln(G_1(x_i)+G_{i2}(x))-(k+1)$$

$$\times\left(\frac{I_0(G_1(x))^c\ln G_1(x)}{1+(G_1(x))^c}\right.$$

$$\left.+\frac{I_1(G_1(x_i)+G_{i2}(x))^c\ln(G_1(x_i)+G_{i2}(x))}{1+(G_1(x_i)+G_{i2}(x))^c}\right)$$

$$-\frac{I_2k(G_1(x_i)+G_{i2}(\eta))^c\ln(G_1(x_i)+G_{i2}(\eta))}{1+(G_1(x_i)+G_{i2}(\eta))^c}. \tag{A.51}$$

The second-order partial derivatives are

$$\frac{\partial^2 L}{\partial\gamma_0^2}$$

$$=-\left\{(k+1)c^2\left(\frac{I_0(G_1(x))^c}{(1+(G_1(x))^c)^2}+\frac{I_1(G_1(x_i)+G_{i2}(x))^c}{(1+(G_1(x_i)+G_{i2}(x))^c)^2}\right)\right.$$

$$\left.+\frac{I_2kc^2(G_1(x_i)+G_{i2}(\eta))^c}{(1+(G_1(x_i)+G_{i2}(\eta))^c)^2}\right\}, \tag{A.52}$$

$$\frac{\partial^2 L}{\partial\gamma_1\partial\gamma_0}$$

$$=-(k+1)c^2\left\{\frac{I_0(G_1(x))^c}{(1+(G_1(x))^c)^2}\left[\ln s_0+\frac{1}{1+\gamma_1}+R_1(x)\right]\right.$$

$$+\frac{I_1(G_1(x_i)+G_{i2}(x))^c}{(1+(G_1(x_i)+G_{i2}(x))^c)^2}\left[\left(\frac{1}{1+\gamma_1}+R_1(x_i)\right)\right.$$

$$\times\left(\frac{G_1(x_i)}{G_1(x_i)+G_{i2}(x)}\right)+\ln s_0-\frac{G_{i2}(x)\ln s_i}{G_1(x_i)+G_{i2}(x)}\Bigg]\Bigg\}$$

$$-\frac{I_2kc^2(G_1(x_i)+G_{i2}(\eta))^c}{(1+(G_1(x_i)+G_{i2}(\eta))^c)^2}\left[\left(\frac{1}{1+\gamma_1}\right)\left(\frac{G_1(x_i)}{(G_1(x_i)+G_{i2}(\eta))}\right)\right.$$

$$\left.+\ln s_0+\frac{R_1(x_i)G_1(x_i)-G_{i2}(\eta)\ln s_i}{(G_1(x_i)+G_{i2}(\eta))}\right], \tag{A.53}$$

$$\frac{\partial^2 L}{\partial c\partial\gamma_0}$$

$$=\frac{1}{c}\left(\frac{\partial L}{\partial\gamma_0}\right)+(k+1)c\left(I_0\left(\frac{(G_1(x))^c\ln G_1(x)}{(1+(G_1(x))^c)^2}\right)\right.$$

$$\left.+I_1\left(\frac{(G_1(x_i)+G_{i2}(x))^c\ln(G_1(x_i)+G_{i2}(x))}{(1+(G_1(x_i)+G_{i2}(x))^c)^2}\right)\right)$$

$$+I_2\left(\frac{kc(G_1(x_i)+G_{i2}(\eta))^c\ln(G_1(x_i)+G_{i2}(\eta))}{(1+(G_1(x_i)+G_{i2}(\eta))^c)^2}\right), \tag{A.54}$$

$$\frac{\partial^2 L}{\partial\gamma_1^2}$$

$$=-I_0\left\{\left[\ln s_0+\frac{1}{1+\gamma_1}+R_1(x)\right]^2\left(\frac{(k+1)c^2(G_1(x))^c}{(1+(G_1(x))^c)^2}\right)\right.$$

$$\left.\times\left(\left(\frac{(k+1)c(G_1(x))^c}{1+(G_1(x))^c}\right)-(c-1)\right)A(t)\right\}$$

$$-I_1\left\{\left[\ln s_0+\left(\frac{1}{1+\gamma_1}+R_1(x_i)\right)\left(\frac{G_1(x_i)}{G_1(x_i)+G_{i2}(x)}\right)\right.\right.$$

$$\left.-\frac{G_{i2}(x)\ln s_i}{G_1(x_i)+G_{i2}(x)}\right]^2\left(\frac{(k+1)c^2(G_1(x_i)+G_{i2}(x))^c}{(1+(G_1(x_i)+G_{i2}(x))^c)^2}+(c-1)\right.$$

$$\left.-\frac{(k+1)c(G_1(x_i)+G_{i2}(x))^c}{1+(G_1(x_i)+G_{i2}(x))^c}\right)$$

$$+\left(\frac{(k+1)c(G_1(x_i)+G_{i2}(x))^c}{1+(G_1(x_i)+G_{i2}(x))^c}-(c-1)\right)$$

$$\times \left(\frac{K(x)}{G_1(x_i)+G_{i2}(x)}\right)\Bigg\} - I_2 \Bigg\{\left[\ln s_0 + \left(\frac{G_1(x_i)}{G_1(x_i)+G_{i2}(\eta)}\right)\right.$$

$$\left.\times \left(\frac{1}{1+\gamma_1}\right) + \frac{R_1(x_i)G_1(x_i) - G_{i2}(\eta)\ln s_i}{G_1(x_i)+G_{i2}(\eta)}\right]^2$$

$$\left(\frac{kc^2(G_1(x_i)+G_{i2}(\eta))^c}{(1+(G_1(x_i)+G_{i2}(\eta))^c)^2} - \frac{kc(G_1(x_i)+G_{i2}(\eta))^c}{1+(G_1(x_i)+G_{i2}(\eta))^c}\right)$$

$$+ \left(\frac{kc(G_1(x_i)+G_{i2}(\eta))^c}{1+(G_1(x_i)+G_{i2}(\eta))^c}\right)\left(\frac{K(\eta)}{G_1(x_i)+G_{i2}(\eta)}\right)\Bigg\}, \tag{A.55}$$

$$\frac{\partial^2 L}{\partial \gamma_1 \partial c}$$

$$= I_0 \left[\ln s_0 + \frac{1}{1+\gamma_1} + R_1(x)\right] \left(\frac{(k+1)c(G_1(x))^c \ln G_1(x)}{(1+(G_1(x))^c)^2}\right.$$

$$\left. - 1 + \frac{(k+1)(G_1(x))^c}{1+(G_1(x))^c}\right) + I_1 \left[\ln s_0 + \left(\frac{1}{1+\gamma_1} + R_1(x_i)\right)\right.$$

$$\left.\times \left(\frac{G_1(x_i)}{G_1(x_i)+G_{i2}(x)}\right) - \frac{G_{i2}(x)\ln s_i}{G_1(x_i)+G_{i2}(x)}\right]$$

$$\times \left(\frac{(k+1)c(G_1(x_i)+G_{i2}(x))^c \ln(G_1(x_i)+G_{i2}(t))}{(1+(G_1(x_i)+G_{i2}(x))^c)^2}\right.$$

$$\left. + \frac{(k+1)(G_1(x_i)+G_{i2}(x))^c}{1+(G_1(x_i)+G_{i2}(x))^c} - 1\right) + I_2 \left[\left(\frac{G_1(x_i)}{G_1(x_i)+G_{i2}(\eta)}\right)\right.$$

$$\left.\times \left(\frac{1}{1+\gamma_1}\right) + \ln s_0 + \frac{R_1(x_i)G_1(x_i) - G_{i2}(\eta)\ln s_i}{G_1(x_i)+G_{i2}(\eta)}\right]$$

$$\times \left(\frac{k(G_1(x_i)+G_{i2}(\eta))^c}{1+(G_1(x_i)+G_{i2}(\eta))^c}\right.$$

$$\left. + \frac{kc(G_1(x_i)+G_{i2}(\eta))^c \ln(G_1(x_i)+G_{i2}(\eta))}{(1+(G_1(x_i)+G_{i2}(\eta))^c)^2}\right) \tag{A.56}$$

and

$$\frac{\partial^2 L}{\partial c^2} = -\sum_{j=0}^{1} \frac{I_j}{c^2} - (k+1)\left(\frac{I_0 (G_1(x))^c (\ln G_1(x))^2}{(1+(G_1(x))^c)^2} + \frac{I_1 (G_1(x_i)+G_{i2}(x))^c (\ln(G_1(x_i)+G_{i2}(x)))^2}{(1+(G_1(x_i)+G_{i2}(x))^c)^2}\right) - \frac{I_2 k (G_1(x_i)+G_{i2}(\eta))^c (\ln(G_1(x_i)+G_{i2}(\eta)))^2}{(1+(G_1(x_i)+G_{i2}(\eta))^c)^2}. \tag{A.57}$$

The elements of the Fisher information matrix for an observation are the negative expectations of these second-order partial derivatives. Using (A.52)–(A.57), these are as follows:

$$E\left[-\frac{\partial^2 L}{\partial \gamma_0^2}\right] = \frac{kc^2 (G_1(x_i)+G_{i2}(\eta))^c}{(1+(G_1(x_i)+G_{i2}(\eta))^c)^{2+k}} + \left(\int_0^{x_i} \left(\frac{(G_1(x))^c}{(1+(G_1(x))^c)^2}\right) f_1(x)dx + \int_{x_i}^{\eta} \left(\frac{(G_1(x_i)+G_{i2}(x))^c}{(1+(G_1(x_i)+G_{i2}(x))^c)^2}\right) f_{i2}(x)dx\right)(k+1)c^2, \tag{A.58}$$

$$E\left[-\frac{\partial^2 L}{\partial \gamma_1 \partial \gamma_0}\right] = (k+1)c^2 \left\{\int_0^{x_i} \left(\frac{(G_1(x))^c}{(1+(G_1(x))^c)^2}\right) \times \left[\ln s_0 + \frac{1}{1+\gamma_1} + R_1(x)\right] f_1(x)dx + \int_{x_i}^{\eta} \left(\frac{(G_1(x_i)+G_{i2}(x))^c}{(1+(G_1(x_i)+G_{i2}(x))^c)^2}\right)\left[\ln s_0 + \left(\frac{1}{1+\gamma_1} + R_1(x_i)\right)\right.\right.$$

$$\times \left(\frac{G_1(x_i)}{G_1(x_i)+G_{i2}(x)}\right) - \frac{G_{i2}(x)\ln s_i}{G_1(x_i)+G_{i2}(x)}\Bigg] f_{i2}(x)dx\Bigg\}$$

$$+ \frac{kc^2(G_1(x_i)+G_{i2}(\eta))^c}{(1+(G_1(x_i)+G_{i2}(\eta))^c)^{2+k}} \left[\left(\frac{G_1(x_i)}{G_1(x_i)+G_{i2}(\eta)}\right)\left(\frac{1}{1+\gamma_1}\right)\right.$$

$$\left. + \ln s_0 + \frac{R_1(x_i)G_1(x_i) - G_{i2}(\eta)\ln s_i}{G_1(x_i)+G_{i2}(\eta)}\right], \tag{A.59}$$

$$E\left[-\frac{\partial^2 L}{\partial c\partial\gamma_0}\right]$$

$$= -(k+1)c\left(\int_0^{x_i}\left(\frac{(G_1(x))^c\ln G_1(x)}{(1+(G_1(x))^c)^2}\right) f_1(x)dx\right.$$

$$\left. + \int_{x_i}^{\eta}\left(\frac{(G_1(x_i)+G_{i2}(x))^c\ln(G_1(x_i)+G_{i2}(x))}{(1+(G_1(x_i)+G_{i2}(x))^c)^2}\right) f_{i2}(x)dx\right)$$

$$- \frac{kc(G_1(x_i)+G_{i2}(\eta))^c\ln(G_1(x_i)+G_{i2}(\eta))}{(1+(G_1(x_i)+G_{i2}(\eta))^c)^{k+2}}, \tag{A.60}$$

$$E\left[-\frac{\partial^2 L}{\partial\gamma_1^2}\right]$$

$$= \int_0^{x_i}\left\{\left[\ln s_0 + \frac{1}{1+\gamma_1} + R_1(x)\right]^2\left(\frac{(k+1)c^2(G_1(x))^c}{(1+(G_1(x))^c)^2}\right)\right.$$

$$\left. + \left(-(c-1)\left(\frac{(k+1)c(G_1(x))^c}{1+(G_1(x))^c}\right)\right)A(x)\right\} f_1(x)dx$$

$$+ \int_{x_i}^{\eta}\left\{\left[\ln s_0 + \left(\frac{1}{1+\gamma_1} + R_1(x_i)\right)\left(\frac{G_1(x_i)}{G_1(x_i)+G_{i2}(x)}\right)\right.\right.$$

$$\left. - \frac{G_{i2}(x)\ln s_i}{G_1(x_i)+G_{i2}(x)}\right]^2\left((c-1)+(k+1)\right.$$

$$\left.\left(\frac{c^2(G_1(x_i)+G_{i2}(x))^c}{(1+(G_1(x_i)+G_{i2}(x))^c)^2}\right) - \frac{(k+1)c(G_1(x_i)+G_{i2}(x))^c}{1+(G_1(x_i)+G_{i2}(x))^c}\right)$$

$$+ \left((k+1)\left(\frac{c(G_1(x_i)+G_{i2}(x))^c}{1+(G_1(x_i)+G_{i2}(x))^c}\right) - (c-1)\right)$$

$$\left.\times\left(\frac{K(x)}{G_1(x_i)+G_{i2}(x)}\right)\right\} f_{i2}(x)dx + (1+(G_1(x_i)+G_{i2}(\eta))^c)^{-k}$$

$$\times \left\{ \left[\left(\frac{G_1(x_i)}{G_1(x_i) + G_{i2}(\eta)} \right) \left(\frac{1}{1+\gamma_1} \right) + \ln s_0 \right. \right.$$

$$\left. + \frac{R_1(x_i)G_1(x_i) - G_{i2}(\eta)\ln s_i}{G_1(x_i) + G_{i2}(\eta)} \right]^2 \left(\frac{kc^2(G_1(x_i) + G_{i2}(\eta))^c}{(1 + (G_1(x_i) + G_{i2}(\eta))^c)^2} \right.$$

$$\left. - \frac{kc(G_1(x_i) + G_{i2}(\eta))^c}{1 + (G_1(x_i) + G_{i2}(\eta))^c} \right) + \left(\frac{kc(G_1(x_i) + G_{i2}(\eta))^c}{1 + (G_1(x_i) + G_{i2}(\eta)))^c} \right)$$

$$\left. \times \left(\frac{K(\eta)}{G_1(x_i) + G_{i2}(\eta)} \right) \right\}, \tag{A.61}$$

$$E\left[-\frac{\partial^2 L}{\partial \gamma_1 \partial c} \right]$$

$$= - \int_0^{x_i} \left[\ln s_0 + \frac{1}{1+\gamma_1} + R_1(x) \right] \left(\frac{(k+1)c(G_1(x))^c \ln G_1(x)}{(1 + (G_1(x))^c)^2} \right.$$

$$\left. - 1 + \frac{(k+1)(G_1(x))^c}{1 + (G_1(x))^c} \right) f_1(x)dx$$

$$- \int_{x_i}^{\eta} \left[\ln s_0 + \left(\frac{1}{1+\gamma_1} + R_1(x_i) \right) \left(\frac{G_1(x_i)}{G_1(x_i) + G_{i2}(x)} \right) \right.$$

$$\left. - \frac{G_{i2}(x)\ln s_i}{G_1(x_i) + G_{i2}(x)} \right] \left(\frac{(k+1)(G_1(x_i) + G_{i2}(x))^c}{1 + (G_1(x_i) + G_{i2}(x))^c} \right.$$

$$\left. + \frac{(k+1)c(G_1(x_i) + G_{i2}(x))^c \ln(G_1(x_i) + G_{i2}(x))}{(1 + (G_1(x_i) + G_{i2}(x))^c)^2} - 1 \right)$$

$$\times f_{i2}(x)dx - (1 + (G_1(x_i) + G_{i2}(\eta))^c)^{-k} \left[\left(\frac{G_1(x_i)}{G_1(x_i) + G_{i2}(\eta)} \right) \right.$$

$$\left. \times \left(\frac{1}{1+\gamma_1} \right) + \ln s_0 + \frac{R_1(x_i)G_1(x_i) - G_{i2}(\eta)\ln s_i}{G_1(x_i) + G_{i2}(\eta)} \right]$$

$$\times \left(\frac{k(G_1(x_i) + G_{i2}(\eta))^c}{1 + (G_1(x_i) + G_{i2}(\eta))^c} \right.$$

$$\left. + \frac{kc(G_1(x_i) + G_{i2}(\eta))^c \ln(G_1(x_i) + G_{i2}(\eta))}{(1 + (G_1(x_i) + G_{i2}(\eta))^c)^2} \right), \tag{A.62}$$

$$E\left[-\frac{\partial^2 L}{\partial c^2}\right]$$

$$= \frac{k(G_1(x_i) + G_{i2}(\eta))^c(\ln(G_1(x_i) + G_{i2}(\eta)))^2}{(1 + (G_1(x_i) + G_{i2}(\eta))^c)^{2+k}}$$

$$+ \int_0^{x_i} \left(\frac{f_1(x)}{c^2}\right) dx + \int_{x_i}^{\eta} \left(\frac{f_{i2}(x)}{c^2}\right) dx$$

$$+ (k+1)\left(\int_0^{x_i} \left(\frac{(G_1(x))^c(\ln G_1(x))^2}{(1+(G_1(x))^c)^2}\right) f_1(x)dx\right.$$

$$\left. + \int_{x_i}^{\eta} \left(\frac{(G_1(x_i) + G_{i2}(x))^c(\ln(G_1(x_i) + G_{i2}(x)))^2}{(1 + (G_1(x_i) + G_{i2}(x))^c)^2}\right) f_{i2}(x)dx\right), \tag{A.63}$$

where

$$E[I_4(t)] = 1 - F_4(\eta),$$

$$E\left[\frac{\partial L}{\partial \gamma_i}\right] = 0, \quad \text{for } i = 0, 1, \quad \text{and}$$

$$E\left[\frac{\partial L}{\partial c}\right] = 0.$$

Thus, elements of Fisher Information Matrix are given in (A.58)–(A.63).

APPENDIX B

B.1 CALCULATIONS OF THE ELEMENTS OF THE FISHER INFORMATION MATRIX GIVEN IN SECTION 8.6.2

The second-order partial derivatives are:

$$\frac{\partial^2 \mathrm{L}}{\partial \gamma_0^2} = \alpha^2 \cdot \left(I(z)\left(\frac{-2e^{-z}}{(1+e^{-z})^2}\right) - \bar{I}(z)\frac{e^{-\tau'}}{(1+e^{-\tau'})^2}\right), \tag{B.1}$$

$$\frac{\partial^2 \mathrm{L}}{\partial \gamma_0 \partial \gamma_1} = \frac{1}{(1+\gamma_1)} \cdot \left(T_1 \frac{\partial^2 L}{\partial \gamma_0^2} + \alpha \frac{\partial^2 L}{\partial \alpha \partial \gamma_0} - 2\frac{\partial L}{\partial \gamma_0}\right), \tag{B.2}$$

$$\frac{\partial^2 \mathrm{L}}{\partial \gamma_1 \partial \gamma_0} = \frac{\partial^2 \mathrm{L}}{\partial \gamma_0 \partial \gamma_1}, \tag{B.3}$$

$$\frac{\partial^2 \mathrm{L}}{\partial \alpha^2} = -\left(\frac{1}{\alpha^2}\right) \cdot \left(I(z)\left(1 + \frac{2z^2 e^{-z}}{(1+e^{-z})^2}\right) + \bar{I}(z)\left(\frac{\tau'^2 e^{-\tau'}}{(1+e^{-\tau'})^2}\right)\right), \tag{B.4}$$

$$\frac{\partial^2 \mathrm{L}}{\partial \gamma_0 \partial \alpha} = \frac{1}{\alpha}\frac{\partial \mathrm{L}}{\partial \gamma_0} + I(z)\left(\frac{2ze^{-z}}{(1+e^{-z})^2}\right) + \bar{I}(z)\left(\frac{\tau' e^{-\tau'}}{(1+e^{-\tau'})^2}\right), \tag{B.5}$$

$$\frac{\partial^2 \mathrm{L}}{\partial \alpha \partial \gamma_0} = \frac{\partial^2 \mathrm{L}}{\partial \gamma_0 \partial \alpha}, \tag{B.6}$$

$$\frac{\partial^2 \mathrm{L}}{\partial \gamma_1^2} = \frac{1}{(1+\gamma_1)} \cdot \left(-\frac{\partial \mathrm{L}}{\partial \gamma_1} - \frac{1}{(1+\gamma_1)} \cdot \left(\alpha \frac{\partial \mathrm{L}}{\partial \alpha} + \frac{\partial \mathrm{L}}{\partial \gamma_0} \right) + \left(\alpha \frac{\partial^2 \mathrm{L}}{\partial \gamma_1 \partial \alpha} + T_1 \frac{\partial^2 \mathrm{L}}{\partial \gamma_1 \partial \gamma_0} \right) \right), \tag{B.7}$$

$$\frac{\partial^2 \mathrm{L}}{\partial \alpha \partial \gamma_1} = \frac{1}{(1+\gamma_1)} \cdot \left(\frac{\partial \mathrm{L}}{\partial \alpha} + \alpha \frac{\partial^2 \mathrm{L}}{\partial \alpha^2} + T_1 \frac{\partial^2 \mathrm{L}}{\partial \alpha \partial \gamma_0} \right), \tag{B.8}$$

$$\frac{\partial^2 \mathrm{L}}{\partial \alpha \partial \gamma_1} = \frac{\partial^2 \mathrm{L}}{\partial \gamma_1 \partial \alpha}. \tag{B.9}$$

The elements of the Fisher information matrix for an observation at $\xi_i, i = 1, 2$, are the negative expectations of these second-order partial derivatives. Using (B.1)–(B.9), the elements of Fisher information matrix are as follows:

$$E\left[\frac{-\partial^2 L}{\partial \gamma_0^2}\right] = \alpha^2 \cdot \left(\int_{-\infty}^{\tau'} \frac{-2e^{-2z}}{(1+e^{-z})^4} dz + \frac{e^{-2\tau'}}{(1+e^{-\tau'})^3} \right), \tag{B.10}$$

$$E\left[\frac{-\partial^2 L}{\partial \gamma_0 \partial \gamma_1}\right] = \frac{1}{(1+\gamma_1)} \cdot \left(\alpha E\left[\frac{-\partial^2 L}{\partial \gamma_0 \partial \alpha}\right] + T_1 E\left[\frac{-\partial^2 L}{\partial \gamma_0^2}\right] \right), \tag{B.11}$$

$$E\left[\frac{-\partial^2 L}{\partial \gamma_1 \partial \gamma_0}\right] = E\left[\frac{-\partial^2 L}{\partial \gamma_0 \partial \gamma_1}\right], \tag{B.12}$$

$$E\left[\frac{-\partial^2 L}{\partial \gamma_0 \partial \alpha}\right] = -\left(\int_{-\infty}^{\tau'} \frac{2ze^{-2z}}{(1+e^{-z})^4} dz + \frac{\tau' e^{-2\tau'}}{(1+e^{-\tau'})^3} \right), \tag{B.13}$$

$$E\left[\frac{-\partial^2 L}{\partial \alpha \partial \gamma_0}\right] = E\left[\frac{-\partial^2 L}{\partial \gamma_0 \partial \alpha}\right], \tag{B.14}$$

$$E\left[\frac{-\partial^2 L}{\partial \gamma_1^2}\right] = \frac{1}{(1+\gamma_1)} \cdot \left(\alpha E\left[\frac{-\partial^2 L}{\partial \gamma_1 \partial \alpha}\right] + T_1 E\left[\frac{-\partial^2 L}{\partial \gamma_0 \partial \gamma_1}\right] \right) \tag{B.15}$$

$$E\left[\frac{-\partial^2 L}{\partial \gamma_1 \partial \alpha}\right] = \frac{1}{(1+\gamma_1)} \cdot \left(\alpha E\left[\frac{-\partial^2 L}{\partial \alpha^2}\right] + T_1 E\left[\frac{-\partial^2 L}{\partial \gamma_0 \partial \alpha}\right] \right), \tag{B.16}$$

$$E\left[\frac{-\partial^2 L}{\partial \alpha \partial \gamma_1}\right] = E\left[\frac{-\partial^2 L}{\partial \gamma_1 \partial \alpha}\right], \tag{B.17}$$

$$E\left[\frac{-\partial^2 L}{\partial\,\alpha^2}\right] = \left(\frac{1}{\alpha}\right)^2 \cdot \left(\int_{-\infty}^{\tau'} \left(\frac{e^{-z}}{(1+e^{-z})^2} + \frac{2z^2 e^{-2z}}{(1+e^{-z})^4}\right) dz \right.$$

$$\left. + \frac{\tau'^2 e^{-2\tau'}}{(1+e^{-\tau'})^3}\right), \tag{B.18}$$

where $T_1 = 1 - \gamma_0 - \log(1+\gamma_1) + \log(s_0/\zeta)$.

These elements have been obtained using the following results:

(i) $E[I(z)] = \dfrac{e^{-\tau'}}{1+e^{-\tau'}}$. (B.19)

(ii) $E[\bar{I}(z)] = \dfrac{1}{1+e^{-\tau'}}$. (B.20)

(iii) $E\left[\dfrac{\partial L}{\partial \gamma_j}\right] = 0, (j = 0, 1)$ and $E\left[\dfrac{\partial L}{\partial \alpha}\right] = 0.$ (B.21)

B.2 CALCULATIONS OF THE ASYMPTOTIC VARIANCE OF THE TEST STATISTIC GIVEN IN SECTION 8.6.3 FOR TIME-CENSORED CASE

Let $H' = nF^{-1}$, and h_{ij} be the (i,j)th element of matrix H' of order 3×3.

Using delta method, V in (8.25) has been obtained as follows:

$$\begin{aligned}&\text{AsVar}(\hat{\mu}_0)\\ &\quad = \text{AsVar}(\ln(\eta)) = \text{AsVar}(g_1(\hat{\gamma}_0, \hat{\gamma}_1))\\ &\quad = \left(\frac{\partial g_1}{\partial\,\hat{\gamma}_0}\right)^2 h'_{11} + \left(\frac{\partial g_1}{\partial\,\hat{\gamma}_1}\right)^2 h'_{22} + 2\left(\frac{\partial g_1}{\partial\,\hat{\gamma}_0}\right)\left(\frac{\partial g_1}{\partial\,\hat{\gamma}_1}\right) h'_{12} = \frac{V_1}{n},\end{aligned} \tag{B.22}$$

$$\begin{aligned}&\text{AsVar}(\hat{\sigma}_0)\\ &\quad = \text{AsVar}((\alpha')^{-1}) = \text{AsVar}(g_2(\hat{\gamma}_0, \hat{\gamma}_1))\\ &\quad = \left(\frac{\partial g_2}{\partial\,\hat{\alpha}}\right)^2 h_{33} + \left(\frac{\partial g_2}{\partial\,\hat{\gamma}_1}\right)^2 h_{22} + 2\left(\frac{\partial g_2}{\partial\,\hat{\alpha}}\right)\left(\frac{\partial g_2}{\partial\,\hat{\gamma}_1}\right) h_{32} = \frac{V_3}{n},\end{aligned} \tag{B.23}$$

$$
\begin{aligned}
&\text{Ascov}(\hat{\mu}_0, \hat{\sigma}_0) \\
&\quad = \text{Ascov}(g_1(\hat{\gamma}_0, \hat{\gamma}_1),\ g_2(\hat{\gamma}_0, \hat{\gamma}_1)) \\
&\quad = \left(\frac{\partial g_1}{\partial \hat{\gamma}_0}\right)\left(\frac{\partial g_2}{\partial \hat{\gamma}_0}\right) h'_{11} + \left(\frac{\partial g_1}{\partial \hat{\gamma}_1}\right)\left(\frac{\partial g_2}{\partial \hat{\gamma}_1}\right) h'_{22} \\
&\qquad + \left(\frac{\partial g_1}{\partial \hat{\alpha}}\right)\left(\frac{\partial g_2}{\partial \hat{\alpha}}\right) h'_{33} + \left(\left(\frac{\partial g_1}{\partial \hat{\gamma}_0}\right)\left(\frac{\partial g_2}{\partial \hat{\gamma}_1}\right) + \left(\frac{\partial g_1}{\partial \hat{\gamma}_1}\right)\left(\frac{\partial g_2}{\partial \hat{\gamma}_0}\right)\right) h'_{12} \\
&\qquad + \left(\left(\frac{\partial g_1}{\partial \hat{\gamma}_0}\right)\left(\frac{\partial g_2}{\partial \hat{\alpha}}\right) + \left(\frac{\partial g_1}{\partial \hat{\alpha}}\right)\left(\frac{\partial g_2}{\partial \hat{\gamma}_0}\right)\right) h'_{13} \\
&\qquad + \left(\left(\frac{\partial g_1}{\partial \hat{\gamma}_1}\right)\left(\frac{\partial g_2}{\partial \hat{\alpha}}\right) + \left(\frac{\partial g_1}{\partial \hat{\alpha}}\right)\left(\frac{\partial g_2}{\partial \hat{\gamma}_1}\right)\right) h'_{23} = \frac{V_2}{n},
\end{aligned}
\tag{B.24}
$$

$$
\text{AsVar}(\text{T}') = \frac{(V_1 - 2kV_2 + k^2 V_3)}{n} = \frac{V}{n}. \tag{B.25}
$$

Here, $g_1 \equiv g_1(\hat{\gamma}_0, \hat{\gamma}_1)$ and $g_2 \equiv g_2(\hat{\gamma}_0, \hat{\gamma}_1)$.

B.3 CALCULATIONS OF THE ELEMENTS OF THE FISHER INFORMATION MATRIX GIVEN IN SECTION 9.4

The first-order partial derivatives of (9.13) with respect to A, α, and c are as follows:

$$
\begin{aligned}
\frac{\partial \ln L}{\partial A} &= \frac{n_a}{A} - k \sum_{i=1}^{n} \delta_{2i} \ln(1 + e^{z_i}) + k n_a \ln(1 + e^{\tau'}) \\
&\quad + k \sum_{i=1}^{n} \bar{\delta}_{1i} \bar{\delta}_{2i} \ln(1 + e^{\tau'}) - k \sum_{i=1}^{n} \bar{\delta}_{1i} \bar{\delta}_{2i} \ln(1 + e^{\eta'}),
\end{aligned}
\tag{B.26}
$$

$$
\begin{aligned}
\frac{\partial \ln L}{\partial \alpha} &= \left(\frac{c}{\alpha}\right) \cdot \left[-(n_a + n_u) + (k+1) \sum_{i=1}^{n} \delta_{1i} \left(\frac{e^{z_i}}{(1 + e^{z_i})}\right) \right. \\
&\quad + (kA + 1) + \sum_{i=1}^{n} \delta_{2i} \left(\frac{e^{z_i}}{(1 + e^{z_i})}\right) - k(A-1) n_a \left(\frac{e^{\tau'}}{(1 + e^{\tau'})}\right)
\end{aligned}
$$

$$- k(A-1)\sum_{i=1}^{n}\bar{\delta}_{1i}\bar{\delta}_{2i}\left(\frac{e^{\tau'}}{(1+e^{\tau'})}\right)$$

$$+ kA\sum_{i=1}^{n}\bar{\delta}_{1i}\bar{\delta}_{2i}\left(\frac{e^{\eta'}}{(1+e^{\eta'})}\right)\Bigg], \tag{B.27}$$

$$\frac{\partial \ln L}{\partial c} = \left(\frac{1}{c}\right)\cdot\Bigg[(n_a+n_u) + \sum_{i=1}^{n} z_i\delta_{1i} + \sum_{i=1}^{n} z_i\delta_{2i}$$

$$- (k+1)\sum_{i=1}^{n}\delta_{1i}\left(\frac{z_i e^{z_i}}{(1+e^{z_i})}\right) - (kA+1)\sum_{i=1}^{n}\delta_{2i}\left(\frac{z_i e^{z_i}}{(1+e^{z_i})}\right)$$

$$+ k(A-1)n_a\left(\frac{\tau' e^{\tau'}}{(1+e^{\tau'})}\right) + k(A-1)\sum_{i=1}^{n}\bar{\delta}_{1i}\bar{\delta}_{2i}\left(\frac{\tau' e^{\tau'}}{(1+e^{\tau'})}\right)$$

$$-kA\sum_{i=1}^{n}\bar{\delta}_{1i}\bar{\delta}_{2i}\left(\frac{\eta' e^{\eta'}}{(1+e^{\eta'})}\right)\Bigg]. \tag{B.28}$$

The second-order partial derivatives are

$$\frac{\partial^2 \ln L}{\partial A^2} = -\frac{n_a}{A^2}, \tag{B.29}$$

$$\frac{\partial^2 \ln L}{\partial A \partial \alpha} = -\left(\frac{kc}{\alpha}\right)\cdot\Bigg[-\sum_{i=1}^{n}\delta_{2i}\left(\frac{e^{z_i}}{(1+e^{z_i})}\right) + (n-n_u)\left(\frac{e^{\tau'}}{(1+e^{\tau'})}\right)$$

$$- n_c\left(\frac{e^{\eta'}}{(1+e^{\eta'})}\right)\Bigg], \tag{B.30}$$

$$\frac{\partial^2 \ln L}{\partial A \partial c} = -\left(\frac{k}{c}\right)\cdot\Bigg[\sum_{i=1}^{n}\delta_{2i}\left(\frac{z_i e^{z_i}}{(1+e^{z_i})}\right) - (n-n_u)\left(\frac{\tau' e^{\tau'}}{(1+e^{\tau'})}\right)$$

$$+ n_c\left(\frac{\eta' e^{\eta'}}{(1+e^{\eta'})}\right)\Bigg], \tag{B.31}$$

$$\frac{\partial^2 \ln L}{\partial \alpha \partial c} = \left(\frac{1}{c}\right) \cdot \frac{\partial \ln L}{\partial \alpha} + \left(\frac{1}{\alpha}\right) \cdot \left[(k+1) \sum_{i=1}^{n} \delta_{1i} \left(\frac{z_i e^{z_i}}{(1+e^{z_i})^2}\right) + kAn_c \left(\frac{\eta' e^{\eta'}}{(1+e^{\eta'})^2}\right) - (n-n_u)k(A-1) \left(\frac{\tau' e^{\tau'}}{(1+e^{\tau'})^2}\right) + (kA+1) \sum_{i=1}^{n} \delta_{2i} \left(\frac{z_i e^{z_i}}{(1+e^{z_i})^2}\right)\right], \tag{B.32}$$

$$\frac{\partial^2 \ln L}{\partial \alpha^2} = -\left(\frac{1}{\alpha}\right) \cdot \frac{\partial \ln L}{\partial \alpha} - \left(\frac{c}{\alpha}\right)^2 \cdot \left[(k+1) \sum_{i=1}^{n} \delta_{1i} \left(\frac{e^{z_i}}{(1+e^{z_i})^2}\right) + kAn_c \left(\frac{e^{\eta'}}{(1+e^{\eta'})^2}\right) - (n-n_u)k(A-1) \left(\frac{e^{\tau'}}{(1+e^{\tau'})^2}\right) + (kA+1) \sum_{i=1}^{n} \delta_{2i} \left(\frac{e^{z_i}}{(1+e^{z_i})^2}\right)\right], \tag{B.33}$$

$$\frac{\partial^2 \ln L}{\partial c^2} = -\left(\frac{1}{c}\right)^2 \cdot \left[(n_u + n_a) + (k+1) \sum_{i=1}^{n} \delta_{1i} \left(\frac{z_i^2 e^{z_i}}{(1+e^{z_i})^2}\right) + kAn_c \left(\frac{\eta'^2 e^{\eta'}}{(1+e^{\eta'})^2}\right) - (n-n_u)k(A-1) \left(\frac{\tau'^2 e^{\tau'}}{(1+e^{\tau'})^2}\right) + (kA+1) \sum_{i=1}^{n} \delta_{2i} \left(\frac{z_i^2 e^{z_i}}{(1+e^{z_i})^2}\right)\right]. \tag{B.34}$$

The elements of the Fisher information matrix $\mathrm{F}(A, \alpha, c)$ for an observation are the negative expectations of the second-order partial derivatives (B.29)–(B.34). These are as follows:

$$E\left[-\frac{\partial^2 \ln L}{\partial A^2}\right] = \frac{np_a}{A^2}, \tag{B.35}$$

$$E\left[-\frac{\partial^2 \ln L}{\partial A \partial \alpha}\right] = \left(\frac{kc}{\alpha}\right) \cdot \left[-\int_{\tau'}^{\eta'} \frac{nkAe^{2z}}{(1+e^{\tau'})^{-k(A-1)}(1+e^z)^{(kA+2)}}dz + (n-np_u)\left(\frac{e^{\tau'}}{(1+e^{\tau'})}\right) - np_c\left(\frac{e^{\eta'}}{(1+e^{\eta'})}\right)\right], \tag{B.36}$$

$$E\left[-\frac{\partial^2 \ln L}{\partial \alpha \partial A}\right] = E\left[-\frac{\partial^2 \ln L}{\partial A \partial \alpha}\right], \tag{B.37}$$

$$E\left[-\frac{\partial^2 \ln L}{\partial A \partial c}\right] = \left(\frac{k}{c}\right) \cdot \left[\int_{\tau'}^{\eta'} \frac{nkAze^{2z}}{(1+e^{\tau'})^{-k(A-1)}(1+e^z)^{(kA+2)}}dz - (n-np_u)\left(\frac{\tau' e^{\tau'}}{(1+e^{\tau'})}\right) + np_c\left(\frac{\eta' e^{\eta'}}{(1+e^{\eta'})}\right)\right] \tag{B.38}$$

$$E\left[-\frac{\partial^2 \ln L}{\partial c \partial A}\right] = E\left[-\frac{\partial^2 \ln L}{\partial A \partial c}\right], \tag{B.39}$$

$$E\left[-\frac{\partial^2 \ln L}{\partial \alpha^2}\right] = \left(\frac{c}{\alpha}\right)^2 \cdot \left[n(k+1)\int_{-\infty}^{\tau'} \frac{ke^{2z}}{(1+e^z)^{(k+3)}}dz - (n-np_u)k(A-1)\left(\frac{e^{\tau'}}{(1+e^{\tau'})}\right)^2 + n(kA+1)\int_{\tau'}^{\eta'} \frac{kAe^{2z}}{(1+e^{\tau'})^{-k(A-1)}(1+e^z)^{(kA+3)}}dz + kAnp_c\left(\frac{e^{\eta'}}{(1+e^{\eta'})}\right)^2\right], \tag{B.40}$$

$$E\left[-\frac{\partial^2 \ln L}{\partial \alpha \partial c}\right] = \left(-\frac{1}{\alpha}\right) \cdot \left[n(k+1)\int_{-\infty}^{\tau'} \frac{kze^{2z}}{(1+e^z)^{(k+3)}}dz - (n-np_u)k(A-1)\left(\frac{\tau' e^{\tau'}}{(1+e^{\tau'})^2}\right)\right.$$

$$+ kAnp_c\left(\frac{\eta' e^{\eta'}}{(1+e^{\eta'})^2}\right) + n(kA+1)$$

$$\left.\times \int_{\tau'}^{\eta'} \frac{kAze^{2z}}{(1+e^{\tau'})^{-k(A-1)}(1+e^z)^{(kA+3)}}dz\right], \tag{B.41}$$

$$E\left[-\frac{\partial^2 \ln L}{\partial c \partial \alpha}\right] = E\left[-\frac{\partial^2 \ln L}{\partial \alpha \partial c}\right], \tag{B.42}$$

$$E\left[-\frac{\partial^2 \ln L}{\partial c^2}\right] = \left(\frac{1}{c}\right)^2 \cdot \left[(np_u + np_a) + n(k+1)\right.$$

$$\times \int_{-\infty}^{\tau'} \frac{kz^2e^{2z}}{(1+e^z)^{(k+3)}}dz - (n - np_u)k(A-1)$$

$$\times \left(\frac{\tau'^2 e^{\tau'}}{(1+e^{\tau'})^2}\right) + kAnp_c\left(\frac{\eta'^2 e^{\eta'}}{(1+e^{\eta'})^2}\right) + n(kA+1)$$

$$\left.\times \int_{\tau'}^{\eta'} \frac{kAz^2e^{2z}}{(1+e^{\tau'})^{-k(A-1)}(1+e^z)^{(kA+3)}}dz\right] \tag{B.43}$$

These expectations are calculated with the aid of,

$$E\left[\frac{\partial L}{\partial A}\right] = 0, \quad E\left[\frac{\partial L}{\partial \alpha}\right] = 0 \quad \text{and} \quad E\left[\frac{\partial L}{\partial c}\right] = 0.$$

B.4 CALCULATION OF THE ASYMPTOTIC VARIANCE OF THE TEST STATISTIC GIVEN IN SECTION 9.5

Let $H = nF^{-1}$, and h_{ij} be the (i, j)th element of matrix H of order 3×3. Using delta method, V in (9.17) has been obtained as follows:

$$\begin{aligned}\text{AsVar}(\hat{\mu}) &= \text{AsVar}(g_1(\hat{A}, \hat{\alpha}, \hat{c})) \\ &= \text{AsVar}(\ln(\hat{\alpha})) = \left(\frac{\partial g_1}{\partial \hat{\alpha}}\right)^2 h_{22} = \frac{V_1}{n},\end{aligned} \tag{B.44}$$

$$\begin{aligned}\text{AsVar}(\hat{\sigma}) &= \text{AsVar}(g_2(\hat{A}, \hat{\alpha}, \hat{c}))\\ &= \text{AsVar}((\hat{c})^{-1}) = \left(\frac{\partial g_2}{\partial \hat{c}}\right)^2 h_{33} = \frac{V_3}{n},\end{aligned} \tag{B.45}$$

$$\begin{aligned}\text{Ascov}(\hat{\mu}, \hat{\sigma}) &= \text{Ascov}(g_1(\hat{A}, \hat{\alpha}, \hat{c}),\ \ g_2(\hat{A}, \hat{\alpha}, \hat{c}))\\ &= \left(\left(\frac{\partial g_1}{\partial \hat{\alpha}} \times \frac{\partial g_2}{\partial \hat{c}}\right) + \left(\frac{\partial g_1}{\partial \hat{c}} \times \frac{\partial g_2}{\partial \hat{\alpha}}\right)\right) h_{23} = \frac{V_2}{n},\end{aligned} \tag{B.46}$$

$$\therefore \quad \text{AsVar}(T_1) = \frac{(V_1 - 2k_1 V_2 + k_1^2 V_1)}{n} = \frac{V}{n}. \tag{B.47}$$

Here $g_1 \equiv g_1(\hat{A}, \hat{\alpha}, \hat{c})$ and $g_2 \equiv g_2(\hat{A}, \hat{\alpha}, \hat{c})$.

B.5 CALCULATIONS OF THE SECOND PARTIAL DERIVATIVES AND ELEMENTS OF THE FISHER INFORMATION MATRIX GIVEN IN SECTION 10.5

The second partial derivatives are given below:

$$\begin{aligned}\frac{\partial^2 \log L}{\partial \theta_j^2} &= \frac{(n_{11} + n_{21})}{\theta_j^2} - \frac{2}{\theta_j^3} \cdot \left((n_{2\bullet} + n_c)\tau^\delta + \sum_{i=1}^{n_{11}} w_i^\delta + \sum_{i=1}^{n_{12}} w_i^\delta\right)\\ &\quad - \frac{2}{\theta_j^3} \cdot \left(A\left(\sum_{i=1}^{n_{21}} (w_i^\delta - \tau^\delta) + \sum_{i=1}^{n_{22}} (w_i^\delta - \tau^\delta)\right.\right.\\ &\qquad \left.\left. + \, n_c(\eta^\delta - \tau^\delta)\right)\right); \quad j = 1, 2,\end{aligned} \tag{B.48}$$

$$\frac{\partial^2 \log L}{\partial \theta_1 \partial \theta_2} = 0 = \frac{\partial^2 \log L}{\partial \theta_2 \partial \theta_1}, \tag{B.49}$$

$$\begin{aligned}\frac{\partial^2 \log L}{\partial \theta_j \partial A} &= \frac{1}{\theta_j^2} \cdot \left(\sum_{i=1}^{n_{21}} (w_i^\delta - \tau^\delta) + \sum_{i=1}^{n_{22}} (w_i^\delta - \tau^\delta)\right.\\ &\qquad \left. + \, n_c(\eta^\delta - \tau^\delta)\right); \quad j = 1, 2,\end{aligned} \tag{B.50}$$

$$\frac{\partial^2 \log L}{\partial A \partial \theta_j} = \frac{\partial^2 \log L}{\partial \theta_j \partial A} \tag{B.51}$$

$$\frac{\partial^2 \log L}{\partial A^2} = -\frac{n_{2\bullet}}{A^2}. \tag{B.52}$$

The elements of Fisher Information Matrix are obtained using the following results with $n_1. \equiv n_1$ and $n_2. \equiv n_2$:

(I) $n = n_1 + n_2 + n_c$, n is fixed, where (B.53a)

(i) $n_1 \sim \text{Bin}(n, p_{n_1})$, (B.53b)

(ii) $n_2 \sim \text{Bin}(n, p_{n_2})$, (B.53c)

(iii) $n_c \sim \text{Bin}(n, p_{n3})$, (B.53d)

(iv) $p_{n_1} = P[0 < W < \tau]$

$$= \int_0^\tau \frac{\delta w^{\delta-1}}{\theta_g} e^{-w^\delta/\theta_g} dt = 1 - e^{-\tau^\delta/\theta_g}, \tag{B.53e}$$

(v) $p_{n_2} = P[\tau < W < T]$

$$= \int_\tau^T \frac{\delta A w^{\delta-1}}{\theta_g} e^{-A(w^\delta - \tau^\delta)/\theta_g - \tau^\delta/\theta_{1g}} dw$$

$$= e^{-\tau^\delta/\theta_g}(1 - e^{-A(T^\delta - \tau^\delta)/\theta_g}), \tag{B.53f}$$

(vi) $p_{n_3} = 1 - p_{n_1} - p_{n_2}$. (B.53g)

(II) $n_{11} \sim \text{Bin}(n_1, \pi_{11})$, where

$$E[n_{11}] = E[E[n_{11}|n_1]]$$

$$= E\left[n_1 \frac{\theta_1^{-1}}{\theta_1^{-1} + \theta_2^{-1}}\right] = n\frac{\theta_1^{-1}}{\theta_g^{-1}} p_{n_1} = n\frac{\theta_1^{-1}}{\theta_g^{-1}}(1 - e^{-\tau^\delta/\theta_g}).$$

(using (i) and (iv)). (B.54)

(III) $n_{12} \sim \text{Bin}(n_2, \pi_{12})$, where

$$E[n_{12}] = E[n_1 - n_{11}]$$

$$= n\left(1 - \frac{\theta_1^{-1}}{\theta_g^{-1}}\right) p_{n_1} = n\left(1 - \frac{\theta_1^{-1}}{\theta_g^{-1}}\right)(1 - e^{-\tau^\delta/\theta_g}).$$

(using (II), (i) and (iv)). (B.55)

(IV) $n_{21} \sim \text{Bin}(n_1, \pi_{21})$, where

$$E[n_{21}] = E[E[n_{21}|n_2]] = E\left[n_2 \frac{\theta_1^{-1}}{\theta_1^{-1} + \theta_2^{-1}}\right] = \frac{\theta_1^{-1}}{\theta_1^{-1} + \theta_2^{-1}} E[n_2]$$

$$= n \frac{\theta_1^{-1}}{\theta_1^{-1} + \theta_2^{-1}} p_{n_2} = n \frac{\theta_1^{-1}}{\theta_g^{-1}} e^{-\tau^\delta/\theta_g}(1 - e^{-A(T^\delta - \tau^\delta)/\theta_g}).$$

(using (ii) and (v)). (B.56)

(V) $n_{22} \sim \text{Bin}(n_2, \pi_{22})$

$$E[n_{22}] = E[n_2 - n_{21}] = E[n_2 - (n_{21}|n_2)] = E[n_2 - E[n_{21}|n_2]]$$

$$= E\left[n_2 - n_2 \frac{\theta_1^{-1}}{\theta_g^{-1}}\right] = n\left(1 - \frac{\theta_1^{-1}}{\theta_g^{-1}}\right) p_{n_2}$$

$$= n\left(1 - \frac{\theta_1^{-1}}{\theta_g^{-1}}\right) e^{-\tau^\delta/\theta_g}(1 - e^{-A(T^\delta - \tau^\delta)/\theta_g}).$$

(using (IV), (ii), and (v)). (B.57)

(VI)

$$E\left[\sum_{i=1}^{n_{11}} w_i^\delta\right] = E\left[E\left[\sum_{i=1}^{n_{11}} w_i^\delta | n_{11}\right]\right]$$

$$= E\left[\sum_{i=1}^{n_{11}} \int_0^\tau \frac{w_i^\delta \frac{1}{\theta_1} \delta w_i^{\delta-1} e^{-w_i^\delta/\theta_g}}{\int_0^\tau \frac{1}{\theta_1} \delta w_i^{\delta-1} e^{-w_i^\delta/\theta_g} dw_i} dw_i\right]$$

$$= E\left[\sum_{i=1}^{n_{11}} \int_0^\tau \frac{w_i^\delta \frac{1}{\theta_1} \delta w_i^{\delta-1} e^{-w_i^\delta/\theta_g}}{\frac{\theta_g}{\theta_1}(1 - e^{-\tau^\delta/\theta_g})} dw_i\right]$$

$$= n \frac{\theta_1^{-1}}{\theta_g^{-1}} (\theta_g - e^{-\tau^\delta/\theta_g}(\tau^\delta + \theta_g)). \quad \text{(B.58)}$$

(VII)

$$E\left[\sum_{i=1}^{n_{12}} w_i^\delta\right] = E\left[E\left[\sum_{i=1}^{n_{12}} w_i^\delta | n_{12}\right]\right]$$

$$= n\left(1 - \frac{\theta_1^{-1}}{\theta_g^{-1}}\right) (\theta_g - e^{-\tau^\delta/\theta_g}(\tau^\delta + \theta_g)). \quad \text{(B.59)}$$

(VIII)

$$E\left[\sum_{i=1}^{n_{21}} w_i^\delta\right]$$

$$= E\left[E\left[\sum_{i=1}^{n_{21}} w_i^\delta | n_{21}\right]\right]$$

$$= E\left[\sum_{i=1}^{n_{21}} \int_\tau^T \frac{w_i^\delta f(w_i,1)dw_i}{\frac{\delta A}{\theta_1}\int_\tau^T w_i^{\delta-1} w_i^\delta e^{-A(w_i^\delta-\tau^\delta)/\theta_g-\tau^\delta/\theta_g} dw_i}\right]$$

$$= E\left[\sum_{i=1}^{n_{21}} \int_\tau^T \frac{w_i^\delta f(w_i,1)dw_i}{\left(\frac{\theta_g}{\theta_1}e^{-\tau^\delta/\theta_i}(1-e^{-A(T^\delta-\tau^\delta)/\theta_g})\right)}\right]$$

$$= \frac{n\theta_1^{-1}}{\theta_1^{-1}+\theta_2^{-1}}((\tau^\delta + A^{-1}\theta_g)e^{-\tau^\delta/\theta_g}$$

$$- (T^\delta + A^{-1}\theta_g)e^{-A(T^\delta-\tau^\delta)/\theta_g-\tau^\delta/\theta_g}). \qquad \text{(B.60)}$$

(IX)

$$E\left[\sum_{i=1}^{n_{22}} w_i^\delta\right] = E\left[E\left[\sum_{i=1}^{n_{22}} w_i^\delta | n_{22}\right]\right]$$

$$= n\left(1-\frac{\theta_1^{-1}}{\theta_1^{-1}+\theta_2^{-1}}\right)((\tau^\delta + A^{-1}\theta_g)e^{-\tau^\delta/\theta_g}$$

$$- (T^\delta + A^{-1}\theta_g)e^{-A(T^\delta-\tau^\delta)/\theta_g-\tau^\delta/\theta_g}). \qquad \text{(B.61)}$$

The elements of the Fisher information matrix for an observation are the negative expectations of these second-order partial derivatives of $\log L$ with respect to θ_1, θ_2 and A. Using (B.48)–(B.61) these are obtained as:

$$E\left[-\frac{\partial^2 \log L}{\partial \theta_j^2}\right] = -\frac{(E[n_{11}]+E[n_{21}])}{\theta_j^2} + \frac{2}{\theta_j^3}\cdot((E[n_{2\bullet}]+E[n_c])\tau^\delta$$

$$+ A_{11}(\tau) + A_{12}(\tau)) + \frac{2}{\theta_j^3}\cdot(A(A_{21}(\tau)+A_{22}(\tau)$$

$$+ E[n_c](\eta^\delta - \tau^\delta))); \quad j = 1,2, \qquad \text{(B.62)}$$

$$E\left[-\frac{\partial^2 \log L}{\partial \theta_1 \partial \theta_2}\right] = 0 = E\left[-\frac{\partial^2 \log L}{\partial \theta_2 \partial \theta_1}\right], \qquad \text{(B.63)}$$

$$E\left[-\frac{\partial^2 \log L}{\partial\theta_j \partial A}\right] = -\left(\frac{1}{\theta_j^2}\right)(A_{21}(\tau) + A_{22}(\tau)$$
$$+E[n_c](\eta^\delta - \tau^\delta)); \quad j = 1, 2, \tag{B.64}$$

$$E\left[-\frac{\partial^2 \log L}{\partial A \partial\theta_j}\right] = E\left[-\frac{\partial^2 \log L}{\partial\theta_j \partial A}\right] \tag{B.65}$$

$$E\left[-\frac{\partial^2 \log L}{\partial A^2}\right] = \frac{E[n_{2\bullet}]}{A^2}, \tag{B.66}$$

where

$$E[n_{2\bullet}] = E[n_{21}] + E[n_{22}], \quad A_{11}(\tau) = E\left[\sum_{i=1}^{n_{11}} w_i^\delta\right],$$

$$A_{12}(\tau) = E\left[\sum_{i=1}^{n_{12}} w_i^\delta\right], \quad A_{21}(\tau) = E\left[\sum_{i=1}^{n_{21}} (w_i^\delta - \tau^\delta)\right] \quad \text{and}$$

$$A_{22}(\tau) = E\left[\sum_{i=1}^{n_{22}} (w_i^\delta - \tau^\delta)\right].$$

B.6 CALCULATION OF THE ASYMPTOTIC VARIANCE OF THE ESTIMATED VALUE OF MEAN LIFETIME, $\hat{\mu}$, GIVEN IN SECTIONS 10.3.1 AND 10.3.2

Using delta method, asymptotic variance of mean lifetime of $\hat{\mu}$ ($\text{Var}(\nu)$) has been obtained as follows:

$$\begin{aligned}\text{Var}(\nu) &= \text{AsVar}(\hat{\mu})\\ &= \text{AsVar}\left(\frac{\Gamma(1/\delta)\hat{\theta}_\bullet^{1/\delta}}{\delta}\right) = \text{AsVar}[\hat{h}(\hat{\theta}_1, \hat{\theta}_2, \hat{A})] = \underset{\sim}{\hat{H}}'F^{-1}\underset{\sim}{\hat{H}}, S\end{aligned}$$

where $\underset{\sim}{\hat{H}}'$ is a transpose of

$$\underset{\sim}{\hat{H}}, \underset{\sim}{\hat{H}}' = \left[\frac{\partial\hat{h}}{\partial\hat{\theta}_1} \quad \frac{\partial\hat{h}}{\partial\hat{\theta}_2} \quad \frac{\partial\hat{h}}{\partial\hat{A}}\right]' \quad \text{and} \quad \hat{h} \equiv \hat{h}(\hat{\theta}_1, \hat{\theta}_2, \hat{A}).$$

APPENDIX C

C.1 BILEVEL PROGRAMMING

In cost minimization problems, the optimal plans may be obtained using **Bilevel programming** approach.

Bilevel programming problems are mathematical optimization problems where the set of all variables is partitioned between two vectors x and y, and x is to be chosen as an optimal solution of a second mathematical programming problem parameterized in y. Thus, the bilevel programming problem is hierarchical in the sense that its constraints are defined in part by a second optimization problem. The bilevel programs facilitate the formulation of several practical problems that involve an hierarchical decision-making process like engineering design, analysis of competitive economies, transport system planning, signal optimization, network design, strategic offensive and defensive force problems, government regulation, management, etc.

Let, this second problem be introduced first as follows:

$$\underset{x}{\text{Min}}\{f(x,y) : g(x,y) \leq 0, h(x,y) = 0\}, \tag{C.1}$$

where

$$\begin{aligned} f &: \mathbb{R}^n \times \mathbb{R}^m \to \mathbb{R}, \\ g &: \mathbb{R}^n \times \mathbb{R}^m \to \mathbb{R}^p, \\ h &: \mathbb{R}^n \times \mathbb{R}^m \to \mathbb{R}^q, \end{aligned}$$

$$g(x,y) = (g_1(x,y),\ldots,g_p(x,y))^T,$$
$$h(x,y) = (h_1(x,y),\ldots,h_q(x,y))^T.$$

This problem will also be referred to as the **lower level** or the **follower's problem**. Let $\Psi(y)$ denote the solution set of problem (C.1) for fixed $y \in \mathbb{R}^m$. Then is a so-called **point-to-set mapping** from $\mathbb{R}^m$ into the power set of $\mathbb{R}^n$, denoted by $\Psi : \mathbb{R}^m \to 2^{\mathbb{R}^n}$.

Denote some element of $\Psi(y)$ by $x(y)$ and assume for the moment that this choice is unique for all possible y. Then, the aim of the bilevel programming problem is to select that parameter vector y describing the "environmental data" for the lower level problem which is the optimal one in a certain sense. To be more precise, this selection of y is conducted so that certain (nonlinear) equality and/or inequality constraints

$$G(x(y),y) \le 0, \quad H(x(y),y) = 0 \tag{C.2}$$

are satisfied and an objective function $Z(x(y),y)$ is minimized, where

$$Z : \mathbb{R}^n \times \mathbb{R}^m \to \mathbb{R}, \quad G : \mathbb{R}^n \times \mathbb{R}^m \to \mathbb{R}^k, \quad H : \mathbb{R}^n \times \mathbb{R}^m \to \mathbb{R}^l.$$

It is assumed that all functions Z, G, H, are sufficiently smooth, i.e., that all the gradients and Hessian matrices of these functions exist and are smooth. Clearly these assumptions can be weakened at many places. The problem of determining a best solution y^* can thus be described as that of finding a vector y^* of parameters for the parametric optimization problem (C.1) which together with the response $x(y) \in \Psi(y)$ proves to satisfy the constraints (C.2) and to give the best possible function value for $F(x(y),y)$. That is

$$\operatorname*{Min}_{y}\{Z(x(y),y) : G(x(y),y) \le 0, H(x(y),y) = 0, x(y) \in \Psi(y)\}. \tag{C.3}$$

This problem is the bilevel programming problem or the **leader's** problem. The function Z is called **the upper level objective** and the functions G and H are called **the upper level constraint functions**. Strongly speaking, this definition of the bilevel programming problem is valid only in the case when the lower level solution is uniquely determined for each possible y. See for reference Dempe (2002).

Formulations (C.1)–(C.3) can also be written as:

$$\begin{aligned}
&\underset{y}{\text{Min}}\, Z(x,y)\\
&\text{s.t}\\
&\quad G(x,y) \leq 0\\
&\quad H(x,y) = 0\\
&\underset{x}{\text{Min}}\, f(x,y)\\
&\text{s.t}\\
&\quad g(x,y) \leq 0\\
&\quad h(x,y) = 0.
\end{aligned}$$

In this work, the second problem is referred to as the lower level problem; optimum values of the experimental variables associated with the stress loading scheme being used are found using optimality criterion based on variance optimality, D-optimality, or A-optimality. In the upper level, optimal values found in the lower level are used to find out the experimental variables related with life test sampling plans, for example, optimum number of test units subject to producer's and consumer's risks, and optimum expected cost per lot.

APPENDIX D

D.1 TAMPERED FAILURE RATE (TFR) MODEL UNDER CONSTANT-STRESS SETUP

Consider a constant-stress PALT wherein a test unit is tested either under use condition or under accelerated conditions till the censoring time. Let there be $(m+1)$ test chambers in all including the one in which items are tested under normal operating condition. The TFR model assumes that the effect of changing acceleration factor α_{i-1} in $(i-1)^{\text{th}}$ test chamber to α_i in i^{th} test chamber is to multiply the failure rate function $h_{i-1}(y)$ by α_i $i = 1, 2, \ldots, m$. Thus, the effect of changing stress through acceleration factor in different test chambers has multiplicative effect on initial failure rate function. Hence, the tampered failure rate model under constant-stress loading is given by

$$h^*(y) = \begin{cases} h(y) & \text{under used condition} \\ h_1(y) = \alpha_1 h(y) & \text{at } 1^{\text{st}} \text{ stress level} \\ h_2(y) = \alpha_2 h_1(y) = \alpha_1 \alpha_2 h(y) & \text{at } 2^{\text{nd}} \text{ stress level} \\ h_3(y) = \alpha_3 h_2(y) = \prod_{i=1}^{3} \alpha_i \cdot h(y) & \text{at } 3^{\text{rd}} \text{ stress level} \\ \vdots & \vdots \\ h_j(y) = \alpha_j h_{j-1}(y) = \prod_{i=1}^{j} \alpha_i \cdot h(y) & \text{at } j^{\text{th}} \text{ stress level} \\ \vdots & \vdots \\ h_m(y) = \alpha_m h_{m-1}(y) = \prod_{i=1}^{m} \alpha_i \cdot h(y) & \text{at } m^{\text{th}} \text{ stress level,} \end{cases} \tag{D.1}$$

where $\alpha_i > 1$, $i = 1, 2, \ldots, m$.

Unlike fully accelerated life testing wherein a regression structure on the stress factor (e.g., temperature, voltage) is specified, the acceleration factor α_i, $i = 1, 2, \ldots, m$ is assumed to be a parameter of the model in PALT.

Let $\alpha_i^* = \prod_{k=1}^{i} \alpha_k$, then (D.1) can be re-written as

$$h^*(y) = \begin{cases} h(y) & \text{under normal condition} \\ h_1(y) = \alpha_1^* h(y) & \text{at } 1^{\text{st}} \text{ stress level} \\ h_2(y) = \alpha_2^* h(y) & \text{at } 2^{\text{nd}} \text{ stress level} \\ h_3(y) = \alpha_3^* h(y) & \text{at } 3^{\text{rd}} \text{ stress level} \\ \quad\vdots & \\ h_j(y) = \alpha_j^* h(y) & \text{at } j^{\text{th}} \text{ stress level} \\ \quad\vdots & \\ h_m(y) = \alpha_m^* h(y) & \text{at } m^{\text{th}} \text{ stress level,} \end{cases} \tag{D.2}$$

where $\alpha_m^* > \alpha_{m-1}^* > \cdots > \alpha_1^* > 1$. Thus, $h(y) < h_1(y) < h_2(y) < \cdots < h_m(y)$.

REFERENCES

Aal, A. (2009). fWLR supported process development based on V- and J-ramp stress tests. *Integrated Reliability Workshop Final Report, 2009. IRW '09. IEEE International*, 141–144.

Abd-Elfattah, A.M., Hassan, A.S., and Nassr, S.G. (2008). Estimation in step-stress partially accelerated life tests for the Burr type-XII distribution using type–censoring. *Statistical Methodology*, **5**(6), 502–514.

Abdel-Ghaly, A.A., Amin, Z.H., and Omar, D.A. (2008). Estimation of the Burr type-XII distribution for partially accelerated life tests using censored data. *Model Assisted Statistics and Applications*, **3**(4), 317–334.

Abdel-Ghaly, A.A., Attia, A.F., and Abdel-Ghani, M.M. (1996). The estimation problem of partially accelerated life tests for the Weibull distribution by maximum likelihood method with censored data. *Proceedings of the 31st Annual Conference of Statistics, Computer Science and Operation Research*, Cairo University, Egypt, 128–138.

Abdel-Ghaly, A.A., Attia, A.F., and Abdel-Ghani, M.M. (1997). The Bayesian estimation of weibull parameters in step partially accelerated life tests with censored data. *Proceedings of the 32th Annual Conference of Statistics, Computer Sciences and Operation Research ISSR*, Cairo, University, 45–59.

Abdel-Ghaly, A.A., Attia, A.F., and Abdel-Ghani, M.M. (2002). The maximum likelihood estimates in step partially accelerated life tests for the Weibull parameters in censored data. *Communications Statistics — Theory and Methods*, **31**(4), 551–573.

Abdel-Ghaly, A.A., El-Khodary, E.H., and Ismail, A.A. (2002). Maximum likelihood estimation and optimal design in step partially accelerated life tests for the Pareto distribution with type-I censoring. *Proceedings of the 14th Annual Conference on Statistics and Computer Modelling in Human and Social Sciences, Faculty of Economics and Political Science*, Cairo University, Egypt, 16–29.

Abdel-Ghaly, A.A., El-Khodary, E.H., and Ismail, A.A. (2003a). Estimation and optimum constant-stress partially accelerated life test plans for a Pareto distribution with type-I censoring. *Proceedings of the 38th Annual Conference of Statistics, Computer Sciences and Operation Research, ISSR*, Cairo University, Egypt, 49–65.

Abdel-Ghaly, A.A., El-Khodary, E.H., and Ismail, A.A. (2003b). Estimation and optimal design in step partially accelerated life tests for the pareto distribution using type-II censoring. *Proceedings of the 15th annual conference on Statistics and Computer Modelling in Human and Social Sciences, Faculty of Economics and Political Science*, Cairo, University, Egypt, 35–47.

Abdel-Ghani, M.M. (1998). *Investigation of some lifetime models under partially accelerated life tests. Ph.D. thesis*, Department of Statistics, Faculty of Economics and Political Science, Cairo University, Egypt.

Abdel-Ghani, M.M. (2004). The estimation problem of the log-logistic parameters in step partially accelerated life tests using type-I censored data. *The National Review of Social Sciences*, **41**(2), 1–19.

Abdel-Hamid, A.H. (2009). Constant-partially accelerated life tests for the Burr type-XII distribution with progressive type-II censoring. *Computational Statistics and Data Analysis*, **53**(7), 2511–2523.

Aggarwala, R. (2001a). Progressive interval censoring: Some mathematical results with applications to inference. *Communication in Statistics — Theory and Methods*, **30**(Iss. 8–9), 1921–1935.

Aggarwala, R. (2001b). Progressive censoring: A review. In *Handbook of Statistics* (N. Balakrishnan and C.R. Rao eds.), **20**, 373–429, Elsevier Science B.V., Netherlands.

Ahmad, N. and Islam, A. (1996). Optimal accelerated life test designs for burr type XII distributions under periodic inspection and type I censoring. *Naval Research Logistics*, **43**(8), 1049–1077.

Ahmad, N., Islam, A., Kumar, R., and Tuteja, R.K. (1994). Optimal design of accelerated life test plans under periodic inspection and type I censoring: The case of Rayleigh failure law. *South African Statistical Journal*, **28**(2), 27–35.

Ahmad, N., Islam, A., and Salam, A. (2006). Analysis of optimal accelerated life test plans for periodic inspection: The case of exponentiated Weibull failure model. *International Journal of Reliability and Quality Management*, **23**(8), 1090–1046.

Ahmad, N., Quadri, S.M.K., and Choudhary, N. (2008). Design of accelerated Life tests for periodic inspection with Burr Type III distributions: Models, assumptions and applications. In *Trends in Applied Statistics Research*, (M. Ahsamullah ed.), 27–46, Nova Science Publishers Inc.

Al-Ghamdi, A.S. and Hassan, A.S. (2009). Optimum step stress accelerated life testing for Lomax distribution. *Journal of Applied Sciences Research*, **5**(12), 2153–2164.

Al-Masri, A.Q. and Al-Haj Ebrahem, M. (2009). Optimum times for step-stress cumulative exposure model using log-logistic distribution with known scale parameter. *Austrian Journal of Statistics*, **38**(1), 59–66.

Al-Nasser, A.D. and Al-Omari, A.I. (2013). Acceptance sampling plan based on truncated life tests for exponentiated Fréchet distribution. *Journal of Statistics and Management Systems*, **16**(1), 13–24.

Alhadeed, A.A. and Yang, S.S. (2002). Optimal simple step-stress plan for Khamis-Higgins Model. *IEEE Transactions on Reliability*, **51**(2), 212–215.

Alhadeed A.A. and Yang, S.S. (2005). Optimal simple step-stress plan for cumulative exposure model using log-normal distribution. *IEEE Transactions on Reliability*, **54**(1), 64–68.

Allen, W.R. (1959). Inference from tests with continuously increasing stress. *Operations Research*, **17**(3), 303–312.

Aly, H.M. and Bleed, S.O. (2013). Estimating and planning step stress accelerated life test for generalized logistic distribution under type-I censoring. *International Journal of Applied Mathematics & Statistical Sciences*, **2**(2), 1–16.

Aly, H.M. and Ismail, A.A. (2008). Optimum simple time-step-stress plans for partially accelerated life testing with censoring. *Far East Journal of Theoretical Statistics*, **24**(2), 175–200.

American Society for Quality Control (1980). *American National Standard — Sampling procedures and tables for inspection by variables for percent nonconforming* — ANSI/ASQC:Z1.9-1980, Milwaukee, Wis.

American Society for Quality Control (1987). *American National Standard-Terms, Symbols, and Definitions for Acceptance Sampling.* ANSI/ASQC A2-1987. Milwaukee, Wis.

American Society for Quality Control (2002). ANSI/ISO/ASQ-A3534-2. *Statistics-Vocabulary and Symbols — Statistical Quality Control*, American Society for Quality, Milwaukee, WI.

Aroian, L.A. (1964). Some comments on truncated sequential tests for the exponential distribution. *Industrial Quality Control*, **21**, 309–312.

Aslam, M. (2007). A double acceptance sampling plan based on truncated life testing Rayleigh distribution. *European Journal of Scientific Research*, **17**(4), 605–611.

Aslam, M., Balamurali, Jun, C.-H., and Hussain, B. (2015). Design of SkSP-R Variables Sampling Plans. *Revista Colombiana de Estadistica*, **38**(2), 413–429.

Aslam, M. and Jun, C.-H. (2009a). A group acceptance sampling plan for truncated life test having Weibull distribution. *Journal of Applied Statistics*, **36**(9), 1021–1027.

Aslam, M. and Jun, C.-H. (2009b). A group acceptance sampling plans for truncated life tests based on the inverse Rayleigh and log-logistic distributions. *Pakistan Journal of Statistics*, **25**(2), 1–13.

Aslam, M. and Jun, C.-H. (2010). A double acceptance sampling plan for generalized log-logistic distribution with known shape parameters. *Journal of Applied Statistics*, **37**(3), 405–414.

Aslam, M., Jun, H.C., and Ahmad, M. (2009). A double acceptance sampling plan based on the truncated life tests in the Weibull model. *Journal of Statistical Theory and Applications*, **8**(2), 191–206.

Aslam, M., Jun, H.C., and Ahmad, M. (2010). Design of a time truncated double sampling plan for a general life distribution. *Journal of Applied Statistics*, **37**(8), 1369–1379.

Aslam, M., Mahmood, Y., Lio, Y.L., Tsai, T.-R., and Khan, M.A. (2011). Double acceptance sampling plans for Burr Type XII distribution percentiles under the truncated life test. *Journal of the Operational Research Society*, **63**(7), 1010–1017.

Aslam, M. and Shahbaz, M.Q. (2007). Economic reliability test plans using the generalized exponential distributions. *Journal of Statistics*, **14**, 53–60.

Attia, A.F., Aly, H.M., and Bleed, S.O. (2011). Estimating and planning accelerated life test using constant stress for generalized logistic distribution under type-I censoring. *ISRN Applied Mathematics*, **203618**, 15 (pages).

Bai, D.S., Cha, M.S., and Chung, S. (1992) Optimum simple ramp tests for the Weibull distribution and type-I censoring. *IEEE Transactions on Reliability*, **41**(3), 407–413.

Bai, D.S. and Chun, Y.R. (1991). Optimum simple step-stress accelerated life tests with competing causes of failure. *IEEE Transactions on Reliability*, **40**(5), 622–627.

Bai, D.S. and Chun, Y.R. (1993). Non-parametric inferences for ramp-stress test under random censoring. *Reliability Engineering and System Safety*, **41**(3), 217–223.

Bai, D.S., Chun, Y.R., and Cha, M.S. (1997). Time-censored ramp tests with stress bound for the Weibull life distribution. *IEEE Transactions on Reliability*, **46**(1), 99–107.

Bai, D.S., Chun, Y.R., and Chung, S.W. (1993). Optimal design of partially accelerated life tests for lognormal distribution under type-I censoring. *Reliability Engineering and System Safety*, **40**(1), 85–92.

Bai, D.S., Chun, Y.R., and Kim, J.G. (1995). Failure-censored accelerated life tests sampling plans for Weibull distribution under expected test time constraint. *Reliability Engineering and System Safety*, **50**(1), 61–68.

Bai, D.S. and Chung, S.W. (1992). Optimal design of partially accelerated life tests for the exponential distribution under type-I censoring. *IEEE Transactions on Reliability*, **41**(3), 400–406.

Bai, D.S. and Chung, S.W. (1998). Optimal designs of simple step-stress ALTs for lognormal lifetime distributions. *International Journal of Reliability Quality and Safety Engineering*, **5**(4), 315–336.

Bai, D.S., Chung, S.W., and Cha, M.S. (1992). Optimum simple ramp tests for the Weibull distribution and type-I censoring. *IEEE Transactions on Reliability*, **41**(3), 407–413.

Bai, D.S. and Kim, M.S. (1993). Optimum simple step-stress accelerated life tests for Weibull distribution and type-I censoring. *Naval Research Logistics*, **40**(2), 193–210.

Bai, D.S., Kim, J.G., and Chun, Y.R. (1993). Design of failure-censored accelerated life test sampling plans for lognormal and Weibull distributions. *Engineering Optimization*, **21**(3), 197–212.

Bai, D.S., Kim M.S., and Lee, S.H. (1989). Optimum simple step-stress accelerated life tests with censoring. *IEEE Transactions on Reliability*, **38**(5), 528–532.

Bai, D.S. and Yun, H.J. (1996). Accelerated life tests for products of unequal size. *IEEE Transactions on Reliability*, **45**(4), 611–618.

Baklizi, A. (2003). Acceptance sampling based on truncated life tests in the Pareto distribution of the second kind. *Advances and Applications in Statistics*, **3**(1), 33–48.

Baklizi, A. and El-Masri, A.E.K. (2004). Acceptance sampling based on truncated life tests in the Birnbaum–Saunders model. *Risk Analysis*, **24**(6), 1453–1457.

Balakrishnan, N. (2009). A synthesis of exact inferential results for exponential step-stress models and associated optimal accelerated life-tests. *Metrika*, **69**, 351–396.

Balakrishnan, N., Beutner, E., and Kateri, M. (2009). Order restricted inference for exponential step-stress models. *IEEE Transactions on Reliability*, **58**(1), 132–142.

Balakrishnan, N. and Han, D. (2008). Exact inference for simple step-stress model with competing risks for failure from exponential distribution under type-II censoring. *Journal of Statistical Planning and Inference*, **138**(12), 4172–4186.

Balakrishnan, N. and Han, D. (2009). Optimal step-stress testing for progressively type-I censored data from exponential distribution. *Journal of Statistical Planning and Inference*, **139**(5), 1782–1798.

Balakrishnan, N., Kundu, D., Ng, H.K.T., and Kannan, N. (2007). Point and interval estimation for a simple step-stress model with type-II censoring. *Journal of Quality Technology*, **39**(1), 35–47.

Balakrishnan, N., Leiva, V., and López, J. (2007). Acceptance sampling plans from truncated life tests based on the generalized Birnbaum–Saunders distribution. *Communications in Statistics-Simulatoin and Computation*, **36**(3), 643–656.

Balakrishnan, N., Xie, Q., and Kundu, D. (2009). Exact inference for a simple step-stress model from the exponential distribution under time constraint. *Annals of the Institute of Statistical Mathematics*, **61**, 251–274.

Balakrishnan, N., Xie, Q., and Zhang, L. (2009). Inference for a simple step-stress model with type-I censoring and log-normally distributed lifetimes. *Communications in Statistics — Theory and Methods*, **38**(10), 1690–1709.

Balamurali, S. and Jun, C.-H. (2006). Repetitive group sampling procedure for variables inspection. *Journal of Applied Statistics*, **33**(3), 327–338.

Balamurali, S. and Usha, M. (2013). A New Failure Censored Sampling System for Weibull Distribution. *International Journal of Performability Engineering*, **9**(1), 3–12.

Balasooriya, U. (1995). Failure-censored reliability sampling plans for the exponential distribution. *Journal of the Statistical Communication and Simulation*, **52**(4), 337–349.

Balasooriya, U. and Low, C.K. (2004). Competing causes of failure and reliability tests for Weibull lifetimes under type I progressive censoring. *IEEE Transactions on Reliability*, **53**(1), 29–36.

Balasooriya, U. and Saw, L.C. (1998). Reliability sampling plans for the two-parameter exponential distribution under progressive censoring. *Journal of Applied Statistics*, **25**(5), 707–714.

Balasooriya, U., Saw, L.C., and Gadag, V. (2000). Progressively censored reliability sampling plans for the Weibull. *Technometrics*, **42**(2), 160–167.

Bessler, S., Chernoff, H. and Marshall, A.W. (1962). An optimal sequential accelerated life test. *Technometrics*, **4**(3), 367–379.

Bhattacharyya, G.K. and Soejoeti, Z. (1989). A tampered failure rate model for step-stress accelerated life test. *Communications in Statistics — Theory and Methods*, **18**(5), 1627–1643.

Blischke, W.R. and Murthy, D.N.P. (1994). *Warranty Cost Analysis.* Marcel Dekker, Inc., New York.

Blischke, W.R. and Murthy, D.N.P. (1996). *Product Warranty Handbook*, Marcel, Dekker, Inc., New York.

Brass, W. (1971). On the scale of mortality. In *Biological aspects of Mortality, Symposia of the Society for the Study of Human Biology* (W. Brass, ed.), Taylor & Francis Ltd., London.

Bulgren, W. and Hewette, J. (1973). Double sample tests for hypotheses about the mean of an exponential distribution. *Technometrics*, **15**(1), 187–190.

Cacciari, M., Montanari, G.C., and Barry, C.P. (1992). Thermal endurance of electrical insulating materials studied by the use of the Kalman filter. *European Transactions on Electrical Power Engineering*, **6**(2), 187–194.

Chan, C.K. (1990). A proportional hazard approach to correlate SiO_2-breakdown voltage and time distributions. *IEEE Transactions on Reliability*, **39**(2), 147–150.

Chernoff, H. (1962). Optimal accelerated life designs for estimation. *Technometrics*, **4**(3), 381–408.

Chiodo, E. and Mazzanti, G. (2004). The log-logistic model for reliability characterization of power system components subjected to random stress. *SPEEDAM, CAPRI*, Italy, 239–244.

Chiodo, E. and Mazzanti, G. (2006). Indirect reliability estimation for electric devices via a dynamic stress-strength model. *Proceedings of IEEE SPEEDAM*, Taormina, S31_19-S31_24.

Chung, S.W., Seo, Y.S., and Yun, W.Y. (2006). Acceptance sampling plans based on failure-censored step-stress accelerated tests for Weibull distributions. *Journal of Quality in Mechanical Engineering*, **12**(4), 373–396.

Cook, R.D. and Jhonson, M.E. (1986). Generalized Burr–Pareto-Logistic distribution with application to a uranium exploration data set. *Technometrics*, **28**(2), 123–131.

Cox, D. (1972). Regression models and life tables. *Journal of the Royal Statistical Society Series B*, **34**(2), 187–220.

Cox, D.R. and Oakes, D. (1984). *Analysis of Survival Data*, Chapman and Hall, London.

Craiu, R.V. and Lee, T.C.M. (2005). Model selection for the competing risks model with and without masking. *Technometrics*, **47**(4), 457–467.

David, H.A. (1981). *Order Statistics*, John Wiley & Sons, New York.

David, H.A. and Moeschberger, M.L. (1971). Life tests under competing risks of failure and the theory of competing risk. *Biometrics*, **27**, 909–933.

David, H.A. and Moeschberger, M.L. (1978). *The Theory of Competing Risks*, Macmillan Publishing, Co., Inc.

David, P.K. and Montanari, G.C. (1992). Compensation effect in thermal aging investigated according to Eyring and Arrhenius models. *European Transactions on Electrical Power Engineering*, **2**(3), 187–194.

DeGroot, M.H. and Goel, P.K. (1979). Bayesian estimation and optimal designs in partially accelerated life testing. *Naval Research Logistics Quarterty*, **26**(2), 223–235.

Dempe, S. (2002). *Foundations of Bilevel Programming*, Springer, US.

Ding, C., Yang, C., and Tse, S.-K. (2010). Accelerated life test sampling plans for the Weibull distribution under Type I progressive interval censoring with random removals. *Journal of Statistical Computation and Simulation*, **80**(8), 903–914.

Dodge, H.F. (1955a). Skip-lot sampling plan. *Industrial Quality Control*, **11**(5), 3–5.

Dodge, H.F. (1955b). Chain sampling inspection plans. *Industrial Quality Control*, **11**(4), 10–13.

Donghoon, H. and Balakrishnan, N. (2010). Inference for a simple step-stress model with competing risks for failure from the exponential distribution under time constraint. *Computational Statistics & Data Analysis*, **54**(9), 2066–2081.

Duncan, A.J. (1986). *Quality Control and Industrial Statistics*, 5th edition, Richard D. Irwin, Homewood, Illinois.

Efron, B. (1982). The Jackknife, the Bootstrap and Other Re-sampling Plans, CBMS/NSF Regional Conference Series in Applied Mathematics Philadelphia, PA.

Elsayed, E.A. (2003). Accelerated life testing. In *Handbook of Reliability Engineering* (H. Pham ed.), 415–428, Springer, New York.

Elsayed, E.A. (2012). *Reliability Engineering*, 2nd edition, John Wiley & Sons Inc., Hoboken, New Jersey.

Elsayed, E.A. (2013). Design of reliability test plans. In *Stochastic Reliability and Maintenance Modeling, Essays in Honor of Professor Shuji Osaki on his 70th Birthday*, (T. Dohi and T. Nakagawa, eds.), 7–40, Springer Series in Reliability Engineering, New York.

Elsayed, E.A. and Zhang, H. (2006). Optimum multiple-stress-type accelerated life testing plans based on proportional odds model with simple step-stress loading. *Journal Europeen des Systemes Automatises*, **40**(7), 745–762.

Elsayed, E.A. and Zhang, H. (2007). Design of PH-based accelerated life testing plans under multiple-stress-type. *Reliability Engineering and System Safety*, **92**, 286–292.

Endicott, H.S., Hatch, B.D., and Schmer, R.G. (1965). Application of the Eyring model to capacitor aging data, *IEEE Transactions on Component Parts*, **12**(1), 34–42.

Epstein, B. (1954). Truncated life tests in the exponential case. *Annals of Mathematics and Statistics*, **25**(3), 555–564.

Epstein, B. and Sobel, M. (1953). Life testing. *Journal of the American Statistical Association*, **48**(263), 486–502.

Epstein, B. and Sobel, M. (1955). Sequential life test in the exponential case. *Annals of Mathematics and Statistics*, **26**(1), 82–93.

Fairbanks, K. (1988). A two-stage life test for the exponential parameter. *Technometrics*, **30**(2), 175–180.

Fan, T.H., Wang, W.L., and Balakrishnan, N. (2008). Exponential progressive step-stress life testing with link function based on box-cox transformation. *Journal of Statistical Planning and Inference*, **138**, 2340–2354.

Fard, N. and Li, C. (2009). Optimum simple step-stress accelerated life test design for reliability prediction. *Journal of Statistical Planning and Inference*, **139**(5), 1799–1808.

Fei, H. (2000). The statistical analysis of combined data from the progressively and constantly life tests under power-Weibull model. *Journal of Mathematical Applications*, **13**(3), 102–106.

Fei, H., Leng, S., and Shu, D. (1991). Progressive stress accelerated life test of solid tantalum electrolytic capacitor, *Chinese Journal of Applied Probability and Statistics*, **7**(3), 330–335.

Fei, H. and Xun, X. (1996). Statistical analysis for progressive stress accelerated life testing in the exponential case, ICRMS'96. Guangzhou, China, No 5.12–15. *Proceedings of the Third International Conference on R. Maintainability and Safety*, 420–425, Publishing House of Electronics Industry.

Fertig, K.W. and Mann, N.R. (1980). Life-test sampling plans for two-parameter Weibull populations. *Technometrics*, **22**(2), 165–177.

Freund, R.A. (1957). Acceptance control charts. *Industrial Quality Control*, **14**(4), 13–23.

Gao, N., Hu, J.M., Shi, Y.M., and Qin, X.Q. (2008). Statistical analysis of the mixed accelerated life test for the type-II progressively censored sample. *Journal of Physical Sciences*, **12**, 23–32.

Gaylor, D.W. and Sweeny, H.C. (1965). Design for optimal prediction in simple linear regression. *Journal of the American Statistical Association*, **60**(309), 205–216.

Goba, F.A. (1969). Bibliography on thermal aging of electrical insulation. *IEEE Transactions on Electrical Insulation*, **EI-4**(2), 31–58.

Goel, P.K. (1971). Some Estimation Problems in the Study of Tampered Random Variables. Technical Report No. 50, Department of statistics, Carnegiemellon university, Pittspurgh, Pennsylvania.

Grosh, D.L. (1989). *A Primer of Reliability Theory*, John Wiley & Sons, Inc., New York.

Goode, H.P. and Kao, J.H.K. (1961). Sampling plans based on the Weibull distribution. *Proceeding of the Seventh National Symposiumon Reliability and Quality Control*, Philadelphia, 24–40.

Gouno, E. (2000). A Bayesian method for inference on step-stress testing. *Second International Conference on Mathematical Methods Abstracts' Book*, **2**, 473–476.

Gouno, E. (2001). An inference method for temperature step-stress accelerated life testing. *Quality and Reliability Engineering International*, **17**(1), 11–18.

Gouno, E. (2007). Optimum step-stress for temperature accelerated life testing. *Quality and Reliability Engineering International*, **23**(8), 915–924.

Gouno, E. and Balakrishnan, N. (2001). Step-stress accelerated life test. In *Handbook of Statistics-20*: *Advances in Reliability*, (N. Balakrishnan and C.R. Rao, eds.), 623–639, North-Holland, Amsterdam.

Gouno, E., Sen, A., and Balakrishnan, N. (2004). Optimal step-stress test under progressive type-I censoring. *IEEE Transactions on Reliability*, **53**(3), 388–406.

Govindaraju, K. and Balamurali, S. (1998). Chain sampling plan for variables inspection. *Journal of Applied Statistics*, **25**(1), 103–109.

Guenther, W.C. (1974). Sample size formulas for some binomial problems. Technometrics, **16**(3), 465–467.

Gui, W. and Zhang, S. (2014). Acceptance sampling plans based on truncated life tests for Gompertz distribution. *Journal of Industrial Mathematics*, **Article ID** 391728, doi: 10.1155/2014/391728.

Gupta, S.S. (1962). Life test sampling plans for normal and lognormal distributions. *Technometrics*, **4**(2), 151–175.

Gupta, S.S. and Groll, P.A. (1961). Gamma distribution in acceptance sampling based on life tests. *Journal of the American Statistical Association*, **56**, 942–970.

Hahn, J.G. and Nelson, W. (1974). A comparison of methods for analyzing censored life data to estimate relationships between stress and product life. *IEEE Transactions on Reliability*, **R-23**, 2–11

Hamaker, H.C. (1979). Acceptance sampling for percent defective by variables. *Journal of Quality Technology*, **11**(3), 139–148.

Han, D. and Balakrishnan, N. (2010). Inference for a simple step-stress model with competing risks for failure from the exponential distribution under time constraint. *Computational Statistics and Data Analysis*, **54**(9), 2066–2081.

Han, D. and Ng, H.K.T. (2013). Comparison between constant-stress and step-stress accelerated life tests under time constraint. *Naval Research Logistics*, **60**(7), 541–556.

Hassan, A.S. and Al-Thobety, A.K. (2012). Optimal design of failure step-stress PALTs with type-II censored inverted Weibull data. *International Journal of Engineering Research and Applications*, **2**(3), 3242–3253.

Hassan, A.S., Assar, S.M., and Shelbaia, A. (2014). Multiple-step stress accelerated life for Weibull Poisson distribution with type I censoring. *International Journal of Basic and Applied Sciences*, **3**(3), 180–189.

Hatter, M. and Moore, A.H. (1980). An evaluation of exponential and Weibull test plans. *IEEE Transactions on Reliability*, **R-25**(2), 100–104.

Herman, R.J. and Patell, K.N.R. (1971). Maximum likelihood estimation for multi-risk model. *Technometrics*, **13**(2), 385–396.

Herzberg, A.M., and Cox, D.R. (1972). Some optimal designs for interpolation and extrapolation. *Biometrika*, **59**(3), 551–561.

Hong, Y., Ma, H., and Meeker, W.Q. (2010). A tool for evaluating time-varying-stress accelerated life test plans with log-location-scale distributions. *IEEE Transactions on Reliability*, **59**(4), 620–627.

Hφyland, A. and Rausand, M. (1994). *System Reliability Theory: Models and Statistical Methods*, John Wiley & Sons, New York.

Hsieh, H.K. (1994). Accelerated life test sampling plans for exponential distribution. *Communication in Statistics-Simulation*, **23**(1), 27–41.

Hsu, J.S.I. (1980). An economic design of skip-lot sampling plans. *Journal of Quality Technology*, **12**(3), 144–149.

Huang, Y.S., Hseish, C.H., and Ho, J.W. (2008). Decisions on an Optimal Life Test Sampling Plan with Warranty Considerations. *IEEE Transactions on Reliability*, **57**(4), 643–649.

International Organization for Standardization. (1981). ISO 3951 *Sampling Procedures and Charts for Inspection by Variables*, Geneva, Switzerland.

International Organization for Standardization. (1989). ISO 2859-1. *Sampling Procedures for Inspection by Attributes* — Part I: *Sampling Plans Indexed by Acceptance Quality Limit (AQL) for Lot-by-Lot Inspection*, ISO, Geneva, Switzerland.

International Organization for Standardization. (1991). ISO 2859-3. *Sampling Procedures for Inspection by Attributes* — Part-3: *Skip-lot Sampling Procedures*, ISO, Geneva, Switzerland.

International Organization for Standardization. (2002). ISO 3534-2. *Statistic-Vocabulary and Symbols* — Part-2: *Applied Statistics*, Geneva, Switzerland.

Islam, A. and Ahmad, N. (1994). Optimal design of accelerated life tests for the Weibull distribution under periodic inspection and type I censoring. *Microelectronics and Reliability*, **34**(9), 1459–1468.

Ismail, A.A. (2004). The test design and parameter estimation of Pareto lifetime distribution under partially accelerated life tests. *Ph.D. Thesis*, Department of Statistics, Faculty of Economics and Political Science, Cairo University, Egypt.

Ismail, A.A. (2009). Optimum constant-stress partially accelerated life tests with type-II censoring: The case of Weibull failure distribution. *International Journal of Statistics and Economics*, **3**(509), 39–46.

Ismail, A.A. (2010). Bayes estimation of Gompertz distribution parameters and acceleration factor under partially accelerated life tests with type-I censoring. *Journal of Statistical Computation and Simulation*, **80**(11), 1253–1264.

Ismail, A.A. (2011). Planning step-stress life tests with type-II censored data. *Scientific Research and Essays*, **6**(19), 4021–4028.

Ismail, A.A. and Aly, H.M. (2009). Optimal planning of failure step-stress partially accelerated life tests under type-II censoring. *International Journal of Mathematical Analysis*, **3**(31), 1509–1523.

Ismail, A.A. and Aly, H.M. (2010). Optimal planning of failure step-stress partially accelerated life tests under type-II censoring. *Journal of Statistical Computation and Simulation*, **80**(12), 1335–1348.

Ismail, A.A. and Sarhan, A.M. (2009). Optimal design of step-stress life test with progressively type-II censored exponential data. *International Mathematical Forum*, **4**(40), 1963–1976.

Johnson, N.L., Kotz, S., and Balakrishnan, N. (1995). *Continuous Univariate Distributions*, 2nd edition. Vol. 1, Wiley, Chichester.

Jun, C.-H., Lee, H., Lee, S.-H., and Balamurali, S. (2010). A variable group sampling plan under failure-censored reliability for Weibull distribution. *Journal of Applied Statistics*, **37**(3), 453–460.

Kamal, M., Zarrin, S., and Arif-ul-Islam (2013). Step stress accelerated life testing plan for two parameter pareto distribution. *RT&A # 01 (28)*, **8**, 30–40.

Kantam, R.R.L. and Rosaiah, K. (1998). Half logistic distribution in acceptance sampling based on life tests. *IAPQR Transactions*, **23**(2), 117–125.

Kantam, R.R.L., Rosaiah, K., and Rao, G.S. (2001). Acceptance sampling based on life tests: Log-logistic models. *Journal of Applied Statistics*, **28**(1), 121–128.

Kateri, M., Kamps, U., and Balakrishnan, N. (2009). A meta-analysis approach for step-stress experiments. *Journal of Statistical Planning and Inference*, **139**(9), 2907–2919.

Kateri, M., Kamps, U., and Balakrishnan, N. (2011). Optimal allocation of change points in simple step-stress experiments under type-II censoring. *Computational Statistics and Data Analysis*, **55**(1), 236–247.

Kateri, M., Kamps, U., and Balakrishnan, N. (2012). A sequential order statistics approach to step-stress testing. *Annals of the Institute of Statistical Mathematical*, **64**(2), 303–318.

Khamis, I.H. (1997). Optimum M-step design with K stresses variables. *Communications in Statistics — Simulations and Computation*, **26**(4), 1301–1313.

Khamis, I.H. and Higgins, J.J. (1996). Optimum 3-step stress tests. *IEEE Transactions on Reliability*, **45**(2), 341–345.

Khamis, I.H. and Higgins, J.J. (1998). A new model for step-stress testing. *IEEE Transactions on Reliability*, **47**(2), 131–134.

Khamis, I.H. and Higgins, J.J. (1999). An alternative to the Weibull step-stress model. *International Journal of Quality & Reliability Management*, **16**(2), 158–165.

Kielpinski, T.J. and Nelson, W. (1975). Optimum censored accelerated life tests for normal and lognormal life distributions, *IEEE Transactions on Reliability*, **R-24** pp. 310–320.

Kim, C.M. and Bai, D.S. (2002). Analysis of accelerated life test data under two failure modes. *International Journal of Reliability, Quality and Safety Engineering*, **9**(2), 111–125.

Kim, M. and Yum, B.J. (2009). Reliability acceptance sampling plans for the weibull distribution under accelerated type-1 censoring. *Journal of Applied Statistics*, **36**(1), 11–20.

Kim, M. and Yum, B.J. (2011). Life test sampling plans for weibull distributed lifetimes under accelerated hybrid censoring. *Statistical Papers*, **52**(2), 327–342.

Klein, J.P. and Basu, A.P. (1981). Weibull accelerated life tests when there are competing causes of failure. *Communications in Statistics Theory and Methods*, **10**(20), 2073–2100.

Klein, J.P. and Basu, A.P. (1982a). Accelerated life testing under competing exponential failure distribution. *IAPQR Transactions*, **7**(1), 1–20.

Klein, J.P. and Basu, A.P. (1982b). Accelerated life tests under competing Weibull causes of failure. *Communications in Statistics — Theory and Methods*, **11**(20), 2271–2286.

Kockerlakota, S. and Balakrishnan, N. (1986). One- and two-sided sampling plans based on the exponential distribution. *Naval Research Logistics Quarterly*, **33**(3), 513–522.

Kwon, Y.I. (1996). A Bayesian life test sampling plan for non-repairable products sold under warranty. *International Journal of Quality & Reliability Management*, **13**(5), 40–49.

Lakshmi, S. and Sankari, G. (2012). Mathematical model for hypoestrogenism of oligomenorrhoea case in exercising women. *American Journal of Mathematics and Mathematical Sciences*, **1**, 47–53.

Lawless, J.F. (1982). *Statistical Models and Methods for Life Data*. John Wiley and Sons, New York.

Lee, J. and Pan, R. (2010). Analyzing step-stress accelerated life testing data using generalized linear models. IIE Transactions (Institute of Industrial Engineers), **42**(8), 589–598.

Lehmann, E.L. (1986). *Testing Statistical Hypotheses*, 2nd edition, John Wiley and Sons Inc., New York.

Lewis, A.W. (1981). The Burr distribution as a general parametric family in survivorship and reliability applications. Ph.D. Dissertation, Department of Biostatistics, North Carolina University.

Li, C. and Fard, N. (2007). Optimum bivariate step-stress accelerated life test for censored data. *IEEE Transactions on Reliability*, **56**(1), 77–84.

Liao, H. and Li, Z. (2008). Multi-objective design of equivalent accelerated life testing plans. *International Journal of Reliability, Quality and Safety Engineering*, **15**(6), 515–538.

Lieberman, G.J. and Resnikoff, G.J. (1955). Sampling plans for inspection by variables. *Journal of the American Statistical Association*, **50**(270), 457–516.

Lin, D. and Chiu, W.K. (1995). An economic accelerated life test acceptance sampling plan. *International Journal of Reliability, Quality and Safety Engineering*, **2**(1), 49–59.

Lin, Z. and Fei, H. (1987). Statistical inference from progressive stress accelerated life tests. *Proceedings of China–Japan Reliability Symposium*, Shanghai, 229–236.

Ling, L., Xu, W., and Li, M. (2009). Parametric inference for progressive type-I hybrid censored data on a simple step-stress ALT model. *Mathematics and Computers in Simulation*, **79**(10), 3110–3121.

Lio, Y.L., Tsai T.-R., and Wu, S.-J. (2010a). Acceptance sampling plans from truncated life tests based on the Birnbaum Saunders distribution for percentiles. *Communications in Statistics — Simulation and Computation*, **39**(1), 119–136.

Lio, Y.L., Tsai T.-R., and Wu, S.-J. (2010b). Acceptance sampling plans for truncated life tests based on the Burr Type XII percentiles. *Journal of Chinese Institute of Industrial Engineers*, **27**(4), 270–280.

Little, R. and Jebe, E.H. (1969). A note on the gain in precision for optimal allocation in regression as applied to extrapolation in S-N fatigue testing. *Technometrics*, **11**(2), 389–392.

Liu, X. and Tang, L.C. (2009). A sequential constant-stress ALT scheme and its Bayesian inference. *Quality and Reliability Engineering International*, **25**(1), 91–109.

Liu, X. and Qiu, W.S. (2011). Modeling and planning of step-stress accelerated life tests with independent competing risks. *IEEE Transactions on Reliability*, **60**(4), 712–720.

Lu, M.W. and Rudy, R.J. (2002). Step-stress accelerated test. *International Journal of Materials & Product Technology*, **17**(Iss. 5–6), 425–434.

Ma, H. and Meeker, W.Q. (2008). Optimum step-stress ALT plans for log-location-scale distributions. *Naval Research Logistics*, **55**(6), 551–562.

Madi, M.T. (1993). Multiple step-stress accelerated life test: The tampered failure rate model, *Communications in Statistical — Theory and Methods*, **22**(9), 2631–2639.

Madi, M.T. (1997). Bayesian inference for partially accelerated life tests using Gibbs sampling. *Microelectronics and Reliability*, **37**(8), 1165–1168.

Mamer, J.W. (1982). Cost Analysis of Pro-rata and Free Replacement Warranties. *Naval Res. Logist. Quart.*, **29**, 345–356.

Mann, N.R. (1972). Design of overstress life-test experiments when failure times have a two-parameter Weibull distribution. *Technometrics*, **14**(2), 437–451.

Mann, N.R., Schafer, R.E., and Singpurwalla, N.D. (1974). *Methods for Statistical Analysis of Reliability and Life Data*. John Wiley and Sons.

McCool, J.I. (1978). Competing risk and multiple comparison analysis for bearing fatigue tests. *ASLE Transactions*, **21**(4), 271–284.

McSorley, E.O., Lu, J.C., and Li, C.S. (2002). Performance of parameter-estimates in step-stress accelerated life-tests with various sample-sizes. *IEEE Transactions on Reliability*, **51**(3), 271–277.

Meeker, W.Q. and Escobar, L.A. (1993). A review of recent research and current issues in accelerated testing. *International Statistical Review*, **61**(1), 147–168.

Meeker, W.Q. and Escobar, L.A. (1998). *Statistical Methods for Reliability, and Data*, John Wiley & Sons, New York.

Meeker, W.Q. and Hahn, G.J. (1985). How to plan an accelerated life test some practical guidelines. *(ASQC Basic References in Quality Control) Statistical Techniques*, American Society for Quality Control, 310W. Wisconsin Ave., Milwaukee, WI 53203, (800), 952–6587.

Meeker, W.Q. and Nelson, W.B. (1975). Optimum accelerated life-tests for the Weibull and extreme value distributions. *IEEE Transactions on Reliability*, **R-24**(5), 321–332.

Meeker, W.Q. and Nelson, W. (1978). Theory for optimum censored life tests for Weibull and extreme value distribution. *Technometrics*, **20**(2), 171–177.

Miller, R. and Nelson, W. (1983). Optimum Simple Step-Stress Plans for Accelerated Life Testing. *IEEE Transactions on Reliability*, **R-32**(1), 59–65.

Mitra, A. (2005). *Fundamentals of Quality Control and Improvement*, 2nd edition, Prentice-Hall of India Private Ltd, New Delhi-1.

Mitra, S., Ganguly, A., Samanta, D., and Kundu, D. (2013). On the simple step-stress model for two-parameter exponential distribution. *Statistical Methodology*, **15**, 95–114.

Mood, A.M., Graybill, F.A., and Boes, D.C. (1974). *Introduction to the Theory of Statistics*, 3rd edition, McGraw-Hill Book Co., Singapore.

Murthy, D.N.P. and Blischke, W.R. (2000). Strategic warranty management: A life-cycle approach. *IEEE Transactions on Engineering Management*, **47**(1), 40–54.

Nachalas, J.A. (1986). A general model for age acceleration during thermal cycling. *Quality and Reliability Engineering International*, **2**, 3–6.

Nechwal, K.N., Nechwal, N.A., Purgailis, M., Rozevskis, U., Strelchonok, V.F., Krast, J., Berzinsh, G., and Moldovan, M. (2009). Optimization of life test sampling plans. *Proceedings of the 9th International Conference, Reliability and Statistics in Transportation and Communication (RelStat'09)*, 21–24 October 2009, Transport and Telecommunication Institute, Lomonosova 1, LV-1019, Riga, Latvia, 108–116.

Nelson, W. (1975). Graphical analysis of accelerated life test data with a mix of failure modes. *IEEE Transactions on Reliability*, **R-24**(4), 230–237.

Nelson, W. (1980). Accelerated life testing: Step-stress models and data analysis. *IEEE Transactions on Reliability*, **R-29**(2), 103–108.

Nelson, W. (1982). *Applied Life Data Analysis*, John Wiley & Sons, New York.

Nelson, W. (1990). *Accelerated Testing: Statistical Models, Test Plans, and Data Analysis*, John Wiley & Sons, New York.

Nelson, W. (2005a). A bibliography of accelerated test plans, Part I: Overview. *IEEE Transactions on Reliability*, **54**(2), 194–197.

Nelson, W. (2005b). A bibliography of accelerated test plans, Part II — References. *IEEE Transactions on Reliability*, **54**(3), 370–373.

Nelson, W. and Kielpinski, T.J. (1976). Theory for optimum censored accelerated tests for normal and lognormal life distribution. *Technometrics*, **18**(1), 105–114.

Nilsson, L. (1985). Estimation from accelerated life tests with increasing stress. Ph.D. Dissertation, Institute of Mathematical Statistics, University of Umea in Sweden.

Park, S.J. and Yum, B.J. (1998). Optimal design of accelerated life tests under modified stress loading methods. *Journal of Applied Statistics*, **25**, 41–62.

Pascual, F.G. (2007). Accelerated life test planning with independent Weibull competing risks with known shape parameter. *IEEE Transactions on Reliability*, **56**(1), 85–93.

Prot, E.M. (1948). Fatigue testing under progressive loading; A new technique for testing materials. *Revue de Metallurgie*, **xiv**, 481–489 (in French), translated by Edward J. Ward, Capt. USAF., WADC TR 52-148, September 1952.

Rezk, H.R., EL Helbawy, A.A., and AL Dayian, G.R. (2014). Statistical inference for a simple step-stress model based on censored data from the Kumaraswamy Weibull distribution. *British Journal of Mathematics & Computer Science*, **4**(11), 1496–1514.

Rosaiah, K. and Kantam, R.R.L. (2005). Acceptance sampling based on the inverse Rayleigh distribution. *Economic Quality Control*, **20**(2), 277–286.

Rosaiah, K., Kantam, R.R.L., and Reddy, J.P. (2007). Economic reliability test plan with Inverse Rayleigh Variate. *Pakistan Journal of Statistics*, **24**(1), 57–65.

Rosaiah, K., Kantam, R.R.L., and Santosh Kumar, C. (2006). Reliability of test plans for exponentiated log-logistic distribution. *Economic Quality Control*, **21**(2), 279–289.

Rosaiah, K., Kantam, R.R.L., and Santosh Kumar, C. (2007). Exponentiated log-logistic distribution: An economic reliability test plan. *Pakistan Journal of Statistics*, **23**(2), 147–156.

Saxena, S., Zarrin, S., Kamal, M., and Arif-Ul-Islam (2012). Optimum step stress accelerated life testing for rayleigh distribution. *International Journal of Statistics and Applications*, **2**(6), 120–125.

Schilling, E.G. (1982). *Acceptance Sampling in Quality Control*, Marcel Dekker, New York.

Schilling, E.G. and Neubauer, D.V. (2009). *Acceptance Sampling in Quality Control*, 2nd edition, CRC Press, Taylor and Francis Group, New York.

Schneider, H. (1989). Failure-censored variables-sampling plans for lognormal and Weibull distributions. *Technometrics*, **31**(2), 199–206.

Schmoyer, R. (1991). Non-parametric analysis for two-level single-stress accelerated life tests. *Technometrics*, **33**(2), 175–186.

Seo, J.H., Jung, M., and Kim, C.M. (2009). Design of accelerated life test sampling plans with a non constant shape parameter. *European Journal of Operational Research*, **197**(2), 659–666.

Seo, S.K. and Yum, B.J. (1991). Accelerated life test plans under intermittent inspection and type-I censoring: The case of Weibull failure distribution. *Naval Research Logistic*, **38**(1), 1–22.

Shaked, M. and Singpurwalla, N.P. (1983). Inference for step-stress accelerated life tests. *Journal of Statist. Planning and Inference*, **7**(4), 295–306.

Sherman, R.E. (1965). Design and evaluation of repetitive group sampling plan. *Technometrics*, **7**(1), 11–21.

Shi, Y.M. (2011). Statistical Inference for the mixed accelerated life testing with progressive censoring. *International Conference on Multimedia Technology*, 2605–2608.

Shi, Y.M., Jin, L., Wei, C., and Yue, Hong-bo. (2013). Constant-stress accelerated life test with competing risk under progressive Type-II hybrid censoring. *Advanced Materials Research*, Vols. 712–715, 2080–2083.

Sobel, M. and Tischendrof, J. (1959). Acceptance sampling with new life test objectives. *Proceedings of 5th National Symposium on Reliability and Quality Control*, 108–118.

Soliman, A.A. (2005). Estimation of parameters of life from progressively censored data using a Burr type-XII model. *IEEE Transactions on Reliability*, **54**(1), 34–42.

Solomon, P., Klein, N., and Albert, M. (1976). A statistical model for step and ramp voltage breakdown tests in thin insulators. *Thin Solid Films*, **35**(3), 321–326.

Srinivasa Rao, G. (2011). Double acceptance sampling plans based on truncated life tests for the Marshall–Olkin extended exponential distribution. *Austrian Journal of Statistics*, **40**(3), 169–176.

Srinivasa Rao, G., Ghitany, M.E., and Kantam, R.R.L. (2009). Reliability test plans for Marshall-Olkin extended exponential distribution. *Applied Mathematical Sciences*, **3**(55), 2745–2755.

Srivastava, P.W. and Mittal, N. (2010). Optimum step-stress partially accelerated life tests for the truncated logistic distribution with censoring. *Applied Mathematical Modelling*, **34**(10), 3166–3178.

Srivastava, P.W. and Mittal, N. (2012a). Optimum multi-level ramp-stress ALT plan with multiple-objectives for Burr type-XII distribution under type-I censoring. *International Journal of Reliability, Quality and Safety Engineering*, **19**(2), Article ID 1250006, p. 17.

Srivastava, P.W. and Mittal, N. (2012b). Optimum multi-objective ramp-stress accelerated life test with stress upper bound for burr type-XII distribution. *IEEE Transactions on Reliability*, **61**(4), 1030–1038.

Srivastava, P.W. and Mittal, N. (2012c). Optimum failure-censored step-stress partially accelerated life test for the truncated logistic life distribution. *International Journal of Reliability and Applications*, **13**(1), 19–35.

Srivastava, P.W. and Mittal, N. (2013a). Optimum multi-objective modified constant-stress ALT plan for the Burr type-XII distribution with type-I censoring. *Journal of Risk and Reliability*, **227**(2), 132–143.

Srivastava, P.W. and Mittal, N. (2013b). Optimum constant-stress partially accelerated life tests for the truncated logistic distribution under time constraint. *International Journal of Operational Research/Nepal*, **2**, 33–47.

Srivastava, P.W. and Mittal, N. (2013c). Failure-censored optimum constant-stress partially accelerated life tests for the truncated logistic life distribution. *Opinion — An International Journal of Business Management, Special Issue on Role of Statistics in Management and Allied Sciences*, **3**(2), 41–66.

Srivastava, P.W. and Mittal, N. (2014). Optimum multi-objective modified step-stress accelerated life test plan for the burr type-XII distribution. *International Journal of Reliability and Applications*, **15**(1), 23–50.

Srivastava, P.W. and Sharma, D. (2014a). Bilevel Programming Based Time-Censored Ramp-Stress ALTSP with Warranty Cost. *International Journal of Performability Engineering*, **10**(3), 281–290.

Srivastava, P.W. and Sharma, D. (2014b). Bilevel Programming Based Step-Stress PALTSP using Tampered Failure Rate Model and Warranty. *International Journal of Reliability, Quality and Safety Engineering*, **21**(5), **Article ID** 1450022, 23 pages, doi: 10.1142/S0218539314500223.

Srivastava, P.W. and Sharma, D. (2014c). Optimum Time-Censored Constant-Stress PALTSP for the Burr Type XII Distribution using Tampered Failure Rate Model. *Journal of Quality and Reliability Engineering*, **Vol.** 2014, **Article ID** 564049, 13 pages. doi: 10.1155/2014/564049.

Srivastava, P.W. and Sharma, D. (2015a). Optimum Time-Censored Simple Ramp-Stress Accelerated Life Test Sampling Plan for the Log-Logistic Distribution. *Journal of Quality and Maintenance Engineering*, **21**(1), 112–132.

Srivastava, P.W. and Sharma, D. (2015b). Bilevel Programming Based Time-Censored Ramp-Stress ALTSP with Warranty Cost. *International Journal of Performability Engineering*, **11**(1), 71–80.

Srivastava, P.W. and Shukla, R. (2008a). A log-logistic step-stress model. *IEEE Transaction on Reliability*, **57**(3), 431–434.

Srivastava, P.W. and Shukla, R. (2008b). Optimum log-logistic step-stress model with censoring. *International Journal of Quality and Reliability Management*, **25**(9), 968–976.

Srivastava, P.W. and Shukla, R. (2009). Optimum simple ramp test for the log-logistic distribution with censoring. *Journal of Risk and Reliability*, **223**(4), 373–379.

Srivastava, P.W., Shukla, R., and Kanwar Sen (2014). Optimum simple step-stress test with competing risks for failure using Khamis–Higgins model under type-I censoring. *International Journal of Operational Research/Nepal*, **3**, 75–88.

Starr, W.T. and Endicott, H.S. (1961). Progressive-stress: A new accelerated approach to voltage endurance. *Transactions of AIEE (Power Apparatus and Systems)*, **80**, 515–522.

Stephens, K.S. (2001). *The Handbook of Applied Acceptance Sampling: Plans, Procedures and Principles*, ASQ Quality Press, Milwaukee, WI.

Suresh, R.P. and Ramanathan, T.V. (1997). Acceptance Sampling plans by variables for a class of symmetric distributions. *Communications in Statistics: Simulation*, **26**(4), 1379–1391.

Tadikamalla, P.R. (1980). A look at the Burr and related distributions. *International Statistical Review*, **48**(3), 337–344.

Tahir, M. (2003). Estimation of the failure rate in a partially accelerated life test: A sequential approach. *Stochastic Analysis and Applications*, **21**(4), 909–915.

Tang, L.C. (1999). Planning for accelerated life tests. *International Journal of Reliability Quality and Safety Engineerin*, **6**(3), 265–275.

Tang, L.C. (2003). Multiple-steps step-stress accelerated life test. In *Handbook of Reliability Engineering* (H. Pham, ed.), 441–455. Springer, New York.

Tang, Y. (2006). Statistical analysis of progressive stress accelerated life testing of Weibull distributions under CE model. *Journal of Systems Science and Mathematical Sciences*, **26**(3), 342–351.

Tang, L.C., Sun, Y.S., Goh, T.N., and Ong, H.L. (1996). Analysis of step-stress accelerated life test data: A new approach. *IEEE Transactions on Reliability*, **45**(1), 69–74.

Tang, L.C., Tan, A.P., and Ong, S.H. (2002). Planning for accelerated life tests with three constant stress levels. *Computers & Industrial Engineering*, **42**(2–4), 439–446.

Tang, L.C. and Xu, K. (2005). A multiple objective framework for planning accelerated life tests. *IEEE Transactions on Reliability*, **54**(1), 58–63.

Tang, L.C. and Yang, G. (2002). Planning multiple levels constant stress accelerated life tests. *Proceedings of Annual Reliability Maintainability Symposium*, 338–342.

Tang, L.C., Yang, G.Y., and Xie, M. (2004). Planning of Step-Stress Accelerated Degradation Test. *RAMS*, 287–292.

Teng, S.L. and Yeo, K.P. (2002). A least-squares approach to analyzing life–stress relationship in step-stress accelerated life tests. *IEEE Transactions on Reliability*, **51**(2), 177–182.

Thomas, M.U. and Rao, S.S. (1999). Warranty economic decision models: A summary and some suggested directions for future research. *Operations Research*, **47**(6), 807–820.

Tsai, T.-R. and Wu, S.-J. (2006). Acceptance sampling based on truncated life tests for generalized Rayleigh distribution. *Journal of Applied Statistics*, **33**(6), 595–600.

Tseng, S. T., Hamada, M. C., and Chiao, H. (1995). Using degradation data from a factorial experiments to improve fluorescent lamp reliability. *Journal of Quality Technology*, **27**(4), 363–369.

United States Department of Defense. (1957). MIL-STD-414. *Military Standard, Sampling Procedures and Tables for Inspection by Variables for Percent Defective*, U.S. Government Printing Office, Wasington, DC.

United States Department of Defense. (1963). MIL-STD-105D. *Sampling Procedures and Tables for Inspection by Attributes: MIL-STD-105D*, U.S. Government Printing Office, Washington, DC.

United States Department of Defense. (1965). MIL-STD- 105D. Factors and procedures for applying MIL-STD- 105D sampling plans to life and reliability testing, quality control and reliability assurance. Technical Report 7, Office of the Assistant Secretary of Defense (Installations and Logistics), Washington, DC.

United States Department of Defense. (1989). MIL-STD-105E. Military standard, sampling procedures and tables for inspection by attributes, U.S. Government Printing Office, Washington, DC.

Viertl, R. (1988). *Statistical Methods in Accelerated Life Testing*, Vandenhoeck & Ruprecht, Gottingen.

Wadsworth, H.M., Stephens, K.S., and Godfrey, A.B. (2001). *Modern Methods for Quality Control and Improvement*, 2nd edition, John Wiley & Sons, New York.

Wald, A. (1947). *Sequential Analysis*, John Wiley, New York. (Reprinted Dover Publications, New York, 1973).

Wallace, W.E. (1985). Present practice and future plans for MIL-STD-781. *Naval Res. Logistics Quart.*, **32**(1), 21–26.

Wallis, W.A. (1948). *Use of Variables in Acceptance Sampling for Percent Defective, Techniques of Statistical Analysis*, The Statistical Research Group, Columbia University, McGraw-Hill, New York.

Wang, B. (2001). Models and statistical analysis for the mixed accelerated life test. *Applied Mathematics — Journal of Chinese Universities Series A*, **16**(1), 101–106.

Wang, B. (2006). Unbiased estimations for the exponential distribution based on step-stress accelerated life-testing data. *Applied Mathematics and Computation*, **173**(2), 1227–1237.

Wang, B. (2010). Interval estimation for exponential progressive Type-II censored step-stress accelerated life-testing. *Journal of Statistical Planning and Inference*, **140**(9), 2706–2718.

Wang, F.U., Chen, Y., Teng, Z.Y., Yin, L.G., Zhang, K.J., Jiang, C.F., Yan, G., Li, W.X.H., Chou, S.S.K. (2008). Accelerated lifetime test for TMR heads by ramped stress. *IEEE Transactions on Magnetics*, **44**(1), 100–103.

Wang, R. and Fei, H. (2004). Statistical inference of Weibull distribution for tampered failure rate model in progressive stress accelerated life testing. *Journal of Systems Science and Complexity*, **17**(2), 237–243.

Watkins, A.J. (2001). Commentary: Inference in simple step-stress models. *IEEE Transactions on Reliability*, **50**(1), 36–37.

Watkins, A.J. and John, A.M. (2008). On constant-stress ALTs terminated by type-II censoring at one of the stress levels. *Journal of Statistical Planning and Inference*, **138**(3), 768–786.

Wu, S.J., Lin, Y.P., and Chen, Y.J. (2006). Planning step-stress life test with progressively Type-I group-censored exponential data. *Statistica Neerlandica*, **60**(1), 45–56.

Xiong, C. (1998). Inference on a simple step-stress model with type-II censored experimental data. *IEEE Transactions on Reliability*, **47**(2), 142–146.

Xiong, C. and Ji, M. (2004). Analysis of grouped and censored data from step-stress life test. *IEEE Transactions on Reliability*, **53**(1), 22–28.

Xiong, C. and Milliken, G.A. (1999). Step-stress life testing with random stress-change times for exponential data. *IEEE Transactions on Reliability*, **48**(2), 141–148.

Xiong, C., Zhu, K., and Ji, M. (2006). Analysis of a simple step-stress life test with a random stress-change time. *IEEE Transactions on Reliability*, **55**(1), 67–74.

Xu, H. and Fei, H.L. (2007). Planning step-stress ALTs with two experimental variables. *IEEE Transactions on Reliability*, **56**(3), 569–579.

Yang, G. (1994). Optimum constant-stress accelerated life-test plans. *IEEE Transactions on Reliability*, **43**(4), 575–581.

Yang, G. (2007). *Life Cycle Reliability Engineering*, John Wiley & Sons, New York.

Yang, G. and Jin, L. (1994). Best compromise test plans for Weibull distributions with different censoring times. *Qual. Reliab. Eng. Int.*, **10**, 411–415.

Yeo, K.P. and Tang, L.C. (1999). Planning step-stress life-test with a target acceleration-factor. *IEEE Transactions on Reliability*, **48**(1), 61–67.

Yin, X.K. and Sheng, B.Z. (1987). Some aspects of accelerated life testing by progressive-stress. *IEEE Transactions on Reliability*, **R-36**(1), 150–155.

Yum, B.J. and Choi, S.C. (1989). Optimal design of accelerated life tests under periodic inspection. *Naval Research Logistics*, **36**(6), 779–795.

Yum, B.J. and Kim, S.H. (1990). Development of life-test sampling plans for Exponential distribution based on accelerated life testing. *Communications in Statistics-Theory and Methodology*, **19**(7), 2735–2743.

Yurkowski, W., Schafer, R.E., and Finkelstein, J.M. (1967). Accelerated testing technology. *Rome Air Development Center Tec Rep. RADC. TR. 2, 67–420. Griffins AFB, NY.*

Zhang, H. (2007). Modeling and planning accelerated life testing with proportional odds, Dissertation, New Brunswick, New Jersey.

Zhang, J.P., Liu, C., Chen, X., Cheng, G.L., and Zhou, A.X. (2014). Study on constant-step stress accelerated life tests in white organic light-emitting diodes. *Luminescence The Journal of Biological and Chemical Luminescence*, **29**(7), 933–937.

Zhu, Y. and Elsayed, E.A. (2011). Design of equivalent accelerated life testing plans under different stress applications. *Quality Technology and Quality Management*, **8**(4), 463–478.

Zimmer, W.J., Keats, J.B., and Wang, F.K. (1998). The Burr XII distribution in reliability analysis. *Journal of Quality and Technology*, **30**(4), 386–394.

INDEX

A

A-optimality, 45
accelerated life test sampling plans, 196, 225
accelerated tests, 2
  accelerated degradation tests, 2
  accelerated life tests, 2
acceleration factor, 36–37, 58, 61, 90, 97, 104, 108, 113, 115, 224
acceptable quality level, 198
acceptable reliability level, 198
acceptance number, 200, 202
acceptance sampling plan, 195–196, 253, 259
Arrhenius relationship, 21
Arrhenius–Weibull model, 25
attribute, 197
attributes sampling plans, 201
average sample number, 200

B

bilevel programming, 226, 241, 257, 270
Burr Type-XII distribution, 19, 121–122, 129, 145, 147–148, 151, 175, 259, 261–262, 272, 276, 280, 299–300

C

catastrophic failures, 3
chain sampling plans, 206
characteristic life, 12, 178–179
competing risks, 177, 191
conditional failure rate, 12
constant-stress, 4, 24, 34–35, 47, 56, 58, 147
constant-stress PALT, 36, 58–59
consumer's risk, 198
continuous inspection, 47, 58, 223
Cox model, 25
cumulative exposure, 28, 30, 32–33, 63, 65–67, 72, 74, 88, 152, 178–179, 182, 193, 228–229
cyclic-stress, 4–5

D

D-optimality, 44, 103, 107, 111, 241, 271, 289
defect, 197
defective, 197
double sampling plan, 201, 204
double specification limits, 201

E

electrical stress, 8
environmental stress, 9
exponential distribution, 13, 299
extreme value distribution, 15
Eyring relationship, 23

F

failure rate, 12–17
fully ALT, 20, 47, 152, 159, 169, 174

G

general rebate policy (GRP), 219
group acceptance sampling plans, 208

H

hard failures, 3
hazard rate, 12

I

inspection, 58
inspection mode(s), 4
  continuous inspection, 4
  periodic inspection, 4
inverse power law, 160, 228–229
inverse power law relationship, 22

K

k-method, 211–212, 215, 217, 225, 257, 259, 263, 280
Khamis–Higgins model, 178, 180, 193

L

life distribution, 11–12
life test sampling plan, 196
likelihood function, 74, 96, 98–99, 101, 182, 286
likelihood ratio, 190
likelihood ratio statistic, 82, 84, 192
log likelihood function, 67, 75, 96, 98, 101, 117, 125, 127, 131, 142, 156, 161, 232
log-location-scale distribution, 32
log-logistic distribution, 17, 20, 64, 66, 72, 84, 145, 176, 225, 228–229, 238, 257
logistic distribution, 16, 18
lognormal distribution, 14–15, 56
lot tolerance percent defective, 199
lot tolerance reliability level, 199
lower specification limit, 200

M

M-method, 212, 216–217
mean lifetime, 13
mechanical stress, 7
modified constant-stress, 56, 158–159, 161, 165, 168–169, 174
modified step-stress, 56, 151–154, 164, 166, 172
multivariable relationship, 24

N

nominal life, 22
nonconforming, 198
nonconformity, 198
normal distribution, 13

O

OC curve, 199, 201, 234, 236, 240, 243, 246, 268, 272, 289
OC function, 236

P

63.2th percentile, 15–16
100*p*th percentile, 12–17, 19
PALT models, 35
periodic inspection, 62
Power–Weibull model, 25
Power–Weibull ramp-stress, 31
Power–Weibull step-stress model, 30
pro-rata warranty, 219
process average, 198
process parameter, 209
producer's risk, 198
product control, 195
progressive-stress, 4–5, 30, 35, 54–56, 119
proportional hazards, 25
proportional odds, 26
pth quantile, 12

R

ramp-stress, 31, 119–120, 123, 130, 134–136, 138, 141–142, 145, 225–226, 228, 231

random-stress, 4, 6
rejectable quality level, 199
reliability acceptance sampling plan, 196
repetitive group sample, 289

S

sensitivity analysis, 65, 88, 115–118, 140–141, 145, 171, 176, 192–193, 222, 255, 278, 280, 295
sequential probability ratio, 206
sequential sampling plans, 206
single sampling plan, 201, 204, 210, 216, 223, 285
single sampling RASP, 202
single specification limit, 200
skip lot sampling plans, 206
soft failure, 3
specification limits, 200
step-stress, 4–5, 28, 32, 34–35, 49, 57–58, 63, 65, 147, 178, 191
step-stress PALT, 37, 59–61, 89–90, 93, 95, 97, 100, 111, 115–118, 259–260, 263–264
stress upper bound, 120–122, 124, 129–130, 134–136, 138, 141, 143

T

tampered random variable, 37
tampering coefficient, 37
tampering point, 37
temperature–humidity relationship, 23
temperature–nonthermal relationship, 23
TFR model, 37, 55, 260–261, 264, 282
thermal cycling, 6
triangular cyclic-stress, 6
truncated logistic distribution, 18, 89, 93, 95, 97–98, 100, 105–106, 109–110, 115, 118
Type A OC curve, 199
Type B OC curve, 199
typical life, 22

U

unacceptable quality level, 199
upper specification limit, 200
usage rate acceleration, 10

V

variable repetitive group sampling plan, 281
variable single sampling plan, 210, 217, 225, 259, 280
variables, 197
variables sampling plan, 201, 208, 214
variance optimality, 44, 234, 257

W

warranty, 260
Weibull distribution, 15, 30, 32, 35, 55, 61, 63, 119–120, 145, 175, 178–179

X

$\bar{X}$-method, 211